Internet Technologies for
Fixed and Mobile Networks

Internet Technologies for Fixed and Mobile Networks

Toni Janevski

ARTECH HOUSE
BOSTON | LONDON
artechhouse.com

Library of Congress Cataloging-in-Publication Data
A catalog record for this book is available from the U.S. Library of Congress.

British Library Cataloguing in Publication Data
A catalog record for this book is available from the British Library.

ISBN-13: 978-1-60807-921-6

Cover design by John Gomes

© 2016 Artech House
685 Canton Street
Norwood, MA 02062

All rights reserved. Printed and bound in the United States of America. No part of this book may be reproduced or utilized in any form or by any means, electronic or mechanical, including photocopying, recording, or by any information storage and retrieval system, without permission in writing from the publisher.

All terms mentioned in this book that are known to be trademarks or service marks have been appropriately capitalized. Artech House cannot attest to the accuracy of this information. Use of a term in this book should not be regarded as affecting the validity of any trademark or service mark.

10 9 8 7 6 5 4 3 2 1

*To my great sons, Dario and Antonio,
and to the most precious woman in my life, Jasmina*

Contents

CHAPTER 1

Introduction to Internet Technologies   1

1.1 Development of the Internet   2
    1.1.1 The Beginning   2
    1.1.2 Standardization of Fundamental Internet Technologies   4
    1.1.3 Worldwide Growth of the Internet   6
    1.1.4 The Growth toward Broadband Internet and New Applications   8
1.2 Internet Architecture   11
    1.2.1 Basic Internet Protocol Architecture   12
    1.2.2 Internet Network Architecture   14
1.3 Characterization of Internet Traffic   18
1.4 Legacy Telecommunications Versus the Internet   20
1.5 Convergence of Telecommunications to an All-IP World   23
1.6 This Book's Structure   27
    References   29

CHAPTER 2

Internet Protocols   31

2.1 Internet Protocol (IP)   31
    2.1.1 Internet Control Message Protocol (ICMP)   34
    2.1.2 Address Resolution Protocol (ARP)   35
2.2 IPv4 Addressing   36
    2.2.1 Classful IPv4 Addressing   37
    2.2.2 Classless IPv4 Addressing   38
    2.2.3 Public and Private IP Addresses   39
    2.2.4 Network Address Translation (NAT)   41
2.3 IPv6 Fundamentals   42
    2.3.1 Flow Labeling in IPv6   44
    2.3.2 Extension Headers   46
    2.3.3 ICMPv6   48
2.4 IPv6 Addressing   50
    2.4.1 Representation of IPv6 Addresses   50
    2.4.2 Allocation of IPv6 Address Space   51
    2.4.3 IPv6 Address Autoconfiguration   52

	2.4.4	Transition from IPv4 to IPv6	53
2.5	Dynamic Host Configuration Protocol (DHCP)		55
	2.5.1	DHCP for IPv4	55
	2.5.2	DHCP Version 6 (DHCPv6)	56
2.6	Transport Protocols in the Internet		59
	2.6.1	User Datagram Protocol (UDP)	60
	2.6.2	Transmission Control Protocol (TCP)	61
2.7	TCP Mechanisms and Versions		69
	2.7.1	TCP Congestion Control	70
	2.7.2	TCP Versions	73
2.8	Internet Governance		74
		References	76

CHAPTER 3

Internet Networking — 79

3.1	What Is a Socket?		79
3.2	TCP Sockets (Stream Sockets)		83
	3.2.1	Interpretation of Binary Data in Internet Networking	85
	3.2.2	TCP Socket Interface	85
3.3	UDP Datagram Sockets		89
3.4	Client–Server Architectures		90
3.5	Peer-to-Peer Architectures		93
3.6	Internet Access Networks		95
	3.6.1	Ethernet and Wi-Fi as Unified Local Access	95
	3.6.2	Fixed Broadband Access Networks	98
	3.6.3	Mobile Broadband Access Networks	99
3.7	Internet Routing		102
	3.7.1	Unicast Routing	104
	3.7.2	Multicast Routing	108
3.8	Border Gateway Protocol (BGP)		113
	3.8.1	BGP Operations	115
	3.8.2	BGP Discussion	117
		References	119

CHAPTER 4

Fundamental Internet Technologies — 121

4.1	Domain Name System (DNS)		121
	4.1.1	Domain Name Space	122
	4.1.2	Name Servers	124
	4.1.3	Resolvers and Resolution	125
	4.1.4	DNS Discussion	127
4.2	File Transfer Protocol (FTP)		128
	4.2.1	Trivial FTP (TFTP)	129
	4.2.2	FTP Discussion	130
4.3	Electronic Mail		130
	4.3.1	Simple Mail Transfer Protocol (SMTP)	132

		4.3.2 Post Office Protocol Version 3 (POP3)	134
		4.3.3 Internet Message Access Protocol Version 4 (IMAP4)	136
	4.4	World Wide Web (WWW)	136
		4.4.1 Hypertext Transfer Protocol (HTTP)	136
		4.4.2 Web Documents	139
	4.5	Multimedia Streaming	140
		4.5.1 Real-Time Transport Protocol (RTP)	141
		4.5.2 Real Time Streaming Protocol (RTSP)	143
		4.5.3 MPEG-2	144
		4.5.4 MPEG-4	146
	4.6	Session Initiation Protocol (SIP) for Internet Signaling	147
		4.6.1 SIP Messages	147
		4.6.2 SIP Addressing	148
		4.6.3 SIP Network Elements	149
	4.7	Internet Security	152
		4.7.1 Internet Security on Network Layer (IPsec)	153
		4.7.2 Internet Security on the Transport Layer (SSL/TLS)	156
	4.8	AAA in Internet	158
		4.8.1 RADIUS	159
		4.8.2 Diameter	161
		References	166

CHAPTER 5

Internet Standardization for Telecom Sector 169

5.1	Next Generation Networks (NGNs) by ITU		170
	5.1.1 Cooperation for NGN Development		171
	5.1.2 NGN Network Concept		172
5.2	Transport and Service Strata of NGN		173
	5.2.1 Transport Stratum		174
	5.2.2 Service Stratum		176
5.3	NGN Architectures		176
5.4	Next-Generation Broadband Access		178
	5.4.1 ITU's Role in Development of Broadband Internet Access		179
	5.4.2 Fixed Broadband Internet Access Technologies		180
	5.4.3 Wireless and Mobile Broadband Internet Access Technologies		187
5.5	Naming and Addressing		195
5.6	Quality of Service (QoS) Framework		197
	5.6.1 Introduction to QoS, QoE, and Network Performance		198
	5.6.2 ITU-T Standardization on QoS and QoE		199
	5.6.3 Overview of Internet QoS		201
	5.6.4 QoS Parameters and Classes		205
	5.6.5 End-to-End QoS		206
5.7	Fixed-Mobile Convergence (FMC)		209
5.8	IP Multimedia Subsystem (IMS)		210
	5.8.1 Proxy-CSCF (P-CSCF)		212
	5.8.2 Serving-CSCF (S-CSCF)		212

	5.8.3 Interrogating-CSCF (I-CSCF)	212
	5.8.4 User Identities in IMS	213
5.9	Discussion	214
	References	214

CHAPTER 6

Internet Technologies for Mobile Broadband Networks — 217

6.1	Mobile IP	218
	6.1.1 Mobile IPv4	218
	6.1.2 Mobile IPv6	221
6.2	Host Identity Protocol (HIP)	225
6.3	All-IP Packet Core for 4G Mobile Networks	227
	6.3.1 Evolved Packet Core (EPC)	228
	6.3.2 Protocol Stacks for 4G Interfaces	229
6.4	QoS Framework for 4G Mobile Internet	234
	6.4.1 QoS in UMTS	234
	6.4.2 QoS in LTE/LTE-Advanced Mobile Networks	236
	6.4.3 QoS in 4G Mobile WiMAX	239
6.5	4G Mobile VoIP	240
	6.5.1 Carrier-Grade Mobile VoIP Implementation	241
	6.5.2 Over-the-Top Mobile VoIP Consideration	243
6.6	Mobile TV	244
	6.6.1 Mobile IPTV Standardization by ITU-T	244
	6.6.2 Multimedia Broadcast Multicast Service (MBMS)	248
6.7	Mobile Web	249
6.8	5G Mobile Broadband Developments	252
	6.8.1 5G Architectures	253
	6.8.2 5G Services and Business Aspects	255
6.9	Regulation and Business Aspects for Mobile Broadband Internet	257
	6.9.1 Regulation Aspects for Mobile Broadband	257
	6.9.2 Business Aspects for Mobile Broadband	260
	6.9.3 Discussion	261
	References	262

CHAPTER 7

Broadband Internet services — 265

7.1	VoIP as the PSTN/PLMN Replacement	266
7.2	Over-the-Top (OTT) Voice-over-IP	267
	7.2.1 Skype	268
	7.2.2 Other OTT VoIP Applications	270
7.3	IPTV Services	271
	7.3.1 IPTV Content Delivery	273
	7.3.2 Internet Technologies Used for IPTV Service	275
	7.3.3 Traffic Management, QoS, and QoE for IPTV	278
7.4	Over-the-Top Multimedia Streaming	280
	7.4.1 YouTube Technologies	281

7.5	Over-the-Top Peer-to-Peer Services		283
	7.5.1 Napster Technology		284
	7.5.2 BitTorrent		284
7.6	Over-the-Top Social Networks		287
	7.6.1 Social Network Concepts and Technology		288
	7.6.2 Comparison of Existing Social Networks and the WWW		289
	7.6.3 The Future of Social Networks: Decentralization		290
7.7	Internet of Things (IoT)		292
7.8	Web of Things		295
7.9	Regulation and Business Aspects for Broadband Internet Services		298
	7.9.1 Business Aspects		298
	7.9.2 Regulatory Aspects		301
7.10	Discussion		302
	References		302

CHAPTER 8

Cloud Computing 305

8.1	ITU's Framework for Cloud Computing	305
8.2	Cloud Systems and Architectures	308
8.3	Cloud Computing Service Models	310
	8.3.1 Infrastructure as a Service (IaaS)	310
	8.3.2 Platform as a Service (PaaS)	311
	8.3.3 Software as a Service (SaaS)	312
	8.3.4 Network as a Service (NaaS)	312
	8.3.5 Communication as a Service (CaaS)	313
	8.3.6 Intercloud Computing	315
8.4	Cloud Security and Privacy	315
	8.4.1 Application Layer Security for Access to the Cloud	317
	8.4.2 Secure Cloud Access	319
8.5	Over the Top (OTT) Cloud Services	320
8.6	Telecom Cloud Implementations	322
	8.6.1 Desktop as a Service (DaaS)	323
	8.6.2 Cloud Communication Center	323
	8.6.3 Service Delivery Platform as a Service (SDPaaP)	324
	8.6.4 End-to-End Service Management by Cloud Provider	325
8.7	Mobile Cloud Computing (MCC)	326
8.8	Regulation and Business Aspects of Cloud Computing	328
	8.8.1 Regulation Aspects of Cloud Computing	329
	8.8.2 Business Aspects of Cloud Computing	330
	References	331

CHAPTER 9

Future Networks 333

9.1	Future Networks Framework by ITU	333
9.2	Network Virtualization for Future Networks	334
9.3	Software-Defined Networking (SDN)	338

9.4	Big Data		339
	9.4.1	Big Data Definition	340
	9.4.2	Big Data Ecosystem and Reference Architecture	341
	9.4.3	Big Data Technologies and Use Cases	343
	9.4.4	Challenges for Big Data	344
9.5	Over-the-Top Versus Telecom Operator Service Models		345
9.6	Cybersecurity		347
	9.6.1	Cybersecurity Technologies	348
	9.6.2	Security in Future Networks	349
9.7	Impact of M2M		352
	9.7.1	Ubiquitous Sensor Network (USN)	352
	9.7.2	Common M2M Service Layer	354
9.8	Smart Networks and Services		356
	9.8.1	Smart Traffic Control	357
	9.8.2	Smart Sustainable Cities	358
9.9	Business and Regulation Challenges in Future Networks		360
	References		361

CHAPTER 10

Conclusions 363

About the Author 365

Index 367

CHAPTER 1

Introduction to Internet Technologies

In the second decade of the 21st century, information and communication technologies (ICTs) are moving toward creation of an Internet that is a unified global network for all types of services and applications. This represents a major shift from the 150-year-old concept of telecommunications, which has traditionally been targeted to telephony services and to broadcast of video and audio (i.e., television and radio). The Internet has moved that concept toward a completely new approach, in which various types of information and services (e.g., voice, video, web, messaging) become readily available to the end user at anytime and from anywhere (i.e., from every access network to the Internet). Thus, the heterogeneity of networks for different services (telephony, data transmission, image, video, audio, multimedia, messaging, and so forth) is converging to one network (the Internet) that unites all the heterogeneity of the various services, terminal devices (phones, computers, mobile devices, TVs, and so forth), and transmission media (copper pairs or cables, fiber, and wireless/mobile access). In fact, the Internet provides access to various communication services in a manner similar to that used by an electrical distribution network to provide electrical power to various electrical appliances and devices (e.g., toasters, washing machines, computers, TV sets). In an electrical network the plug into the network (i.e., the socket) is the same for all of the different devices. In the case of the Internet as a global network, the Internet access is a "socket" for all types of communications services that provide access to information (e.g., to content on a website) or the exchange of information (e.g., via Internet telephony or messaging). So, what is the Internet? The Internet is a network of networks that are interconnected and use the Internet Protocol (IP) [1] for internetworking connectivity and exchange of information (regardless of its type) between different networks and the users that are attached to them.

All technologies used in the Internet for provision of certain services are called Internet technologies. Generally, Internet communication is independent of the underlying transport networks (optical transport networks, satellite transport networks, and so forth) and access networks [Ethernet as a typical local-area network (LAN), modem-based access, wireless Internet access, and so forth]. Hence, Internet technologies include so-called protocol layers above the underlying transport network protocols, starting from the so-called network layer (where the IP is located) and the upper protocol layers (e.g., various applications, such as web servers and browsers, voice applications, messengers, and video players).

1.1 Development of the Internet

The history of telecommunications starts in the 19th century with telegraphy and later telephony. Telephony provided distant mouth-to-ear voice communication in both directions (which is natural for humans) via technical means and paved the way for building a global telecommunication network infrastructure for connecting people around the world. In the first half of the 20th century, the television added real-time video transmission as a service (radio broadcasts preceded television broadcasts), adding to the types of global telecommunication services. Initially, however, all networks and all communications were based on transfer of analog signals over a distance. The second half of the 20th century saw an analog-to-digital (A/D) transition in telecommunication networks, which was first accomplished in the transport networks. The A/D conversion was then done for the telephony (in the access part) and later for the television. The introduction of digital communications end to end, however, was made possible by the introduction of computers in telecommunication networks, including in network nodes (e.g., switches, routers, databases) and in end-user devices (e.g., personal computers, laptops, smartphones). So, all signals in computers and in networks became digits, mainly "1" and "0" bits because it is simplest to decide on the recipient's side of a given link between two possible values. Such developments in parallel in the telecommunication world and in computer science provided the basis for development of data networks where all information (e.g., audio, video, text, various data) is transferred and processed by using ones and zeros, packed in given formats of messages, packets, and units.

How was the Internet born? The birth of the Internet happened as a result of a research project started in 1960s in the United States by the Advanced Research Projects Agency (ARPA), later renamed the Defense Advanced Research Projects Agency (DARPA). At that time no one could really imagine how big the Internet would grow and the global effect it would have not only on the technology, but on society in general. That is similar to trying to predict what will happen 50 years from now, which is virtually impossible. So, the birth of the Internet came about from having a desire to create a network that would be resistant to link failures (i.e., a robust and fault-tolerant network) by segmenting the data (i.e., the information in a digital form) in small chunks called packets, and then routing each such packet from its source to its destination independently from all other packets. Such an approach was suitable for communications in certain environments that had a higher failure probability (e.g., radio access networks, military networks) than others. That initial concept was seriously developed and standardized over time until eventually it became the biggest worldwide packet network, capable of transferring different types of information over the same infrastructure. Although the roots of the Internet were formed in the late 1960s and 1970s, the main protocols were standardized in the 1980s, some of which are still in force today.

1.1.1 The Beginning

The brief history of the Internet [2] starts in 1962, with a series of letters by J. C. R. Licklider discussing the "Galactic Network" concept, which envisioned a globally interconnected set of computers through which everyone could access data and programs from a computer connected to the network, something that is very similar to the Internet that we have today. Licklider was the first head of the DARPA project from which the Internet was developed. At the same time, the first theoretical foundations of the Internet were published by L. Kleinrock. However, the first connections among computers were made over circuit-switched telecommunication networks by means of dial-up modems. So, the beginning of the Internet as a packet-switched network was performed over circuit-switched networks, in which a dedicated channel is established between the two end points.

The ARPANET, which was the first name for the network, became operational in 1969, connecting several universities in the United States. It was the first large-scale packet-switching network worldwide. The main idea behind the ARPANET, followed later by today's Internet, lay in its key technical approach: use of an open network architecture as opposed to the closed networks (at the time of the ARPANET) operated by telecom operators, which provisioned circuit-switched telephony as the main service. With an open architecture, the particular choice of a given network architecture can be selected freely by the network provider, which connects with other such networks through a meta-level interworking architecture [2]. The first host-to-host protocol in the ARPANET (as the predecessor of today's TCP/IP protocol stack) was called the Network Control Protocol (NCP), finished in December 1970. The implementation of NCP was needed for early developers to start developing first applications to be used on hosts connected to the ARPANET. The first "killer" application was electronic mail. Appearing in 1972, the first basic email software for sending and receiving emails was written by Ray Tomlinson and expanded with additional utilities (to list, selectively read, forward and respond to messages, and so on) by Larry Roberts. With the growth of the Internet, email has had an increasing impact on everyday life and a significant influence on personal and business communications today. For example, traditional fax messaging, which was the standard way of exchanging documents in business environments, has been replaced nowadays by email communications. Also, in 1972 the first successful demonstration of the ARPANET was organized by R. Kahn, who had a major influence on the architectural network design of the ARPANET in the 1970s. A key step in the standardization process of new concepts and protocols was made by Steve Crocker, who introduced the so-called Requests for Comments (RFCs) in 1969 as informal memos; they later became major documents that led the advance toward Internet standardization. Currently RFCs are produced and maintained by the Internet Engineering Task Force (IETF). So, the 1960s and 1970s denoted the initial period in the development of the Internet [3] according to the timeline given in Figure 1.1.

The development in which more hosts began to attach to the network influenced changes in thinking. Before the appearance of LANs, the developers of the ARPANET targeted only a small number of computers for attachment to the network and therefore Internet addressing was defined with 32 bits; the first 8 bits identified the network (i.e., network ID) and the last 24 bits identified the host on the given network (i.e., host ID). With the spread of the Internet, certain changes were implemented by introducing the allocation of bits for the network ID and host ID, with the goal of allowing more networks to exist in the Internet. However, the initial NCP relied on the ARPANET for provision of end-to-end reliability, so error control was not included in it. The open networking ideas of Kahn together with V. Cerf's experience with creating the NCP resulted in development of protocols that

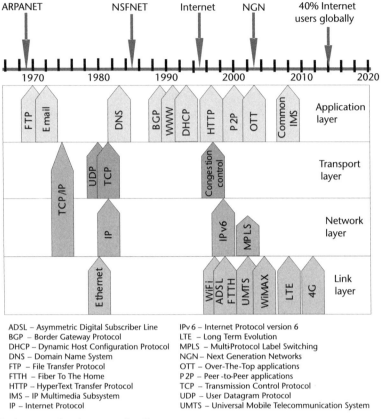

Figure 1.1 Internet development timeline.

fit the open networking approach: the Internet Protocol (IP) and the Transmission Control Protocol (TCP), or simply TCP/IP. Both protocols were first standardized in 1974 as the Internet Transmission Control Program (RFC 675 [4]), which introduced interprocess communication as a basic principle, in which processes were viewed as active elements on all hosts connected to the network. In such an approach, all input/output (I/O) media regardless of its type (e.g., data, audio, video) was viewed as communicating through the use of processes. Because a given process may have to distinguish among several different streams that are originating from that host (or terminating to it), that RFC also introduced the notion of "ports" for a given process through which it communicates with ports of other processes. However, because ports may not be unique, a proposal was made that they be concatenated in each host with a given network identifier (i.e., an IP address on a given network interface) to create so-called "socket" names that would be unique on all interconnected networks. (More details about sockets and Internet networking is given in Chapter 2 of this book.)

1.1.2 Standardization of Fundamental Internet Technologies

The second period of Internet development started at the beginning of the 1980s with the standardization of the main technologies (e.g., protocols, systems) for the Internet and with its transition to a widespread architecture. That was possible at that time due to the appearance and development of personal computers (PCs) and workstations, as well as LANs. Although Internet network design was targeted to provide seamless communication over different underlying transport technologies, there was certainly a needed accommodation for the underlying technology to carry Internet packets. Such technology was the Ethernet, developed by R. Metcalfe in 1973, which later became the dominant access technology to the Internet worldwide, and it is currently being standardized by the Institute of Electrical and Electronics Engineers (IEEE).

So, the main idea regarding the fundamental Internet technologies was established in the 1970s, and the work resulted in the Internet's two main protocol standards, which appeared in 1981, TCP/IP [5]. TCP/IP defines the main "look" of the Internet protocol stack. The first implementation of TCP/IP in a BSD (Berkley Software Distribution) Unix operating system and its introduction were done at the same time for all hosts attached to the ARPANET on January 1, 1983. That day is noted as "Flag Day" for the Internet, when several hundreds of machines switched from NCP (as a single networking protocol) to TCP/IP (as two different protocols, i.e., TCP over IP).

Will there be another flag day in the Internet's future? It will not be possible again because the number of hosts on the Internet grew from several hundred in 1983 to several billion in the second decade of the 21st century, and it is continuing to increase even today at a similar pace. The TCP provided mechanisms for flow control and recovery from lost packets, which were lacking in its predecessor NCP. However, there were applications that did not need TCP mechanisms (e.g., voice transfer over Internet). For such applications the User Datagram Protocol (UDP) was standardized in August 1980 [6], with the goal of providing direct access to IP basic services without any guarantees on packet delivery to the end host. (In the case of UDP it is left to the application to deal with any packet losses and other problems that may appear.) Transition from NCP (which strictly relied on the ARPANET architecture) to TCP/IP provided the possibility of having several "independent" networks for different purposes (e.g., ARPANET for research activities, MILNET for operational defense activities) [2].

Increasing numbers of networks and hosts has led to the need to assign names to hosts instead of numeric addresses. When there was a limited number of hosts in the ARPANET in the 1970s, there was a single table of all hosts and their associated addresses and names. The need to assign names to hosts and mapping (i.e., resolving) host names in IP addresses resulted in development of the Domain Name System (DNS), invented by Paul Mockapetris and standardized with RFCs in November 1983 [7, 8].

Regarding the Internet protocols and their global use, the NSFNET, which was founded by the National Science Foundation (NSF) in the United States and deployed in 1985, has played an important role. The critical decision of the NSFNET was the adoption of the TCP/IP protocol stack as the mandatory protocols on all hosts in the network. One might say that NSFNET was the successor of the initial ARPANET (created by the DARPA project), which further continued the evolution toward the current Internet. The NSFNET extended the network to a wide-area infrastructure by using U.S. federal funding for the task. It started with six backbone nodes connected via 56-Kbps links and until 1995 it grew up to 21 backbone nodes connected with multiples of 45-Mbps links [2]. It expanded the network on all continents, and paved the way toward the global expansion of the Internet that followed afterward. However, it had certain policies in place that prohibited use of the NSFNET for commercial purposes, given that its primary use was for research activities. During the existence of NSFNET (1985–1995), several different commercial networks appeared that were interconnected with the NSFNET and routed traffic over it according to NSFNET so-called "acceptable use" policies. Additionally, NSF and the NSFNET played an important role toward the Internet governance by adopting the previous DARPA approach arranged in a hierarchy under the Internet Architecture Board (IAB) as a governance body (Internet governance is covered in Chapter 2), by ensuring interoperability of ARPANET and NSFNET with a joint authorship of RFC 985 for Internet gateways [9], made by IETF and the Internet Research Task Force (IRTF) with the IAB on one side and the NSF on the other side.

With the growth of the network after 1985 it started to suffer from congestion problems, which initiated work for solving such issues. That resulted in two other important novelties in the network:

- TCP congestion control, introduced in 1988 [10], was targeted to solve network collapses that have appeared in the growing network. It triggered standardization of several TCP algorithms for congestion control [11], as well as different versions of TCP that currently exist, including standardized and proprietary ones.
- The Border Gateway Protocol (BGP) with interdomain routing policies was created in 1989 [12] to support routing policies between different autonomous IP-based networks that were interconnected and exchanging IP traffic. Its version 4, BGP-4 [13], later became the worldwide standard for interconnection of so-called autonomous systems (ASs) in the Internet.

The era of the NSFNET also introduced collaboration between the research community that played a role in building the network from the roots up as the ARPANET (which was decommissioned by 1990) and the growing sector of vendors for commercial network equipment, which knew about the practical problems that were occurring in the network that needed solutions. This collaboration between the research and commercial communities continued and shaped the coming Internet as a collection of communities and technologies that are continuously changing.

1.1.3 Worldwide Growth of the Internet

The 1980s gave to the world the main Internet technologies, such as IP, TCP, UDP, DNS, and BGP, which shaped the evolving Internet. However, the main applications that existed in the Internet in the 1980s were the File Transfer Protocol (FTP) and email, which well suited the scholarly community in the beginning and the business community later. However, to aid the global spread of the Internet toward residential users, an application was needed that would suit their needs. Such an application (or one might refer to it as a technology) became the World Wide Web (WWW), invented by Tim Berners-Lee at the beginning of the 1990s. The WWW has proven to be the most important single Internet application so far, which led to acceptance of the Internet as a single worldwide packet-switching network, by winning over the competition from European packet-switching technology, the Asynchronous Transfer Mode (ATM) in the 1980s and 1990s, as well as the International Telecommunication Union's X.25 standard for packet-switched services, which was set up in most countries during the 1970s and 1980s. Important events that helped the development and growth of the WWW and the Internet itself are as follows [3]:

- The WWW was invented by Berners-Lee at CERN, Switzerland. He got the initial idea in 1989, developed the first web server and web browser in 1990, and set up the first website on November 13, 1990. On April 30, 1993, CERN announced that the WWW would be free to anyone (without royalties), which paved the way for global use and spread of the WWW and the Internet.
- In October 1994 Berners-Lee founded the World Wide Web Consortium (W3C) at the Massachusetts Institute of Technology (MIT), which was supported by DARPA (from the U.S. side) and the European Commission.
- In May 1996 the first version of the application layer communication protocol for the WWW, which was called HTTP (Hypertext Transfer Protocol), version 1.0, was standardized [14]. In January 1997 an improved version of the HTTP, version 1.1 [15], was standardized and is still in use today.

The WWW was the most important technology of the Internet from an application point of view. However, several other Internet technologies also have roots in the 1990s. The growth of the Internet brought to the forefront the issue of a limited number of IP addresses, because it had initially been designed for a limited number of hosts (regarding the IP addressing space with 32-bit-long addresses). Several solutions for solving this problem were standardized during this period:

- The Dynamic Host Configuration Protocol (DHCP) was standardized in October 1993 with RFC 1531 and RFC 1541 [16]. It provided capabilities for automatic allocation of reusable network addresses (i.e., IP addresses) and additional configuration options. The DHCP provided flexibility in the IP addressing and paved the ground toward Internet global expansion.
- The IP Network Address Translator (NAT) was introduced in an informational RFC 1631 in May 1994 [17]. It provided a short-term solution to problems regarding the IP address exhaustion and scaling in routing by using mapping between reusable private IP addresses and globally unique public IP addresses.
- Internet Protocol version 6 (IPv6) was standardized in December 1998 with RFC 2460 [18]. It was designed to be a long-term solution for overcoming the limitations of its predecessor Internet Protocol (IPv4) regarding the size of the IP addressing space. (IPv6 provides 128-bit-long IP addresses, while IPv4 addresses have 32 bits.)

However, the DHCP and NAT has provided even longer sustainability of IPv4 addresses, thus postponing the introduction of IPv6 until the second decade of the

21st century, when it has already started to be implemented on a global scale due to exhaustion of IPv4 address spaces in all regions in the world.

In the middle of the 1990s the commercialization of the Internet began (although it was still the NSFNET at that time). This commercialization culminated in April 1995 with the defunding (i.e., privatization) of the NSFNET backbone network [2]. During the NSFNET era, the Internet grew by up to 50,000 networks, of which more than half were located in the United States. With the decommissioning of the NSFNET by 1995, standardization of HTTP in 1996–1997 as a basic communication protocol for the Web started the exponential growth of a number of Internet hosts in the late 1990s.

Although the word "internet" had appeared before, one informal definition of the Internet was given in a resolution passed by the Federal Networking Council (FNC) on October 24, 1995. FNC is a body that was formed to coordinate sharing of the NSFNET infrastructure. According to that definition, the term "Internet" refers to the global information system that [2]:

1. Is logically linked together by a globally unique address space based on the IP or its successors;
2. Is able to support communications using the TCP/IP suite or its subsequent extensions or new versions, as well as other IP-compatible protocols; and
3. Provides, uses, or makes accessible, either publicly or privately, high-level services layered on the communications and related infrastructure.

So, one may say that in practice 1995 is the year when the Internet started its global expansion and growth as a commercial network targeted to all possible types of information and users, thus going beyond the limits even imagined for it at the beginning.

1.1.4 The Growth toward Broadband Internet and New Applications

The explosion of the WWW by the end of the 1990s made the Web the most used application in terms of the number of transferred packets or bytes. The beginning of the 2000s was marked by the appearance of different proprietary (i.e., not standardized, developed by certain developers or companies) peer-to-peer (P2P) systems and services, such as Napster in 1999 (for music sharing), Gnuttela in 2000 and Kazaa in 2001 (for file sharing), BitTorrent in 2001 (also for file sharing), and so forth. BitTorrent quickly became the most used P2P application due to it ability to simultaneously download different pieces of a single file from multiple computers. Also, there appeared different proprietary P2P video/audio streaming applications (e.g., SopCast), mainly targeted to streaming of live events (e.g., sport games, concerts). However, the development of new applications was triggered by the development of technologies for broadband access to the Internet, because some services require higher bit rates end to end (e.g., video streaming).

The 2000s were characterized by several other important Internet services related to real-time media over the Internet such as voice communications (i.e., telephony over Internet) and video sharing services:

- Skype as a voice-over-IP (VoIP) technology, based on a P2P approach, was released for the first time in 2003. Soon afterward it became the largest best-effort VoIP service on the Internet, taking a significant share of the international telephone traffic from telecom operators due to free Skype-to-Skype calls. However, as is usual for the best-effort Internet services, it lacks QoS (quality of service) support, numbering support in most of the countries worldwide (with the aim of Skype users receiving calls from other non-Skype users), and standardization (Skype is a proprietary application, i.e., not standardized, thus it is limited to be offered only by the company that owns it). However, its ability to be run over different devices connected to the Internet, and especially to provide free Skype-to-Skype calls (including conference calls, video conferences, and messaging), made it a recognizable and favorite best-effort VoIP service for more than a decade.

- YouTube as a video-sharing website appeared in 2005, providing end users with the ability to upload and share their videos or to view videos uploaded by other users. It is the world largest video-sharing website and provides videos with different resolutions (i.e., different image quality). It is used by individual users as well as certain media companies and organizations. In fact, YouTube introduced the user-generated video contents, which has become possible due to availability at that time (2005) of different small handheld cameras, either stand-alone video cameras or cameras integrated in mobile devices.

The 2000s were also characterized by the growth of broadband Internet access to residential users, including fixed broadband and mobile broadband (Figure 1.2). Although broadband access for business users was available even before the 2000s by using LANs (e.g., Ethernet), the "real" expansion of broadband access to the Internet was driven by the standardization of digital subscriber line (DSL) technologies by the International Telecommunication Union (ITU) for access via the existing twisted-pair lines, which were initially deployed in 20th century for the circuit-switched telephony by the telecom operators. The American National Standards Institute (ANSI) first standardized the technology in 1998 [19], and further

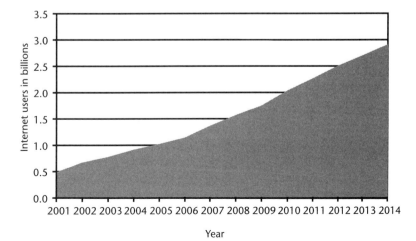

Figure 1.2 Internet growth in the 21st century.

standardization was done by the ITU-T in 1999 and on [20, 21]. ADSL has provided several megabits per second in the downstream direction (from the Internet toward the end user) with asymmetrically lower bit rates in the upstream direction (from the end user toward the Internet), which has resulted in nearly 100 times higher bit rates than the maximum 56 Kbps provided via dial-up modem over a telephone line or the maximum 144 Kbps provided via the Integrated Services Digital Network (ISDN). So, deployment of ADSL globally was significant in the growth of Internet use by residential users and in the development of new applications and increasing use of video-based services (since video is much more demanding regarding the required bit rates end to end) such as video streaming (e.g., live videos) and video on demand (e.g., YouTube videos), which are asymmetrical services in that they demand much higher bit rates in the downstream direction than in the upstream direction.

Fixed broadband Internet penetration on a global scale was also fostered by the standardization of fiber access technologies (e.g., fiber-to-the-home technologies, standardized by the ITU) and their deployment, which provide several tens to several hundreds of megabits per second to residential users [22–24]. Additionally, cable networks, which were initially deployed for TV broadcast to residential users, introduced Internet access through the DOCSIS (Data-Over-Cable Service Interface Specifications) standard [25]. So, all fixed networks, either over twisted pairs, coaxial cables, or fiber, have standardized broadband access to the Internet.

Wireless and mobile technologies developed in parallel with the Internet during the 1990s, leading to integration of mobile networks and the Internet in the 2000s. For example, the most successful digital mobile technology in the world, the European Global System for Mobile (GSM) communications, was standardized in the 1990s for mobile telephony, and was further upgraded by the General Packet Radio System (GPRS) by the end of the 1990s to provide Internet connectivity to mobile terminals (despite GPRS's initial average bit rate of several tens of kilobits per second).

At the same time as the GSM/GPRS system was being standardized, the IEEE developed and standardized the IEEE 802.11 wireless LANs as a wireless extension to the Ethernet, with the legacy IEEE 802.11 standard standardized in 1997 [26]. There have been many amendments and newer versions of the standard since then, such as 802.11a/b/g/n/ad/ac on the physical layer and 802.11e (for QoS), 802.11i (for security), and so forth, on the medium access control (MAC) layer. The IEEE 802.11 LANs are also known widely as Wi-Fi and globally they became the most used wireless local access network (WLAN) for connecting to the Internet. They are used as an extension of wired Ethernet in business environments, as well for home wireless LANs (e.g., as an extension of fixed broadband access), due to their low price (there is no need for synchronization between the sender and receiver, which reduces the price of the equipment) and their initial positioning for use in unlicensed frequency bands (e.g., 2.4 or 5 GHz). The IEEE further extended its broadband wireless portfolio with the IEEE 802.16 wireless metropolitan-area network known as WiMAX in 2004, and Mobile WiMAX in 2005 (first release) [27]. WiMAX, however, has not had the same success as Wi-Fi, mainly due to competition in developed countries from 3GPP mobile broadband technologies, which evolved from the globally successful GSM/GPRS mobile networks. (GSM belongs to the so-called second-generation mobile systems, i.e., 2G, while GSM/GPRS belongs to the so-called 2.5G.)

Mobile broadband Internet started with 3G (third-generation mobile system) technologies, such as UMTS/HSPA (universal mobile telecommunication systems/high speed packet access) as an evolution from GSM/GPRS mobile networks, and Mobile WiMAX, release 1 (based on the IEEE 802.16e-2005 standard in the radio access network), which provided average bit rates in the range of several megabits per second (higher in the downstream direction). At the start of the 2010s, this technology was being standardized and deployed in fourth-generation (4G) systems, including LTE-Advanced from 3GPP (LTE stands for long-term evolution) and WirelessMAN-Advanced (with IEEE 802.16m in the radio access part [28]), which is also known as Mobile WiMAX, release 2, according to the WiMAX Forum. 4G mobile broadband is characterized by its all-IP requirements; that is, all traffic, including user and signaling traffic, should be transferred using IP packets and Internet technologies, as specified in the ITU–Radiocommunications (ITU-R) specification for IMT-Advanced radio interfaces (IMT stands for international mobile telecommunications) [29]. The 4G mobile networks are targeted to provide access to bit rates of from several tens up to several hundreds of megabits per second, thus paving the way for their use in more demanding applications. Because mobile communications are personal (i.e., each mobile user has a separate device) and mobile users usually carry their mobile devices (e.g., smartphones) all the time and everywhere, mobile broadband development is becoming one of the main drivers toward higher Internet penetration worldwide, and vice versa. Emerging Internet applications and services provided to mobile users drive the mobile industry by creating a continuing need for higher bit rates and hence new mobile network equipment as well as more capable mobile devices in terms of processing capabilities, memory capacities, network interfaces, and battery life.

The penetration of broadband Internet access led to the creation of application ecosystems (e.g., Google's ecosystem, Apple's ecosystem) that are used for work, administration, entertainment, socialization, learning, storage, and so forth. Higher bit rates also made possible video services with higher resolutions, such as high-definition (HD) video, as well as emerging cloud computing services.

1.2 Internet Architecture

The Internet has revolutionized telecommunications, including communication architectures and technologies, and the revolution is ongoing. In terms of the development of telecommunication and Internet network architectures, two main periods can be distinguished:

- Integration of connection-oriented circuit-switched telecommunication networks and packet-switched connectionless Internet in the 1990s and 2000s; and
- All-IP based networks and services, that is, the convergence of all networks, services, and devices toward the Internet and Internet technologies in the 2010s and beyond.

The Internet's architectural design was initially based on low-cost networks, which are flexible in terms of their network architecture (e.g., a new IP network can be added on the run) and open to all different types of services (e.g., voice, video, WWW, email, P2P) by having more intelligent end-user devices (with installed operating systems that contain TCP/IP protocols). Such an Internet architecture led to its success in the long term and finally to the convergence of legacy telecommunication services toward the Internet. However, legacy telecommunication services are also influencing the Internet's architecture with requirements for QoS guarantees end to end and for signaling, thus changing/upgrading its initial design.

1.2.1 Basic Internet Protocol Architecture

For easier development of the software for communication between processes on the same or on different machines, the tasks to be performed need to be separated and grouped on different so-called "layers" that have precisely defined interfaces between them. Each such layer typically has a communications protocol, which is, in fact, a system of digital rules for data exchange between same types of protocols on different machines (e.g., hosts, routers) or between the protocols on adjacent layers on the same machine. Each communications protocol uses a defined format for the messages that are exchanged, so it must define syntax and semantics as well as communications algorithms or mechanisms (e.g., congestion control in TCP). The design of the architecture for communication between different processes needed not to be limited to a particular system, but to be applied globally. One such global standard for the protocol architecture in telecommunication systems (including hardware and software) is the Open Systems Interconnection (OSI) model standardized by the International Standardization Organization (ISO) [30] and ITU-T [31]. The OSI protocol reference model has seven layers (Figure 1.3). They are, going from the bottom to the top, the physical layer, data-link layer, network layer, transport layer, session layer, presentation layer, and application layer.

The Internet protocol layering model has evolved over the years. In the early years it was suggested that only three layers needed to be defined, where each of the layers would be able to have multiple protocols [32]. Later, in 1989 with RFC

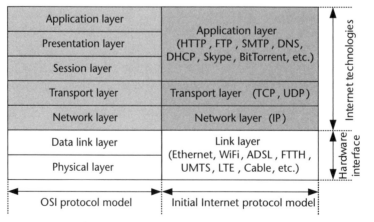

Figure 1.3 Internet protocols mapped to the OSI protocol layering model.

1122 [33], four protocol layers were explicitly designed in the Internet architecture (Figure 1.3):

- Application layer: The top layer in the Internet protocol suite combines the functions of the presentation and application layers of the OSI reference model. The application layer is divided into two sublayers: (1) user protocols to provide service directly to end users, and (2) support protocols that provide common systems functions. The most common Internet user protocols in 1989 were Telnet (for remote login) [34], FTP [35], and Simple Mail Transfer Protocol (SMTP) for email communications [36].
- Transport layer: This layer provides end-to-end services for applications, which is present with two main transport protocols:
  - TCP as a reliable connection-oriented transport protocol, and
  - UDP as a connectionless transport service based on datagram delivery (each datagram is carried by an IP packet).
- Internet layer: This is the layer below the Internet transport protocols, and it uses IP to carry the data from the source host (i.e., the sender's side) to the destination host (i.e., the recipient's side). The IP itself is defined as a connectionless datagram internetwork service (without any guarantees for the packet delivery to the destination host), which provides network addressing, type-of-service (ToS) specification, a segmentation option (and reassembly), and security information.
- Link layer: This layer is a communications protocol at the bottom of the Internet protocol stack. It is used to interface with a given network to which the host is connected. Because there are many different networks, there are many different network interfaces (i.e., link layer communication protocols). However, the Internet was integrated initially with Ethernet, invented by Robert Metcalfe in 1973, with the first published specification coming out in 1980 from a group of companies (Digital, Intel, and Xerox) [37]. At the same time IEEE started to standardize Ethernet and published it as standard IEEE 802.3 in 1983 [38]. The IEEE 802 networks have been a good fit for the Internet protocol stack since the 1980s [39]. The connection between the IEEE 802.3 Ethernet and the IP was realized via the Address Resolution Protocol (ARP) for converting network IP addresses into 48-bit-long Ethernet MAC addresses [40].

So, the Internet protocol stack defines a type of hourglass with the IP protocol in the middle (with both of its existing versions, IPv4 and IPv6), several transport protocols and many applications above, and many underlying transport technologies below IP (i.e., interfaces). Considering that the OSI-1 layer (the physical layer) and the OSI-2 layer (the data-link layer) are defined for telecommunication link technologies (including wired and wireless ones), then the Internet protocol suite has, in fact, five protocol layers, since the Internet does not go into the standardization down to the lowest two OSI layers. However, the link layer must support IP on the network layer by having certain defined message transfer units (MTUs) and by mapping IP addresses to lower link layer addresses (also referred to as physical addresses). In this manner, even Ethernet standardization follows the OSI layering model, which is defined from its first specification [37] and further standardized as the IEEE 802.3 standard. So, in practice, the Internet is considered to have five layers in its protocols suite [38].

The legacy telecommunications technologies are converging toward the Internet architecture and Internet technologies, which include telephony (i.e., VoIP in the Internet environment) and television to be carried over Internet [i.e., IP television (IPTV)]. But, telephony requires signaling and it is provided by appropriate protocols for such a purpose. For example, Signaling System Number 7 (SS7), which is also packet based although not IP based, has been used in legacy telecommunication network since the 1980s until currently. The current globally accepted protocol for signaling in Internet environments is the Session Initiation Protocol (SIP) [41]. For example, for VoIP with standardized signaling, SIP is needed for voice calls (e.g., call initiation, management, termination). So, one can conclude that with the convergence of all services onto the Internet, for certain services (e.g., VoIP, IPTV) the Internet protocol suite is approaching the OSI reference protocol layering model. However, there can always be different point of views regarding the Internet protocol layering (e.g., four layers, five layers, six layers, seven layers).

Generally, in the Internet protocol architecture the network interface contains OSI layers 1 and 2, the network layer is IP, and the transport layer protocols are TCP and UDP (both over IP), whereas OSI layers 5 to 7 are considered to be a single layer (the fifth layer), which is the application layer in the Internet (as shown in Figure 1.3).

Layer splitting also occurs in the Internet protocol architecture. For example, for Internet access over DSL, the Point-to-Point Protocol over Ethernet (PPPoE) is used. It is placed between the link layer and IP (i.e., the network layer). Further, Multi-Protocol Label Switching (MPLS) [42] is implemented in most of the Internet transport networks nowadays between the link layer protocol and IP (therefore MPLS appears as OSI layer 2.5). Additionally, for real-time applications (e.g., VoIP, IPTV) the Real-time Transport Protocol (RTP) [43] is used over UDP, where both protocols (RTP and UDP) belong to the transport layer (i.e., OSI-4 layer).

1.2.2 Internet Network Architecture

The Internet is composed of many IP networks. Each IP network is separated from the other networks through IP routers. The elements (i.e., network nodes) that may exist in an IP network in the Internet are as follows:

- Hosts: Different devices and computers connected to the Internet have applications running on the network (including network-based hosts such as servers, and end-user hosts such as computers). Each host connects to the Internet via a network interface card (e.g., Ethernet, Wi-Fi, ADSL modem). While Ethernet and Wi-Fi were initially designed for the Internet, the modems (modulator–demodulator devices) are used to connect hosts to the Internet via a network that was not initially designed for Internet traffic (e.g., over legacy telecommunication networks for telephony).

- Repeaters and hubs: Repeaters are used for longer Ethernet cables (e.g., over 100m) to regenerate the digital signal (i.e., clean the accumulated noise during the transmission). When a given repeater has multiple ports, it is called a hub. Both repeaters and hubs work on the physical layer (i.e., OSI layer 1). However, today repeaters and hubs are rarely used in access networks such as Ethernet, because they have been replaced by switches.
- Bridges: These enable communication among different LANs. They are also used to divide a single so-called collision domain (in Ethernet LAN) into two while maintaining a single so-called broadcast domain for the packets. They function on the OSI-2 layer, and typically are used to break large congested LANs into smaller and more efficient LANs.
- Switches: These are network devices that filter and forward OSI layer 2 packet frames between different ports, based on Ethernet's 48-bit MAC addresses (also referred to as physical addresses). The switch differs from the hub in that it only forwards the layer 2 packets to physical ports (on the switch) involved in the communication (i.e., a host connected to a given port sends or receives the packet). It associates the MAC addresses of hosts with its ports based on self-learning (using previous forwarding of packets in both directions, to/from connected hosts). Switches typically operate on OSI layer 2, but some, called multilayer switches (e.g., layer 2/3 switches), operate on the network layer (i.e., OSI layer 3) as routers. However, switches (with their typical operation on layer 2) are most commonly used today in building Ethernet LANs and wireless LANs (i.e., Wi-Fi networks) as most typical representatives of the IP access networks.
- Routers: These devices route packets from one IP network to other IP network(s) by using information that is carried in the header of each IP packet transferred over the Internet. So, for routing of IP packets, routers use the network (IP) addresses of the sender and recipient. Since they process information stored in the headers of the IP packets and IP is placed on the network layer (OSI layer 3), routers operate on OSI layer 3. Each router has several to many physical ports (e.g., Ethernet ports) through which it connects to other routers, switches, or hosts. (In Internet access networks the hosts are usually connected to Ethernet switches, which are further connected to the routers.)

In order for network elements to communicate with each other, certain rules must be applied. These rules are defined within the protocols on different layers. Protocols (e.g., TCP, UDP, IP) determine the format and the order of messages exchange and the actions that shall be taken when receiving protocol message units (e.g., datagrams, frames). As an example, Figure 1.4 shows the message structure on different protocol layers for the WWW (for Web browsing).

Generally, a telecommunications network, such as the Internet, is composed of geographically distributed hardware and software components. However, communication among different hosts is always between a peer-protocol (e.g., Web application) on one host with exactly the same peer-protocol on the other corresponding host. (The same statement holds for communication among network nodes, such as routers.) When an application, such as that shown in Figure 1.4, on a given

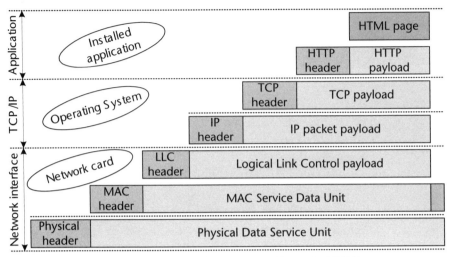

Figure 1.4 Structure of the messages on different protocol layers for WWW.

host (e.g., a web server) sends information (e.g., HTML page) to an application on another host machine (e.g., to a web browser), it has to transmit it through the network interface on the sending host. For that purpose, the information (i.e., the data in HTML) from the application is transferred to the protocol below (that is HTTP, which is used as a communication protocol for the Web), which adds its header and transfers the formed message unit to the transport protocol below the HTTP (which is TCP). The transport protocol adds a transport protocol header to the unit and passes it to the protocol below it, and that protocol is IP. Further, IP adds an IP header and creates the basic message unit in the Internet, called an IP datagram or IP packet (the terms are used interchangeably). The IP packet is then transmitted in that form through the Internet toward its destination. However, on each link the IP packet is placed as payload in an OSI layer 2 frame (e.g., Ethernet frame, Wi-Fi frame) and transmitted over the physical layer of that link (e.g., copper, fiber, or wireless). So, the sender host places the IP packet as payload in data-link frames, which add data-link headers such as a logical link control (LLC) header and MAC header in the case of Ethernet access (in the example given in Figure 1.4). The formed MAC service data unit (MSDU) is relayed within the host toward the physical layer, which adds a physical layer header and transmits the message unit (which has the IP packet embedded in it).

Typically, the IP packet may traverse several switches and routers on the way to the destination host, where the switches process the OSI layer 2 message units (based on MAC addresses) and routers route the packets (based on IP addresses carried in IP headers). OSI layer 2 message units are valid until they reach the next router, which extracts the IP packet, uses IP header information for routing purposes, and puts the IP packet in another OSI layer 2 unit in one of its output ports with the goal of sending it on the next hop. In this manner, the IP packet travels through the Internet as a network of interconnected routers (and switches as layer 2 nodes in the access networks, through which the hosts are connected) until it reaches the destination host (Figure 1.5). At the destination host each header is processed and removed by the same peer-protocol as the one that added the given header on the

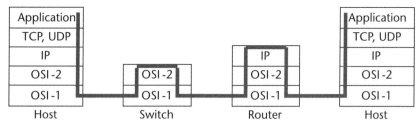

Figure 1.5 Comparison of information processing in an Internet host, switch, and router.

sender's side, going that way from the physical layer up to the application layer (in this example the application is a web browser), which uses the information (e.g., presents the web page).

The Internet is a network of many interconnected networks (Figure 1.6). Based on the size of the geographical area to be covered, these networks are divided into local-area networks, metropolitan-area networks, and wide-area networks. A LAN is the smallest type of network and is used in the home or office (e.g., Ethernet, Wi-Fi). A MAN is a network that covers an entire city or metropolitan area (e.g., Metro Ethernet, WiMAX) and can have many LANs connected to it. However, each MAN and LAN must be connected to a WAN (e.g., national IP backbone network, regional IP backbone network).

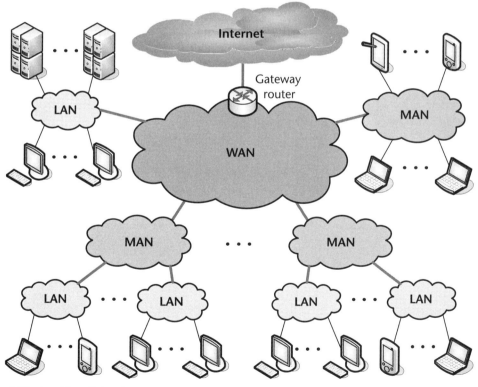

LAN – Local Area Network
MAN– Metropolitan Area Network
WAN – Wide Area Network

Figure 1.6 Internet network architecture.

Each IP-based network consists of interconnected network nodes, such as switches and routers. Internetworking among different types of networks in the Internet is realized by particular routers, called *gateways*, that are placed between two or more separate networks to interconnect them. Forwarding of packets in routers is performed by using the information provided by the Internet Protocol.

Each Internet host and router must have implemented the Internet Protocol on the network layer.

In general, each network connected to the Internet is autonomous. An IP network can function even without being connected to another network. The Internet as a global network is not subject to centralized control, which allows it to be flexible to add new nodes (e.g., routers) and connect to new networks, while still supporting various existing and new applications/services.

The global Internet has a flat architecture that was established during the 1980s. It consists of interconnected autonomous systems (ASs). Each AS consists of many IP networks, where each IP network is characterized by a certain IP address space allocated to it (refer to Chapter 2 for Internet addressing). Each IP network can be divided into several subnets for better network administration and/or management in a process called subnetting. Routing of IP packets between ASs is based on allocated numbers to the ASs [12], while routing within the given AS is based on IP addresses carried in the IP packet [3]. Initially, the Internet has a best-effort architecture, which means that the network gives its "best effort" to transmit every received IP packet to its destination without any guarantees, while leaving all congestion control to TCP at the end hosts. However, for support of real-time applications/services, such as voice and video, some IP networks (within their routers) have implemented functionalities for QoS support and traffic control.

Overall, the creators of the Internet architecture have asked "Is the architecture anything but protocols?" in the discussion "Towards the Future Internet Architecture" (RFC 1287 [44]) at the beginning of the 1990s, in the "doom" of the Internet expansion. And, indeed, the Internet architecture is built on protocols (on various layers) because the Internet is created such that it can use any underlying transport technology that can carry IP packets, that is, having IP on the network layer. Such an architectural design is also the main reason for its unprecedented success.

1.3 Characterization of Internet Traffic

Internet traffic covers a range of multimedia services, which have different traffic characteristics such as required bit rates, burstiness (i.e., peak-to-average bit rates ratio), session duration, as well as different requirements and constraints regarding the performances (e.g., required bandwidth, end-to-end delay, packet losses).

Different applications use different transport protocols depending on their traffic requirements (e.g., TCP, UDP). All control, such as congestion control, is left to the transport protocol (i.e., TCP) or applications (when UDP is used as transport layer protocol) at the end-user hosts.

In general, Internet traffic refers to the amount of data (e.g., volume of data in bytes, number of packets) sent to and received from any network connected to the Internet. However, Internet traffic can be measured on a given port of a router or on a given network interface of a host. Because the IP is present in all hosts and routers on the network layer, Internet traffic is also noted as IP traffic.

Before invention of the WWW the dominant type of traffic on the Internet was FTP traffic, where FTP as an application uses the TCP/IP protocol stack. The other traffic included email applications and Telnet, which also use TCP/IP. After the invention of the WWW at the beginning of 1990s, the Web (which uses TCP/IP) became the single application with the largest volume of traffic on Internet. However, DNS traffic (which accompanies web traffic for resolving website names into IP addresses) is based on the UDP/IP protocol stack. Around 2000 TCP accounted for most of the traffic: 95% or more of the bytes, 85% to 95% of the packets, and 75% to 85% of the flows [45]. However, because most of the TCP traffic was actually web traffic, the WWW accounted for 55% to 90% of the TCP traffic (the smaller shares of the TCP traffic belonged to email and FTP traffic) [45].

From 2000 onward many peer-to-peer applications for file sharing appeared, which contributed significantly to Internet traffic on a global scale. The most used P2P application for file sharing on a global scale is BitTorrent, followed by eDonkey [46, 47]. According to a 2007 Internet study [46], P2P traffic at that time made up between 49% and 83% of all Internet traffic (it varies in different regions in the world). So, P2P traffic moved web traffic to second place during the 2000s. One could expect such a result in the 2000s (when P2P file sharing traffic significantly grew) since the Web is an interactive application that typically requests activity by the end user (of the web services) and hence it is limited in time duration of its usage per user, whereas P2P traffic is typically triggered by the end user (i.e., initiated) and then it is realized without a need for further user interaction.

On the other hand, since its invention Skype has become the most popular best-effort Internet telephony service (i.e., best-effort VoIP, without any guarantees on QoS such as end-to-end delay and required bit rate/bandwidth), as shown by a 2007 Internet study [46]. By the end of the 2000s, the most popular applications on the Internet became the Web, media streaming (e.g., YouTube), P2P file sharing, one-click file hosting (e.g., DropBox), instant messaging, Internet telephony (e.g., Skype), and online gaming (a service type that had started to grow by the end of the 2000s). However, in the mid-2000s social networking web services (e.g., Facebook) appeared that have gained popularity. So, web traffic made its comeback due to the popularity of social networking websites, growing usage of web-based file hosting services, and the higher media richness of web pages [48].

Nowadays all best-effort Internet applications are considered to be over-the-top (OTT) applications. With the penetration of broadband access to the Internet, a significant increase has been recorded for video services, which are offered as video on demand (e.g., YouTube) and as live video streaming (e.g., broadcast of live events, such as sport games, concerts, live news) with different resolutions of the videos. YouTube is a website for video sharing that uses the TCP/IP protocol stack for video provisioning to end users. Similarly, other video-sharing websites that exist as OTT web services also use TCP. In contrast, real-time video and audio in the Skype application use UDP/IP for transmission of video and audio streams. However, TCP/IP is also used by Skype for control information as well as for Skype chat. The TCP and UDP shares in Skype traffic for audio, video and chat services are shown in Figure 1.7.

The 2010s are currently characterized by similar OTT applications and services found in business and residential sectors, but also by many custom applications (including custom TCP-based and custom UDP-based applications), which are playing a significant part in the Internet traffic portfolio.

The most used applications today (according to generated traffic volume in number of bytes) are given in Figure 1.8. Web browsing is at the top of the list of Internet application usage, followed by storage backup (related to cloud computing services), custom TCP application (due to several existing huge applications ecosystems nowadays, where most of the proprietary applications use the TCP/IP suite), and photo and video sharing [48]. One may note that nowadays P2P applications contribute around 6% of the total bandwidth used in the Internet, while social networking has a 1.62% share in the Internet traffic volume worldwide. (This might be due to the smaller sizes of pictures and texts that are typically used by users of such social networking websites.)

Of course, Internet traffic analysis varies in different regions in the world, and among other things it significantly depends on the bit rates (i.e., the bandwidth) with which the users access the Internet, as well as the popularity of certain applications and services in a given region.

1.4 Legacy Telecommunications Versus the Internet

Since the beginning of the 21st century, all legacy telecommunications services, such as telephony and television, have been migrating to the Internet environment by using already standardized Internet technologies by the IETF.

The legacy (i.e., traditional) telephone networks from the 20th century are referred to as the Public Switched Telephone Network (PSTN) [3]. The PSTN

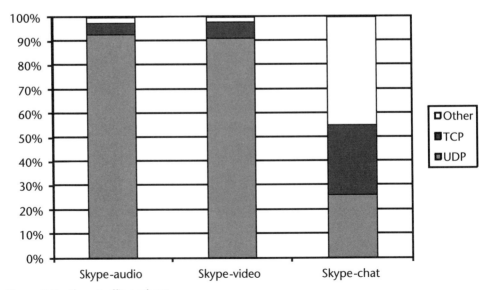

Figure 1.7 Skype traffic analyses.

1.4 Legacy Telecommunications Versus the Internet

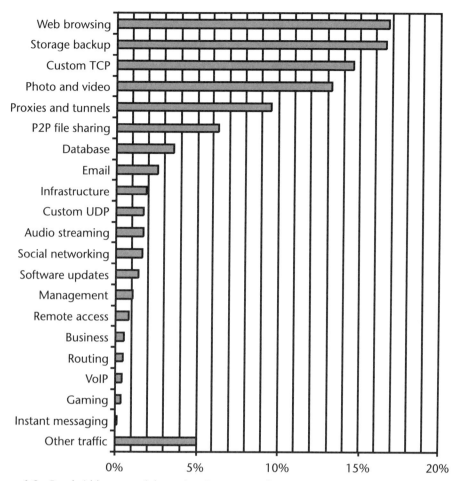

Figure 1.8 Bandwidth usage of the various Internet applications.

provided the technical means for achieving two-way voice communication (in both directions—from caller to the called user, and vice versa) by establishing a direct connection between the telephone devices of the end users by means of switching equipment (i.e., telephone exchanges) in centralized locations. Further, telephone exchanges are interconnected in a given hierarchy by using transmission systems (Figure 1.9). The traditional PSTN is a circuit-switched connection-oriented system, which means that allocated capacity between two end users is dedicated and cannot be shared with other telephone connections, regardless of the intensity of the voice traffic in both directions.

With the goal of allocating network resources (i.e., channels) and managing established telephone connections, PSTN requires standardized signaling (the global standard is SS7, standardized by ITU-T). However, due to the dedication of channels between the end users the end-to-end QoS is guaranteed and hence admission control is used (i.e., a telephone call is accepted only when there are enough free resources to be allocated between the two end points). Telephone networks were built in the 20th century by state-owned national telecom operators in many countries throughout the world. However, each telephone network must be connected with other telephone networks if global connectivity for telephony service is to be

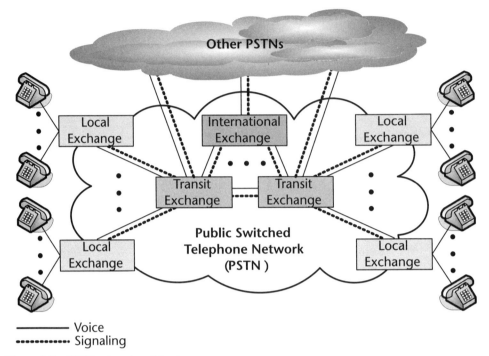

Figure 1.9 PSTN network architecture.

provided. The PSTNs were analog until the 1970s and 1980s, after which they migrated to digital telephone networks. The same PSTN approaches were used for circuit-switched mobile networks in the 1980s and 1990s. For example, GSM is the most typical example of a public land mobile network (PLMN). Both PSTN and PLMN have higher costs compared to the initial Internet network architecture because they require more complex network equipment, such as switching systems and signaling network nodes.

The other legacy telecommunications services, radio and TV, were distributed over separate broadcast networks for radio and TV. Also, these services are strictly standardized regarding the technologies and regulated regarding the contents (standards and regulation aspects vary in different countries or regions in the world).

On the other side, unlike PSTN, the Internet is at the same time a transit network for all IP packets and a network of computing-based devices (e.g., computers, laptops, smartphones) that provides the possibility for anyone from anywhere to access the Internet to retrieve, process, and store all types of information (e.g., voice, video, audio, messages, data). The Internet currently has a flat hierarchy (unlike PSTN) consisting of interconnected ASs, where each AS is a network of interconnected routers. Further, the traditional Internet is based on the best-effort principle, which means that every packet and every connection is admitted in any IP network (PSTN enforces strict admission control). Where the PSTN relies on established point-to-point connections between the telephone devices by using a switch hierarchy, the Internet does not specify such a hierarchy (at least, that is the case with the current design of the Internet for best-effort services, such as WWW and email). However, the most important feature of the Internet is the development of transport and network routing mechanisms that are neutral to the application traffic thus making it a general-purpose network [49]. Another important novelty introduced with the Internet is the change from the traditional telecommunication paradigm where the end users were only consumers of the contents (e.g., watching a TV program or listening to a radio) by allowing end users to create content (e.g., file, photo, video sharing) to upload. Also, end users can invent new proprietary applications over the existing Internet protocol suite (i.e., over the top) and can put an additional network (e.g., LAN) or a server in an office or a home LAN. A comparison of some of the important aspects between the legacy telecommunication networks and legacy Internet is given in Table 1.1.

Generally, on may note that the Internet is also a type of a telecommunication network, which is based on a packet-switching, connectionless approach by default. On the other side, legacy telecommunication networks, such as PSTN/PLMN and TV broadcast networks, are transitioning toward the Internet and hence changing its initial nature.

1.5 Convergence of Telecommunications to an All-IP World

The convergence of legacy telecommunications toward the Internet and Internet technologies is an ongoing process. However, even though the telecom and Internet worlds are converging, telephony and television as main legacy telecommunication

Table 1.1 Comparison Between Legacy Telecommunications and Legacy Internet

	Legacy Telecommunications	*Legacy Internet (Best-Effort)*
Global reach	Yes	Yes
Range of supported applications and services	PSTN/PLMN are optimized for telephony, TV broadcast for TV	General purpose network, open to all types of applications and services (e.g., WWW, email, VoIP)
Innovations	Only telecom and broadcast operators can innovate	Any individuals or organizations can setup a service on the application layer
Content creation	Individual end users generate voice (PSTN/PLMN), companies create content (TV broadcast)	Individuals and companies generate voice, videos, messages, various data files
End-to-end QoS	Yes	No
Standardization	Technology standardization is strictly required	Technology standardization is required, but proprietary solutions are allowed within the networks and on the applications' side
Main standardization organizations globally	ITU (PSTN, TV broadcast), 3GPP (PLMN)	IETF (OSI layer 3 and above), IEEE (OSI layers 1 and 2)
Regulation	Strictly regulated by national regulatory agencies or bodies, including architectures, technologies, numbering, and content	Network neutrality is preserved for all kinds of information (this may differ from country to country)
User devices	Dummy end-user devices, mainly hardware based	Computers and smart end-user devices with hardware and software
Network costs	Higher-cost switches (exchanges)	Lower-cost switches and routers
Pricing	Costs are associated with individual user calls (PSTN/PLMN) or flat fee (TV)	Costs are associated with volume of traffic (in bytes) or access bit rates (in bits per second) or flat fee

services require QoS support end to end. On the other hand, the legacy Internet was created for best-effort services, such as Internet-native services (e.g., FTP, email, WWW, P2P file sharing), which do not demand QoS guarantees. So, the traditional Internet does not have the built-in QoS support that is required for real-time telecommunication services such as telephony and TV. However, the openness of the Internet to various different applications and services (existing and new ones), flexibility in the network architectures, and generally lower costs to run different services over a single network than over a dedicated network for each service (e.g., telephone network, TV broadcast network, data network) have paved the way for the convergence of the whole telecommunication world, including networks and services, toward the Internet by using Internet technologies. Hence, vertical separation of different networks for different services transited to horizontal separation on a converged transport network infrastructure on the bottom of the protocol suite and heterogeneous services/applications on the top, with IP in the middle (Figure 1.10). This transition is referred to as convergence of telecommunications to all-IP.

The convergence of telephony and TV toward the Internet increased the complexity of the Internet due to requirements for QoS support and signaling. For example, PSTN and PLMN can provide end-to-end delay below 150 ms in each direction of the telephony conversation. Delays above 400 ms are unacceptable for telephony [50]. TV transmission over IP requires a high dedicated bandwidth (e.g., several megabits per second per TV channel), which requires QoS support. Additionally, telephony requires signaling for call establishment, because the initial architecture for the Internet was based on a client–server model where a client (typically on the user's side) requests certain information from a server (typically on the network's side). In this Internet approach the client always contacts the server, not the opposite way. Telephony, however, needs a provision for locating the device of the called user.

Convergence of telecommunications toward the Internet is also driven by the development of broadband access to the Internet, including fixed broadband (e.g., Ethernet, ADSL, FTTH, Cable networks) and wireless/mobile broadband (e.g., Wi-Fi, WiMAX, UMTS/HSPA, LTE/LTE-Advanced). In practice, the diffusion of

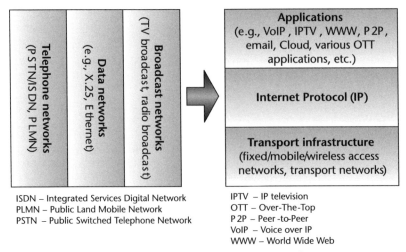

Figure 1.10 Convergence of telecommunication infrastructure to all-IP.

1.5 Convergence of Telecommunications to an All-IP World

broadband is inseparably related to the appearance of the Internet, although there were efforts in the 1980s and 1990s, mainly led by European countries, to standardize Asynchronous Transfer Mode (ATM) as a global packet-switching broadband technology, which finally lost the "battle" with the Internet technologies by the end of the 1990s. The term *broadband* refers to all access technologies (fixed and mobile) that provide enough bandwidth (in megabits per second) end to end for all existing applications to run smoothly. So, broadband Internet access for individual users refers to different bit rates at different times, for example, several megabits per second in the 2000s and several tens of megabits per second in the 2010s.

Convergence of legacy telecommunications toward the Internet required strict standardization of the QoS framework and overlay signaling network (also IP based), which led to standardization of a next-generation network (NGN) framework by the ITU-T. The work on NGN initially started in 2003 [3], with a general overview of NGN specified in 2004 [51]. The NGN was primarily targeted to standardization of QoS-enabled VoIP and IPTV by using standardized Internet technologies while keeping backward compatibility of these two services with legacy end-user equipment. However, NGN also introduced the framework for Internet of Things (IoT) and Web of Things (WoT), as well as further development of NGN concepts for future networks [52] and cloud computing. So, simply said, NGN is based on the Internet principle of separation of services and transport technologies; hence, it generally defines a transport stratum and a service stratum, as shown in Figure 1.11.

The NGN provides the possibility of transiting telephony and television to an all-IP network, with backward compatibility to previous legacy telecommunication services. So, in general, in the late 2000s and in the 2010s triple-play services were offered over single broadband access to the Internet, which included telephony, TV, and best-effort Internet. However, that resulted in offering different bundles of services to the end users (only Internet, telephony and Internet, triple-play, and

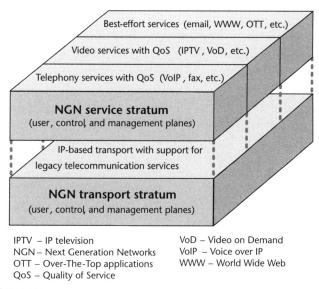

Figure 1.11 NGN strata.

so on). If one adds mobility to the triple-play bundle, it becomes a quadruple-play bundle (e.g., in mobile broadband networks).

Additionally, the development of the NGN is fostering fixed-mobile convergence (FMC) by accepting the use of the IP multimedia subsystem (IMS) as the standard for signaling in all-IP environments, including fixed and mobile networks.

With the convergence of all telecommunications toward the Internet technologies (from the network layer above), two types of service and applications are being offered to end users:

- Services and applications provided by telecom operators (fixed and mobile ones), such as:
  - Telephony (VoIP) and television (IPTV), which are provided with QoS support and signaling. These services, although transferred over IP networks, inherit the capabilities and regulation from their predecessors (i.e., PSTN/PLMN and TV broadcast networks). For example, it is mandatory for emergency calls (e.g., 112 in Europe, 911 in the United States) to be provided in all countries by the telecom operators that provide voice services, but it is not mandatory for the best-effort voice applications such as Skype and Viber.
  - With the convergence to all-IP networks in the 21st century, leased lines for business users, provided via the digital hierarchies of PDH and SDH/SONET in transport networks, are being replaced with virtual private networks (VPNs) as a standardized solution in all-IP transport networks (run by telecom operators) supported by QoS Internet mechanisms and IP traffic engineering (e.g., MPLS) [3].
- OTT services and applications are being offered globally on a best-effort basis (without QoS guarantees end to end) and include web-based services, messaging services, peer-to-peer services, etc. All OTT services are typically available through single access to the Internet, where they are isolated from the QoS-supported VoIP and IPTV services. Hence, network neutrality of the Internet refers to the best-effort access to the Internet and to OTT services.

Convergence of telecommunications toward all-IP has become a reality due to the synergy of the main standardization organizations throughout the world, including ITU, 3GPP, IETF, and IEEE. For example, the ITU has created the NGN framework for transition of all legacy telecommunications services as well as Internet-native services to the Internet as a single global network by using the standardized Internet technologies from IETF (e.g., IP, TCP, UDP, HTTP), as well as 3GPP mobile technologies and LAN and MAN standards from the IEEE. Also, 3GPP has included the Internet network architecture and Internet services with the introduction of GPRS, followed by 3G technologies such as UMTS/HSPA (which were based on a combination of circuit-switched voice and packet-switched Internet access), and further by 4G mobile systems (which were all-IP from the beginning), such as LTE/LTE-Advanced. The IEEE adopted the all-IP approach in their standards for network interfaces from the beginning (i.e., from the 1980s), thus creating Ethernet, Wi-Fi, and WiMAX to best suit the Internet.

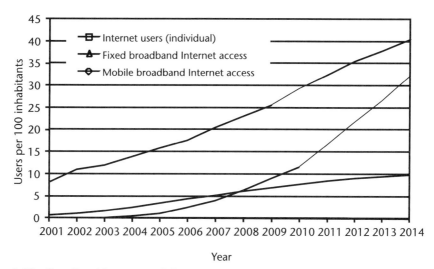

Figure 1.12 Broadband Internet statistics.

In general, convergence of telecommunications towards all-IP contributed to Internet growth in the 21st century by targeting fixed and mobile telecommunication broadband technologies exclusively for Internet access. Figure 1.12 shows that the Internet growth from 2000 onward happened due to the growth of broadband penetration in the world, because broadband access to individual residential users provided the possibility for the Internet to spread outside the initial academic and business environments. Mobile broadband is increasing at a higher pace on a global scale, due to the increasing number of mobile deployments in developing countries, which generally lack a fixed network infrastructure and hence have a lower penetration of fixed broadband Internet access.

Generally, the world of ICT is continuously evolving and changing, including the technologies, regulations, and business aspects. However, such development is currently directed toward all-IP networks and services, which means that all traffic, including user, control, and management information, is carried end to end by using IP packets and Internet technologies.

1.6 This Book's Structure

This book focuses on Internet technologies, in particular, on Internet principles, protocols, and services for fixed and mobile networks, including technologies, regulations, and business aspects.

Introductory chapters cover the Internet fundamentals of today, including the main Internet protocols (IPv4, IPv6, TCP, UDP, DHCP, and so forth) and IP addressing. The book also discusses the basis of Internet networking including the use of sockets for defining end-to-end Internet connections and different Internet architectures (client–server, P2P) accompanied by the main Internet routing principles and protocols. Also, it includes fundamental Internet systems and applications (DNS, WWW, email protocols, Internet multimedia communication, and SIP), which are the legacy Internet technologies.

During the first decades of the 21st century, all legacy telecommunications services from the 20th century, such as telephony and TV, have been migrating to the Internet environment by using already standardized Internet technologies from the IETF. Hence, this book covers Internet standardization for the telecom sector, the NGN, including its main signaling and control system IMS as the standardized approach for FMC.

Further, the book covers Internet technologies for mobile broadband networks, including Mobile IP (MIPv4 and MIPv6), HIP (Host Identity Protocol), all-IP core for 4G networks [i.e., Evolved Packet Core (EPC) and System Architecture Evolution (SAE)], and the QoS framework for 4G mobile Internet. Also, the book incorporates 4G mobile VoIP, mobile TV [IPTV, Multimedia Broadcast Multicast Service (MBMS)], mobile web services [mobile social networks, location-based services (LBS)], as well as business and regulation challenges for mobile broadband Internet.

Broadband access networks, including fixed and mobile ones, are needed to have all Internet services in all environments. In that manner, the book covers VoIP, including its implementation as a PSTN/PLMN replacement, as well as OTT VoIP implementations, such as Skype and Viber. Further, it includes telco-based IPTV services, OTT multimedia streaming (YouTube), OTT peer-to-peer services (BitTorrent, P2P streaming), and OTT social networks (Facebook, Twitter), as well as emerging IoT and WoT, considering the regulation and business aspects for given Internet services.

This book further focuses on cloud computing and future networks from technology, regulation, and business aspects. The book incorporates cloud computing, including ITU's framework, cloud ecosystems, architectures, and cloud service models: Infrastructure as a Service (IaaS), Platform as a Service (PaaS), and Software as a Service (SaaS). Cloud security, OTT cloud services, telco cloud implementations, mobile cloud computing (MCC) services and applications are also covered, as are business and regulation aspects for cloud computing.

It also covers future networks as defined by the ITU, including network virtualization, software defined networking (SDN), and smart networks and services (e.g., smart homes, smart cities). Finally, the book includes big data issues, cybersecurity, OTT versus telco service models, and convergence of regulation toward future networks.

There is no specific prerequisite knowledge for the readers (although some basic pre-knowledge about telecommunications and the Internet in general will be beneficial) because the book is structured in such a way that it covers all required aspects of the Internet technologies in fixed and mobile environments and its network architectures, protocols, and services. So, the book first provides fundamentals of the Internet technologies in the first four chapters and then covers more advanced topics in later chapters.

This book is targeted to managers, engineers, and employees from regulators, government organizations, telecommunication companies, and ICT companies, and to students and professors from academia and anyone else who is interested in understanding, implementing, and regulating Internet technologies in fixed and mobile networks, including technology, regulation, and business aspects.

References

[1] J. Postel, "Internet Protocol," RFC 791, September 1981.

[2] B. M. Leiner et al., "Brief History of Internet," Internet Society, www.internetsociety.com, 2012.

[3] T. Janevski, *NGN Architectures, Protocols and Services*, New York: John Wiley & Sons, April 2014.

[4] V. Cerf, Y. Dalal, C. Sunshine, "Specification of Internet Transmission Control Program," RFC 675, December 1974.

[5] J. Postel, "Transmission Control Protocol," RFC 793, September 1981.

[6] J. Postel, "User Datagram Protocol," RFC 768, August 1980.

[7] P. Mockapetris, "Domain Names—Concepts and Facilities," RFC 822, November 1983.

[8] P. Mockapetris, "Domain Names—Implementation and Specification," RFC 883, November 1983.

[9] Network Technical Advisory Group, National Science Foundation, "Requirements for Internet Gateways—Draft," May 1986.

[10] V. Jacobson, M. Karels, "Congestion Avoidance and Control," *ACM SIGCOMM '88 Symposium*, 18(4): pp. 314–329.

[11] W. Stevens, "TCP Slow Start, Congestion Avoidance, Fast Retransmit, and Fast Recovery Algorithms," January 1997.

[12] K. Lougheed, Y. Rekhter, "A Border Gateway Protocol (BGP)," RFC 1105, June 1989.

[13] Y. Rekhter, T. Li, "A Border Gateway Protocol 4 (BGP-4)," RFC 1654, July 1994.

[14] T. Berners-Lee, R. Fielding, and H. Frystyk, "Hypertext Transfer Protocol—HTTP/1.0," RFC 1945, May 1996.

[15] R. Fielding et al., "Hypertext Transfer Protocol—HTTP/1.1," RFC 2616, June 1999.

[16] R. Droms, "Dynamic Host Configuration Protocol," October 1993.

[17] K. Egevang, P. Francis, "The IP Network Address Translator (NAT)," May 1994.

[18] S. Deering, R. Hinden, "Internet Protocol, Version 6 (IPv6) Specification," RFC 2460, December 1998.

[19] ANSI T1.413, Issue 2, "Network and Customer Installation Interfaces—Asymmetric Digital Subscriber Line (ADSL) Metallic Interface," 1998.

[20] ITU-T Recommendation G.992.1, "Asymmetric Digital Subscriber Line (ADSL) Transceivers," June 1999.

[21] ITU-T Recommendation G.992.3, "Asymmetric Digital Subscriber Line Transceivers 2 (ADSL2)," April 2009.

[22] ITU-T Recommendation ITU-T Q.834.1, "ATM-PON Requirements and Managed Entities for the Network and Network Element Views," June 2004.

[23] ITU-T Recommendation G.983.1, "Broadband Optical Access Systems Based on Passive Optical Networks (PON)," January 2005.

[24] ITU-T Recommendation G.984.1, "Gigabit-Capable Passive Optical Networks (GPON): General Characteristics," March 2008.

[25] CableLabs, "Data-over-Cable Service Interface Specifications—DOCSIS 3.0: MAC and Upper Layer Protocol Interface Specification," 2013.

[26] IEEE 802.11-1997, "Wireless LAN Medium Access Control (MAC) and Physical Layer (PHY) Specifications," June 1997.

[27] IEEE 802.16e-2005, "Part 16: Air Interface for Fixed and Mobile Broadband Wireless Access Systems Amendment 2: Physical and Medium Access Control Layers for Combined Fixed and Mobile Operation in Licensed Bands and Corrigendum 1," February 2006.

[28] IEEE 802.16m-2011, " Part 16: Air Interface for Broadband Wireless Access Systems, Amendment 3: Advanced Air Interface," May 2011.

[29] ITU-R Report M.2134, "Requirements Related to Technical Performance for IMT-Advanced Radio Interface(s)," 2008.

[30] ISO/IEC 7498-1:1994, "Information Technology—Open Systems Interconnection—Basic Reference Model: The Basic Model," November 1994.

[31] ITU-T Recommendation X.200, "Information Technology—Open Systems Interconnection—Basic Reference Model: The Basic Model," July 1994.

[32] M. A. Padlipsky, "Perspective on the ARPANET Reference Model," RFC 871, September 1982.

[33] R. Braden, "Requirements for Internet Hosts—Communication Layers," RFC 1122, October 1989.

[34] J. Postel, J. Reynolds, "Telnet Protocol Specification," RFC, May 1983.

[35] J. Postel, J. Reynolds, "File Transfer Protocol (FTP)," RFC 765, October 1985.

[36] J. Postel, "Simple Mail Transfer Protocol," RFC 821, August 1982.

[37] Digital Equipment Corporation, Intel Corporation, and Xerox Corporation, "The Ethernet, Local Area Network. Data Link Layer and Physical Layer Specifications, Version 1.0," September 1980.

[38] D. E. Comer, *Internetworking with TCP/IP—Principles, Protocols and Architecture*, 4th ed., Upper Saddle River, NJ: Prentice Hall, 2000.

[39] J. Postel, J. Reynolds, "A Standard for the Transmission of IP Datagrams over IEEE 802 Networks," February 1988.

[40] D. C. Plummer, "An Ethernet Address Resolution Protocol," RFC 826, November 1982.

[41] J. Rosenberg et al., "SIP: Session Initiation Protocol," RFC 3261, June 2002.

[42] E. Rosen, A. Viswanathan, R. Callon, "Multiprotocol Label Switching Architecture," RFC 3031, January 2001.

[43] H. Schulzrinne et al., "RTP: A Transport Protocol for Real-Time Applications," July 2003.

[44] D. Clark et al., "Towards the Future Internet Architecture," December 1991.

[45] T. Janevski, *Traffic Analysis and Design of Wireless IP Networks*, Norwood, MA: Artech House, May 2003.

[46] H. Schulze, K. Mochalski, "Internet Study 2007," www.ipoque.com/.

[47] H. Schulze, K. Mochalski, "Internet Study 2008/2009," www.ipoque.com/.

[48] Palo Alto Networks, "Application Usage & Threat Report," http://researchcenter.paloaltonetworks.com/app-usage-risk-report-visualization, June 2014.

[49] Internet Society, "The Internet and the Public Switched Telephone Network," 2012.

[50] ITU Recommendation G.114, "One-Way Transmission Time," 2003.

[51] ITU-T Recommendation Y.2001, "General Overview of NGN," December 2004.

[52] ITU-T Recommendation Y.3001, "Future Networks: Objectives and Design Goals," May 2011.

CHAPTER 2

Internet Protocols

2.1 Internet Protocol (IP)

The Internet network is a global network that was initially designed exclusively for data transmission. However, rapid development of broadband technologies for Internet access has paved the way for many new multimedia services and applications. Contrary to applications, the basic protocols on which the Internet is based have undergone only small changes in the past two or three decades. Although the Internet protocol stack generally consists of protocol layers from the link layer at the bottom up to the application layer on the top, the protocol that makes the Internet a single network is the Internet Protocol (IP), which provides interconnection capabilities between different IP networks.

The IP is a connectionless protocol, which means that there is no requirement for connection establishment before data transmission begins and that each datagram (i.e., IP packet) is independently transferred over the Internet. Also, IP does not guarantee that packets will arrive in the original sequence or will they arrive at their destination (that is left to upper protocols, such as transport protocols and/or applications).

Currently, there are two versions of the IP:

- IP version 4 (IPv4), which is standardized with RFC 791 [1], and
- IP version 6 (IPv6), initially standardized with the RFC 1883 specification in 1995 [2], and finalized in 1998 as RFC 2460 [3].

All functions of the Internet Protocol are defined with different fields in the IP header, which is added to the message unit received from the transport layer protocol (e.g., TCP, UDP) within the source host. On the receiving side, it is extracted by the IP software at the destination host before being sent to the transport protocol. The IP header and IP payload (information received from or sent to the transport protocol entity within a given host) form the IP datagram or, in other words, the IP packet (both terms are used interchangeably). The IP packet is the transfer unit that travels through the Internet, from the sender (source) to the recipient (destination). Each router uses the IP header information (together with additional information, such as operator's policies) to route packets.

Figure 2.1 presents the IPv4 packet header. It consists of the following fields:

- Version field (4 bits): This field indicates the type of IP header. Currently there are only two versions of IP headers, IPv4 and IPv6.
- Internet header length (4 bits): This defines the length of the IP header in number of words, where each word is equal to 32 bits (i.e., 4 bytes). The minimum size of an IPv4 packet header is 20 bytes (i.e., 5 words), while the maximum size is $2^4 - 1 = 15$ words (i.e., 60 bytes).
- Type of Service (ToS) (8 bits): This field provides an indication of the abstract parameters needed for the required QoS support for a given packet. (Each packet is served individually by each router, but packets belonging to the same communication session typically have the same QoS requirements.) In the IPv4 standard [1], bits 0–2 in the ToS field define the type of precedence (i.e., $2^3 = 8$ possible precedence values), bit 3 defines delay requirements ("0" for normal delay, and "1" for low delay), bit 4 defines the throughput requirements ("0" for normal throughput, "1" for high throughput), bit 5 defines the reliability for the IP packet ("0" for normal reliability, "1" for high reliability), and the two remaining bits are left undefined. However, each IP network in a managed system (e.g., telecom operator's network) typically replaces (or maps) the ToS field values for all ingress IP packets (i.e., IP packets entering the IP network) with ToS values that are valid in that IP network.
- Total Length (16 bits): This field defines the length of the IP packet in number of bytes (i.e., octets), including the IP header and packet payload. Hence, the maximum IP packet length is $2^{16} - 1 = 65,535$ bytes. However, in practice IP packets are smaller, typically having sizes up to 1,500 bytes (due to limitations of the MAC transport unit in Ethernet and Wi-Fi networks as the dominant local access network in the Internet).

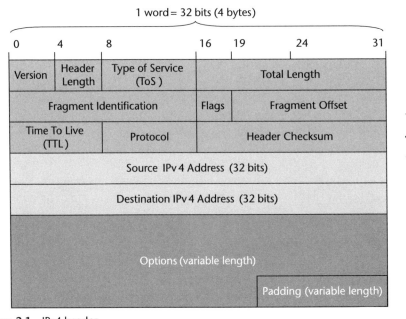

Figure 2.1 IPv4 header

- Identification (16 bits): This field enables fragmentation of larger IP datagrams into smaller ones (on the sender's side) and reassembly of fragments back to the original IP datagram (on the receiver's side) by providing a unique identification for the datagrams from a given sender to a given destination within the maximum datagram lifetime (MDL) for the Internet, which is typically interpreted to be 2 min [4]. However, if it is enforced all the time for a given communication (between a sender and a receiver), it limits the maximum bit rate to approximately 6.6 Mbps (e.g., for common Ethernet access it will be 1,500 bytes × 8 bits/byte × 65,536 datagrams/120 sec), which is not suitable when many individual connections may exceed that bit rate due to existing broadband access technologies on the Internet.
- Flags (3 bits): This field consists of 3 bits that are used to identify whether fragmentation is used and whether it is allowed to be used on the given IP packet. The first bit is reserved and must be 0; the second bit is set to 1 when the IP packet can be fragmented; and the third bit is set if the datagram is not the last fragment.
- Fragment Offset (13 bits): This field specifies the offset of a given fragment in units of 8 bytes (i.e., 64 bits) relative to the first byte of the unfragmented IP packet. The first fragment has an offset equal to zero (also, every unfragmented IP packet is also offset to zero). Because this field consists of 13 bits, $2^{13} = 8,192$ fragments are possible per datagram, but their total length must not exceed 65,535 bytes (which is the maximum size of IPv4 packets).
- Time-to-Live (TTL) (8 bits): This field defines the maximum time the IP packet is allowed to remain in the Internet. Initially it is counted in number of seconds, but each router must decrease its value by at least one (even if it processes the IP packet for less than 1 sec) and therefore the TTL value in practice is used as a maximum number of hops, where a hop is a logical link between two adjacent routers on the datagram's path, that a given datagram may travel before it is destroyed (which is done by the router that decreases its value to zero). It is intended to provide dropping of "lost" packets by the Internet.
- Protocol field (8 bits): This field specifies the protocol that should process the IP payload in the receiving host. Typically that is a transport layer protocol (TCP, UDP, and so on), but it can also be a protocol on the network layer (e.g., IP packet encapsulated within another IP packet, called IP tunneling) and even below the network layer (e.g., layer 2 tunneling). For example, the decimal value of this field is 6 for TCP, 17 for UDP, 4 for IP within IP encapsulation (i.e., tunneling), and so forth.
- Header Checksum (16 bits): This field provides error control of the IP header. Note, however, that this field must be calculated again in each router that decrements the TTL field because the TTL field is part of the IP header, and checksum calculation on a given set of bits (in this case, the IP header) needs to be performed again when any of the bits is changed.
- Source Address (32 bits): This field contains the source IP address of the given datagram. That is typically the IP address of the sender host, but in

certain situations the initial source IP address may be replaced with another IP address (e.g., network address translation [5]).
- Destination Address (32 bits): This field contains the IP address of the destination host (i.e., the IP address to which the datagram is addressed), which is typically specified by the sender host. However, it can also be replaced by an intermediate node with other destination address (e.g., network address translation).
- Options (0 to 40 bytes): This is an optional field. Its use is not mandatory, however, it must be implemented within the IP software in all hosts and nodes in the Internet. The Options field is variable in length. If the length specified in the length field in the IP header is bigger than 5 (words) then the Options field is included, otherwise it is not. However, the total length of the IP header must be an integer number of words (i.e., 32-bits units), so padding or dummy bytes may be used at the end of the Options field. The options may be used for improvement of the Internet protocol by providing additional services for control, debugging, and measurement.

Overall, IPv4 has existed almost without changes for more than three decades. However, certain fields in the IP header have shown to be less scalable than others. Such examples are the TTL value and header checksum combination, as well as the small IPv4 address space with only 32 bits. But at the beginning, no one really knew what the Internet would become in the 21st century.

2.1.1 Internet Control Message Protocol (ICMP)

Several other protocols are necessary for functioning of the IP and for routing and forwarding the datagrams end to end in the Internet. Such protocols are the Internet Control Message Protocol (ICMP) for error reporting and diagnosing on the Internet, as well as the Address Resolution Protocol (ARP), which provides mapping between the network layer identifier (i.e., IP addresses) and the physical addresses (the MAC addresses).

The base IPv4 protocol as a host-to-host network protocol has no error control and reporting messages. Also, IPv4 is not a reliable protocol, so problems can occur during the transport of the datagrams, such as errors in datagram processing in routers, discarding of datagrams due to a TTL value of zero, and so forth. So, occasionally a router or a host needs to communicate with the source host for error reporting. ICMP was created for such purposes [6]. ICMP messages are sent with IP packets in the same manner as transport layer protocol messages. However, ICMP is considered to be an integral part of the IP and therefore must be implemented in every IP software module, that is, in all routers and hosts on the Internet. However, a router cannot detect and report all problems that may appear during datagram processing. So, a certain datagram may not be delivered to its destination without any error reporting with ICMP back to the sender. Therefore, applications that require a high level of reliability (i.e., all sent packets must reach the destination) need a reliable transport protocol (e.g., TCP) or reliable mechanisms implemented on the application layer.

In general, ICMP has two classes of messages: error messages and query messages, the latter of which are based on a request/response principle [7]. Error messages are sent due to different errors in processing of IP packets at routers or end hosts. The query messages are used for different testing or diagnosing network activities (e.g., the echo request/reply is executed after using the "ping" command for testing whether a certain network interface is up and running). Each ICMP message starts with and 8-byte Type followed by an 8-byte Code field (for definition of different messages for given messages type). Base ICMP messages are listed in Table 2.1.

For example, a Destination Unreachable message is generated by a router when it is unable to locate the destination of a datagram. The same message is generated when a datagram with a set Do Not Fragment bit cannot be delivered to its destination because it has to be transferred over a network that requires fragmentation of the datagram.

A Time Exceeded message is sent to the source when the datagram is discarded because the TTL counter went to 0. When a given router often generates such messages, it can be an indication that certain routes that pass that router have a routing loop, or it may be due to traffic congestion in a given node of the network.

2.1.2 Address Resolution Protocol (ARP)

Each network used in the Internet has an underlying addressing scheme for network interfaces. Hence, each network interface has a so-called physical address, which is handled by the link layer protocol (i.e., OSI-2 protocol). In contrast, each interface on each node in the Internet (e.g., host, router) has an assigned IP address as network layer identification that is used for routing and forwarding of IP packets end to end. As the original standard for ARP states [8], the protocol is designed for converting IP addresses into local network addresses (e.g., 48-bit Ethernet addresses).

Each host that accesses the Internet must have one physical address and an IP address for each physical interface. The most deployed IP access networks nowadays are Ethernet and Wi-Fi. (There were more networks in the past such as ATM,

Table 2.1 ICMP Messages

ICMP Message Classification	ICMP Message Type	ICMP Message Description
ICMP error messages	3	Destination Unreachable
	4	Source Quench
	5	Redirect
	11	Time Exceeded
	12	Parameter Problem
ICMP query messages	0	Echo Reply
	8	Echo
	13	Timestamp
	14	Timestamp Reply
	15	Information Request
	16	Information Reply

X.25, Frame Relay, and so forth, but all of them disappeared over the time because, unlike Ethernet, they were not dedicated to the Internet protocol model; Ethernet was initially built for local access to the Internet.) So, the following examples are based on Ethernet access networks.

Each Ethernet is connected to the Internet via a router. When a datagram (addressed to a given host in the Ethernet network) arrives at the boundary router (through which the Ethernet network connects to the Internet), the router should use ARP to determine the appropriate Ethernet address (which corresponds to the datagram's destination IP address) to deliver the datagram framed in an Ethernet frame (i.e., layer 2 unit). So, the router should have cached mapping of the 2-tuple <IP address, Ethernet address> for each machine. Such a mapping table between the IP addresses and the physical (i.e., MAC) addresses of network interfaces in a given local network (e.g., Ethernet) is formed by use of the ARP protocol.

ARP runs encapsulated in the link layer protocol (e.g., Ethernet MAC frame), so it is typically referred to as being between OSI layers 2 and 3. In general, ARP solves the problem of finding the Ethernet address corresponding to a given IP address. But sometimes one needs to solve the opposite problem: Find the IP address corresponding to an Ethernet address. The problem of finding the IP address when the MAC address is known can be resolved with Reverse ARP (RARP) [9]. However, with the standardization of DHCP, which includes more functionality than RARP, RARP has become obsolete.

2.2 IPv4 Addressing

Every network interface on every host and router on the Internet has its own numerical identifier called an IP address. When a given host or router has several network interfaces, then it must have several IP addresses (i.e., each network interface must be assigned its own IP address).

The IP addressing architectures are directly connected with the Internet routing. In this classification of IP routing aspects, the term *domain* is used to identify networking resources (e.g., interconnected routers) under control of a single administration. Therefore, such domains are called administrative domains. Typically, *domain* and *routing domain* are used interchangeably. The domains that have shared resources with other domains are referred to as network service providers or Internet Service Providers (ISPs). However, the Internet is not a homogeneous network, nor does it need to be. In reality, it is a heterogeneous network consisting of many different access and transport networks based on different underlying transport technologies. So, Internet addressing is targeted to help the routing functionalities of the Internet protocols, which are implemented by the network resources (i.e., routers) and partially by the end-user hosts (e.g., computers, smartphones, different devices connected to the Internet).

Each IPv4 address contains two parts: one for identification of the IP network (network number or ID), and the other for identification of the host in an IP network (host number or ID). So, the IP address has a length of 32 bits, consisting of the network ID and the host ID. Therefore, all interfaces that have the same network ID belong to the same IP network. In general, allocation of IP addresses to host and routers is referred to as IP addressing. When IP addresses belong to IP

2.2 IPv4 Addressing

version 4, we refer to it as IPv4 addressing. IPv4 has two addressing architectures: (a) a classful IP addressing architecture and (2) a classless interdomain routing (CIDR) addressing architecture.

In both addressing architectures the IPv4 addresses are canonically represented in so-called decimal-dot notation, in which the 32-bit IP address is segmented into 4 octets and each octet is written separately in decimal number ranging from 0 to 255 (i.e., $2^8 - 1 = 255$). However, decimal-dot notation was created to be used by humans, whereas in Internet machines (e.g., hosts, routers) it is used in its original binary form.

Example of presentation of IPv4 address:
IPv4 address in decimal-dot notation: 192.168.1.0 (four decimal numbers with values between 0 and 255 separated by dots)
IPv4 address in binary notation: 11000000 10101000 00000001 00000000 (32 bits)

2.2.1 Classful IPv4 Addressing

The classful IP addressing architecture was the first global addressing architecture in the Internet since the standardization of IP in 1981 and was used until the introduction of classless IP addressing in 1993. Classful IP addressing was introduced in IPv4 standard [1] and contains the following defined classes (given in Figure 2.2):

- Class A: The highest-order bit is 0, the following 7 bits define the network (overall, a network is identified with 8 bits), and the remaining 24 bits are the host ID.
- Class B: The two highest-order bits are "01," followed by 14 bits for network identification (overall, a network ID consists of the first 16 bits of the IP address) and 16 bits for the host ID.
- Class C: The three highest-order bits are "001" followed by 21 bits for network identification and 8 bits for the host ID.

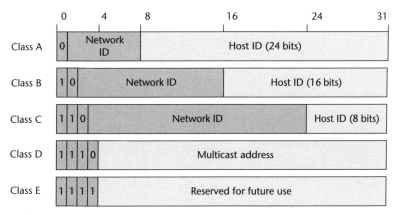

Figure 2.2 Classful IPv4 addresses.

- Class D: This is a multicast type of IP address, in which a single packet is replicated many times and different copies of it travel to different destination hosts that belong to the same multicast group of hosts. These IP addresses always start with "1110" as the highest-order bits. There are no predefined sizes for the network ID and host ID.
- Class E: This class is reserved for future use, and IP addresses start with highest-order bits "1111."

Overall, Class A allows definition of 126 different networks with 16 million hosts each; Class B defines 16,382 networks with 65,534 hosts per network; and Class C includes 2 million networks with 254 hosts per network. Classes A, B and C are dedicated to unicast communications (i.e., communications between a single sender and a single receiver in a given communication direction), whereas class D is dedicated to multicast communications (i.e., one sender sends IP packets to multiple hosts). Besides the single unicast IP address for a given network interface, a host can also obtain one or more multicast addresses of Class D.

2.2.2 Classless IPv4 Addressing

The CIDR addressing architecture was introduced to the Internet in 1993 [10]. Its goal is to provide higher efficiency in the allocation of IP addresses because three classes, A, B and C, do not always fit all different IP networks with a different number of interfaces. It provides the possibility of fine granularity in the IP address allocation process (Figure 2.3). The CIDR also includes the network ID and host ID in the IP address, but contrary to classful addressing it allows allocation of the leftmost continuous significant bits of the IP address as a routing prefix (also referred to as a network prefix or network ID), while the remaining bits identify the host within that IP network (i.e., the host ID). For presentation of the network ID (i.e., routing prefix) CIDR uses an <IP address/IP mask>-tuple, so a logical "and" operation between binary values of the IP address and the IP mask gives the sequence of the leftmost significant bits that form the routing prefix.

So, the mask defines the network part of the IP address. There are two approaches for IP network notation based on CIDR, which is illustrated by the following examples that refer to the same network:

1. *192.168.1.0/255.255.255.0:* In this case the IP mask is given in decimal-dot format, the same as the IP address.
2. *192.168.1.0/24:* In this case the mask is a number that defines the number of leftmost continuous block of "1" bits in the 32-bit mask, so in the given example the first 24 bits of the IP mask are set to "1," while the remaining 8 bits are set to "0."

The CIDR was standardized to solve several problems that were appearing in the 1990s. One of the problems was the exhaustion of the Class B IP addresses, since Class C networks (with up to 254 hosts) were too small for some midsize organizations and the Class B networks (with up to 65,534 hosts) were too big. Another problem was the growth in routing tables in Internet routers, which could not be managed by software, hardware, and people at that time. Finally the third

2.2 IPv4 Addressing

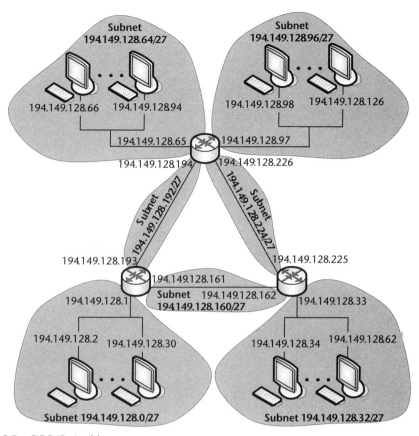

Figure 2.3 CIDR IPv4 addresses.

problem was eventual exhaustion of IPv4 address space with the growth of the Internet. Given the situation, the creation of CIDR provided a short- and mid-term solution to the exhaustion of the Class B address problem and the growth in routing tables. However, the exhaustion of the IPv4 address space (which was slowed down with the CIDR invention) required a long-term solution, which resulted in creation of the next version of the Internet Protocol, that is, Internet Protocol version 6 (IPv6).

2.2.3 Public and Private IP Addresses

Another short- and mid-term solution to prevent exhaustion of the IPv4 address space was the definition of so-called private IP addresses. Such private addresses can be reused in different IP networks (typically access networks to the Internet), but they must be mapped into the public IP addresses if there is a need to communicate over the global Internet. The private IPv4 addresses are defined by the Internet Assigned Numbers Authority (IANA) [11], and they consist of three continuous blocks as shown in Table 2.2.

However, because private IP addresses are reusable, they are not globally unique and hence cannot be used as source and destination addresses in IP packets sent over the public Internet. For routing purposes IP addresses must be public, where public IP addresses are globally unique (i.e., the same unicast IP address

Table 2.2 Private IPv4 Addresses

Private IP Address Block	Prefix	Allocation Size
10.0.0.0–10.255.255.255	10.0.0.0/8	1 Class A network
172.16.0.0–172.31.255.255	172.16.0.0./12	16 Class B networks
192.168.0.0–192.168.255.255	192.168.0.0/16	256 Class C networks

cannot be allocated to two different hosts on the Internet at the same time). In that manner, private IP addresses have no global meaning, so routing information about private networks (IP networks that use private IP addresses) should not propagate on internetwork links, and IP packets carrying private source or destination addresses in their headers should be rejected by the routers in the public Internet. Private networks are used in enterprises, in offices and homes, and so forth. However, for hosts attached to such private networks (and thus having allocated private IP addresses to their network interfaces) to be able to communicate with other hosts anywhere on the global Internet, their packets in both directions (upstream and downstream) must go through a network address translator (NAT) gateway [11], which maps private IP addresses to defined public IP addresses (for the upstream, i.e., outgoing traffic from the private network toward the Internet), and vice versa (in the downstream direction, toward the hosts). Private IP addresses were also used in certain cases in enterprises that required additional security. For example, consider their use in application layer gateways to connect an internal enterprise network to the Internet and control the incoming and outgoing traffic by means of firewalls.

Besides the private IP address blocks (see Table 2.2), there are also defined (by IANA) so-called special-purpose IP address allocations [12], which are given in Table 2.3. For example, link local IP addresses (e.g., 169.254.0.0/16 and 172.16.0.0/12 blocks) are used for self-configuration of IPv4 network interfaces when there is no allocation of an IP address to it (either public or private). Another special-purpose IP address is the Loopback address (127.0.0.0/8), which is an address allocated to the virtual network interface included in the Internet protocol suite through which network applications can communicate when they are running on the same machine. In such cases, all traffic that a given application sends to the

Table 2.3 Special-Purpose IPv4 Address Allocations

IPv4 Address Block	Name	Allocation Date
0.0.0.0/8	"This host on this network"	September 1981 (RFC 791)
100.64.0.0/10	Shared address space	April 2012 (RFC 6598)
127.0.0.0/8	Loopback	September 1981 (RFC 791)
169.254.0.0/16	Link local	May 2005 (RFC 3927)
172.16.0.0/12	Link local	February 1996 (RFC 1918)
192.0.0.0/24	IETF protocol assignments	January 2010 (RFC 6890)
192.0.0.0/29	Dual-stack lite broadband deployments	June 2011
192.0.2.0/24	Documentation	January 2010 (RFC 5737)
192.88.99.0/24	6to4 relay anycast	June 2001 (RFC 3068)
240.0.0.0/4	Reserved	August 1989 (RFC 1112)
255.255.255.255/32	Limited broadcast	October 1984 (RFC 919)

Loopback IP address is processed by the Internet protocol stack as if it has been received from another machine (e.g., host) via a physical network interface.

Recently, the IPv4 address block 100.64.0.0/10 was introduced as shared address space that is intended for use by network providers, which is in that manner different than private address blocks targeted for use by individual end-user hosts. Shared address space is targeted to be used by carrier-grade NAT (CGN) devices, and it is defined mainly to be used to number the interfaces that connect the CGN devices (e.g., routers) to the customer-premises equipment in offices and homes [13]. Like shared address space, the three blocks of private IP addresses (given earlier in Table 2.2) are also considered to be special-purpose IP addresses [12].

2.2.4 Network Address Translation (NAT)

NAT represents mapping among public and private IP addresses that can be performed in a given router placed between the private IP network on one side and the public Internet on the other. NAT is used for two main reasons. The first one is the limited number of public IPv4 addresses. The second one is the security, when one does not want a machine to be directly accessible to the Internet via a public IP address (e.g., billing servers or storage servers).

A public IP address must be globally unique; that is, two different network interfaces cannot be assigned the same public IP address, with the except of multicast IP addresses in which a multicast address is assigned to all hosts that belong to a given multicast group. Only public IP addresses are globally routable over the Internet. Unlike public addresses, private IP addresses can be reused as many times as someone needs (of course, the same private addresses can be reused in different private networks) in different so-called private networks. Mapping between public and private IP addresses must be processed by the router that interconnects the private network and the public Internet. Such mapping is not always 1:1 (i.e., one private IP address to one public IP address). Typically, multiple private IP addresses are mapped into a single public IP address (e.g., for hosts at home that use broadband access to the Internet for residential users).

The procedures for mapping private addresses to public addresses (and vice versa), which is done in the router or a computer with multiple network interfaces (i.e., network cards), are called network address translation. One NAT example is illustrated in Figure 2.4. In the given example, three hosts with private IP addresses (192.168.1.1, 192.168.1.2, and 192.168.1.3) are mapped to one public IP address (the interface of the router) with IP address 194.149.144.1. The router that performs NAT has a so-called NAT table where all mappings between the private addresses on the private IP network and the public address assigned to the router interface through which it is connected to the Internet are stored. Differentiation of packets sent to/from various hosts on the local network (with private IP addresses) is done using different logical numbers, called ports, that are associated with the mapping entries in the NAT table. Therefore, this procedure of mapping is also called network address and port translation (NAPT), which is used interchangeably with NAT.

Overall, the history in the past two decades has shown that NAT (together with allocation of private IP addresses, CIDR addressing, and DHCP standardization)

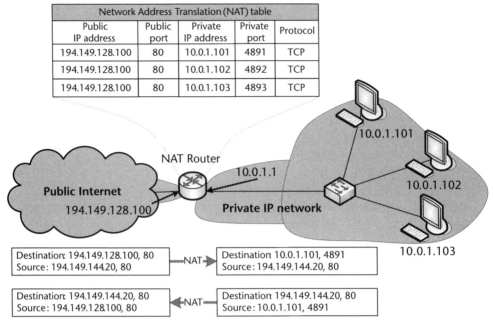

Figure 2.4 NAT example.

has prolonged the time required to reach IPv4 address space exhaustion and hence contributed to the growth of the Internet in the 1990s and 2000s.

2.3 IPv6 Fundamentals

IPv6 was developed as a long-term solution for the IPv4 address exhaustion problem that appeared in the 1990s with the global spread of the Internet. The current IPv6 was standardized with RFC 2460 in 1998 [3]. However, short- and mid-term solutions for prevention of IPv4 exhaustion (e.g., CIDR addressing, special-purpose IPv4 addresses, dynamic allocation of IP addresses, NAT) have prolonged the life of IPv4 until the second decade of the 21st century. Due to the longer lifetime of IPv4, in 2014 the worldwide share of IPv6 was less than 5% [14]. Note, however, that this is an increase of more than 10 times since 2010 when the IPv6 share was below 0.5%. We should expect this percentage to continue to increase significantly in the following years, and after a certain time (e.g., in the 2020s) IPv6 will become the dominant addressing architecture used by the Internet and, hence, in the ICT world.

The IPv6 header is shown in Figure 2.5. Its minimum size is 40 bytes, from which 32 bytes are dedicated to source and destination IPv6 addresses. In particular, IPv6 header includes the following fields [3]:

- Version (4 bits): This field contains the version number (the same as in IPv4 header), and it has value of 6 for IPv6.
- Traffic Class (8 bits): This field has the same size and similar function as the ToS field in IPv4 headers. It is also used to store so-called differentiated services code point (DSCP), which is used for QoS class differentiation.

2.3 IPv6 Fundamentals

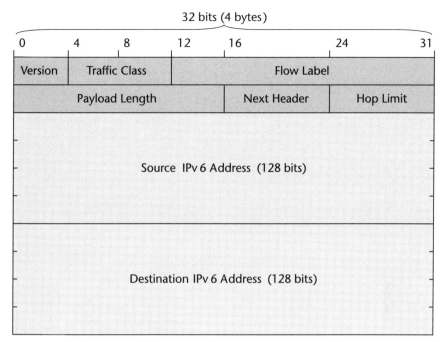

Figure 2.5 IPv6 header.

- Flow Label (20 bits): This field contains the flow label, which is an additional feature of IPv6 (if compared to IPv4). Its goal is to provide different treatment of certain IPv6 packets that belong to a certain flow with specific QoS or other requirements (e.g., security). If all bits in this field are set to zero, the packet is considered unlabeled.
- Payload Length (16 bits): This field specifies the size of the rest of the IPv6 packet after the header, in number of octets (i.e., bytes). If additional headers are included after the given headers by using the Next Header field, they are considered to be part of the packet payload and hence are counted in the packet length.
- Next Header (8 bits): This field selects the type of header that will follow the given IPv6 header, and has a similar function to that of the Protocol field in IPv4 headers [15]. Typically this field specifies the transport layer protocol above the IP (e.g., TCP, UDP), but also it can be used to provide one or more so-called extension headers (e.g., for security support, authentication, mobility support, routing), which are placed between the leftmost IPv6 header and the transport protocol header within its payload.
- Hop Limit (8 bits): This field has the same function as the TTL field in IPv4 headers. It is decremented by one in each network layer node that forwards the IPv6 packet, and finally it is discarded when its value reaches zero, thus preventing "lost" packets from traveling infinitely in the Internet.
- Source Address (128 bits): This field contains the IPv6 address of the sender of the packet (i.e., the packet originator).

- Destination Address (128 bits): This field contains the IPv6 address of the intended recipient of the packet. However, it is not always the ultimate recipient of the packet (e.g., when a Routing Header is present).

To summarize, according to the current IPv6 standard, the main changes from IPv4 to IPv6 were targeted to the following:

- More addressing space is achieved as a result of increasing the 32-bit IPv4 addresses to 128-bits IPv6 addresses.
- The header in IPv6 is simplified by dropping the Header Checksum that existed in IPv4 and resulted in necessary recalculation of the field in each IP packet in each router due to the TTL value decrease. Also, fragmentation and reassembly fields (mandatory in IPv4) are made optional in IPv6. Note, however, that the size of the IPv6 header is larger than the IPv4 header due to larger Source and Destination address fields.
- Novelties in the IPv6 header, which include flow labeling, as well as extension headers (used with the Next Header field) provide for the possibility of transporting additional information (e.g., for authentication, security, QoS, routing) with the IPv6 packets on their way through the Internet.

2.3.1 Flow Labeling in IPv6

Flow labeling capability was introduced in IPv6 with the goal of providing a mechanism for labeling the packets belonging to a certain flow, which was initially considered to be used by senders to request different treatment of certain packets (e.g., QoS support for real-time flows in network routers). However, the development of approaches for QoS in IPv4 networks will probably influence the approaches for the QoS in IPv6-based networks, thus limiting the use of this field (i.e., a gateway router is likely to replace its values with a certain default value according to the policies in the given administrative domain).

By definition, from a network layer point of view [16], a flow is a sequence of IP packets sent from a particular source host to a particular unicast, anycast (as a new type of connection introduced with IPv6), or multicast destination. From the upper-layer protocol viewpoint (e.g., applications), a flow can be considered as all packets in one direction belonging to a certain transport connection (e.g., a TCP connection) or media stream (e.g., voice, video, or multimedia stream). However, the flow is not always mapped 1:1 to the transport connection. For example, several flows can use a single transport connection, and several transport connections can be aggregated in a single flow.

Generally, the sender node selects to label or not label a given flow, while the network nodes may process labeled packets according to their labels or may ignore or rewrite the Flow Label field in IPv6 packets (e.g., in gateway routers at the edge of a given administrative domain). However, the flow label is not protected in any way (even in cases when network layer security solutions are applied, such as IPsec), so any en route change of flow label (e.g., by an attacker) is undetectable.

Even before IPv6 and the introduction of its flow label, a flow between a source and a destination could be distinguished. In such a case (without a flow label), a flow is identified with a so-called 5-tuple, which consists of a source IP address, destination IP address, source port (the port is the identifier in the transport layer protocols, above the IP), destination port, and transport protocol type (e.g., TCP, UDP). Such flow identification is used in IPv4 networks and also can be used in IPv6 networks. However, with flow labeling IPv6 provides the possibility of flow identification by using 3-tuple consisting of a flow label, source IPv6 address, and destination IPv6 address. The 3-tuple provides efficient flow classification in network nodes (e.g., routers) because only IP headers needs to be processed in such a case (since all three elements in the 3-tuple are part of the IP header as the network layer protocol header), while in the case of the 5-tuple there is the need to process the transport layer protocol header to obtain the source port and destination port.

Generally, the flow label can be used in two scenarios:

- Stateful scenario: In this scenario the network nodes (which process the flow labels) need to store the information about the flow, which includes the flow label values. The stateful mechanism may require a certain signaling mechanism [e.g., the standardized Resource reSerVation Protocol (RSVP) [17], or experimental General Internet Signaling Transport (GIST) [18]] to inform other nodes in the downstream direction regarding the way in which the given flow label is used.

- Stateless scenario: In this scenario any node (e.g., a router) that processes flow labels for IPv6 packets does not need to store flow information before or after the packet. In this case the flow label value can be generated randomly. However, if two different flows are assigned the same flow label and have same source and destination addresses, that is, the same 3-tuple, they will receive the same flow treatment through the network, which will not significantly influence the load distribution due to the low probability of such event.

The IPv6 node assigning flow labels must keep track of all triplets (consisting of a flow label, source address, and destination address) in use, with the goal of preventing mixing of different flows. However, a certain programming interface is required for assigning and managing flow labels. Overall, when a flow label is set once to a nonzero value, it must be delivered unchanged to the final destination [16]. However, an exception to this rule (for unchanged delivery of the flow label) has been specified that allows firewalls to rewrite the flow label field for security reasons, such as suspected use of the flow label field as a covert data channel.

Overall, a flow label is useless if it is not actually used. For example, stateful flow labeling is more appropriate for management of aggregated flows in the transport networks in a given administrative domain. In such a case, the flow state is established as a subset for all network nodes (i.e., routers) on the path of the flow. Then, it can be used for flow-specific treatment, which can be signaled or configured (manually, or automatically with defined algorithms).

2.3.2 Extension Headers

Extension headers in IPv6 provide the possibility of additional IPv6 headers carrying optional information. These headers are placed between the IPv6 header and transport layer header in the form of a chain in which each header points to the type of the following header by value in the Next Header field. For example, if the Next Header field contains the value "0" (as a decimal number), then it specifies the next header of type "Hop-by-Hop Option"; if the value is "41," then the next header is also an IPv6 header (used as an extension header in such case); value "43" denotes "IPv6 Routing Header" as the next header, and so forth. The extension headers on the network layer end when the Next Header field contains the value "59" which means "no next header" (i.e., that extension header is followed by the transport protocol header). An example of a chain of extension headers in an IPv6 packet is shown in Figure 2.6.

The base IPv6 standard (RFC 2460 [3]) specifies several new types of extension headers, given as follows:

- Hop-by-Hop Options header: This header is used to carry optional information available to every node (e.g., router) through the packet's path.
- Routing header: This header is used to list one or more network nodes (i.e., routers) through with the packet should pass on its path toward the destination.
- Fragment header: This header is intended to be used by an IPv6 source to fragment an IPv6 packet, which is larger than the MTU in the path toward the destination. In this way the fragmentation and reassembly mechanism that existed in IPv4 via several fields is made optional in IPv6.
- Destination Options header: This header is used to carry optional information targeted for use by the destination host for the given IPv6 packet.
- Authentication header: This header is intended to provide for the integrity of data carried by the IP packet, including certain fields of the header. (Some fields may change on the way such as TTL and Checksum in IPv4 and Hop Limit in IPv6, so their values cannot be protected.) Typically it is used in security architectures in both IPv4 and IPv6 networks [19].
- Encapsulating Security Payload (ESP) header: This header is used for protection of the information in a given IP packet. In fact, it provides encryption of all the information (including the next extension headers, if any) that immediately follows the ESP header, so such information becomes inaccessible for the network nodes on the path toward the destination [20]. The ESP can be provided alone, or with an IP Authentication Header, or in a standardized security architecture for the Internet Protocol [21].

Additionally, several extension headers are dedicated to mobility and multihoming in all-IP heterogeneous network environments (e.g., single host with several different network interfaces connected to different IP-based networks, such as Ethernet, Wi-Fi, or IP-based mobile networks), as shown in Table 2.4.

All extension headers follow the same generic format (as given earlier in Figure 2.6) [22]. It starts with a Next Header value (i.e., an 8-bit selector), followed by the

2.3 IPv6 Fundamentals

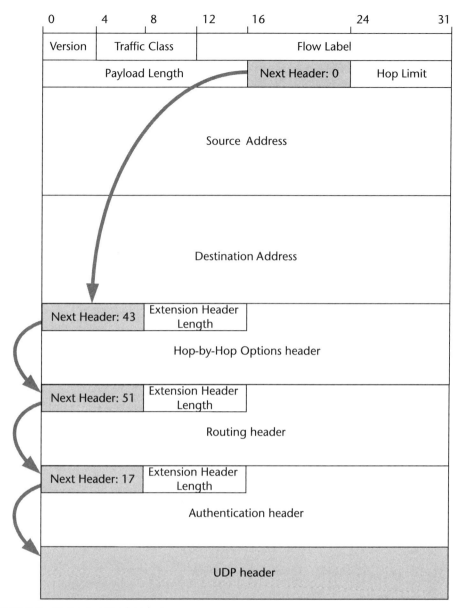

Figure 2.6 IPv6 extension headers.

8-bit Header Extension Length (in number of bytes, not including the first 8 bytes, which contain the Next Header value) and Header Specific Data, which is specific to the type of extension header and hence it can have a variable length. However, the total length of the extension header must be a multiple of 32 bits.

Overall, the IPv6 extension headers bring important advantages to IPv6 over IPv4 by providing for the possibility of single IP packet carrying optional information (e.g., for routing, security, mobility, etc.) that can be processed by the network layer protocol software in all network nodes and hosts. According to IANA [23, 24], all IPv6 extension headers are listed in Table 2.4.

Table 2.4 IPv6 Extension Headers

Protocol Decimal Number	Protocol Name
0	IPv6 Hop-by-Hop Option
43	Routing Header for IPv6
44	Fragment Header for IPv6
50	Encapsulating Security Payload
51	Authentication Header
60	Destination Options for IPv6
135	Mobility Header
139	Experimental use for Host Identity Protocol (HIP)
140	Shim6 Protocol (Level 3 Multihoming Shim Protocol for IPv6)
253	Use for experimentation and testing
254	Use for experimentation and testing

2.3.3 ICMPv6

The Internet Control Message Protocol version 6 (ICMPv6) is an integral part of IPv6 [25], in the same manner as ICMPv4 is an integral part of IPv4. As with ICMP for IPv4, ICMPv6 must be fully implemented by every IPv6 module in the Internet.

In general, ICMPv6 is the ICMP for IPv4 with some changes. The common characteristics of both versions of ICMP are error reporting and diagnosing functions. However, differences do exist between ICMPv6 and ICMPv4. For example, the ICMPv6 header in an IPv6 packet is always preceded by the IPv6 header followed by zero or more extension headers. The ICMPv6 header is identified by a Next Header value of 58, which is different from ICMPv4, which has assigned the protocol number 1 [23]. Also, some protocols that were independent in IPv4, such as ARP and RARP, are integral to ICMPv6 (so, there is no ARP or RARP version 6). For that purpose several new types of messages have been defined in ICMPv6. ICMPv6 messages fall into two classes:

- ICMPv6 error messages: These messages are the same as in ICMPv4 (see Table 2.1), without the Source Quench message (in ICMPv4) and with one new message, the Packet Too Big Message, which is sent by a router in the Internet in response to an IP packet that it cannot forward due to it being a larger packet size than the MTU of the outgoing link.
- ICMPv6 informational messages: Two messages that belong to this class: Echo Request and Echo Reply. They are designed to check whether a given host (i.e., its interface) is connected and accessible on the Internet. Several debugging tools (e.g., ping and traceroute) are based on these messages.

Besides the two main classes of ICMPv6 messages (which mainly correspond to ICMPv4 messages), several other mechanisms are implemented via ICMPv6 in IPv6 (Table 2.5), including the following:

2.3 IPv6 Fundamentals

Table 2.5 Defined ICMPv6 Messages

ICMPv6 Message Classification	ICMP Message Type	ICMP Message Description
Error messages [25]	1	Destination Unreachable
	2	Packet Too Big
	3	Time Exceeded
	4	Parameter Problem
Informational messages [25]	128	Echo Request
	129	Echo Reply
Neighbor discovery messages	133	Router Solicitation [26]
	134	Router Advertisement [26]
	135	Neighbor Solicitation [26]
	136	Neighbor Advertisement [26]
	137	Redirect Message [26]
	141	Inverse Neighbor Discovery Solicitation Message [32]
	142	Inverse Neighbor Discovery Advertisement Message [32]
	148	Certification Path Solicitation Message [33]
	149	Certification Path Advertisement Message [33]
Multicast messages	130	Multicast Listener Query [27]
	131	Multicast Listener Report [27]
	132	Multicast Listener Done [27]
	143	Version 2 Multicast Listener Report (RFC 3810)
	151	Multicast Router Advertisement [28]
	152	Multicast Router Solicitation [28]
	153	Multicast Router Termination [28]
Mobile IPv6 messages	144	Home Agent Address Discovery Request Message [29]
	145	Home Agent Address Discovery Reply Message [29]
	146	Mobile Prefix Solicitation [29]
	147	Mobile Prefix Advertisement [29]
	154	FMIPv6 Messages (Mobile IPv6 Fast Handovers, RFC 5568)

- Neighbor discovery messages: There are five different ICMP packet types: a pair of Router Solicitation and Router Advertisement messages, a pair of Neighbor Solicitation and Neighbor Advertisements messages, and a Redirect message [26].

- Multicast listener and router discovery messages: There are three Multicast Listener Discovery (MLD) messages for IPv6: Query, Report, and Done [27]. The MLD protocol for IPv6 is derived from the Internet Group Management Protocol (IGMP) functionalities for IPv4, but contrary to it the MLD is implemented via ICMPv6 messages. Also, there are three multicast router discovery messages—Advertisement, Solicitation, and Termination [28]—that provide for the possibility of identifying the locations of multicast routers.

- Mobility support messages: ICMPv6 includes several messages, which are targeted to mobility support in all IPv6 networks by using Mobile IPv6 [29]. The messages are a pair of Home Agent Discovery messages (Request and

Response), a Mobile Prefix Solicitation, and a Mobile Prefix Advertisement messages.

Besides the types of messages discussed above, several other ICMPv6 messages have been standardized by IETF and assigned by the IANA [30]. The full list of ICMPv6 messages is given in Table 2.5. So, IPv6, together with its accompanying protocol ICMPv6, provides more compact networking layers than IPv4.

2.4 IPv6 Addressing

The IPv6 addresses are 128-bit identifiers for network interfaces of all hosts and network nodes attached to the Internet. IPv6 defines three types of addresses [31]:

- Unicast address: This type of address identifies a single network interface on a host or router. It is used for unicast Internet communications, in which a single machine (host or router) sends a packet to another single machine attached to the Internet.
- Anycast address: This types of address is an identifier that is assigned to a set of network interfaces that typically belong to different machines. It is called anycast because a packet sent to an anycast address is delivered to any of the interfaces in a given set, typically to the nearest one (according to a measure of distance provided by the routing protocols).
- Multicast address: This is also an IPv6 address assigned to multiple network interfaces. So, a packet that is sent to a multicast address is delivered to all interfaces identified by that multicast address.

Unlike IPv4, the IPv6 addressing architecture does not have broadcast addresses, because their function is replaced by the multicast IPv6 addresses. However, in IPv6 addressing also refers to individual interfaces, not to hosts or nodes. So, a host or a node (e.g., a router) that has multiple network interfaces will have multiple IPv6 addresses. Any interface that is designed to provide Internet communication must have at least one unicast address (link local). In contrast, a single interface may have multiple IPv6 addresses, including unicast, anycast, and multicast addresses. The IPv6 addressing architecture regarding subnetting continues the IPv4 approach, so each link has an associated subnet prefix. However, multiple subnet prefixes may be assigned to the same link.

2.4.1 Representation of IPv6 Addresses

Whereas IPv4 addresses are typically represented in decimal-dot format, IPv6 addresses are usually represented in hexadecimal format (due to their length of 128 bits). However, there are certain accepted conventional forms for representing IPv6 addresses [31].

The typical preferred form of IPv6 representation is "$x1{:}x2{:}x3{:}x4{:}x5{:}x6{:}x7{:}x8$" where the x's have one to four hexadecimal digits (each hexadecimal digit represent

2.4 IPv6 Addressing

4 bits of the IPv6 address), so each *x* field is representing 16 bits of the IPv6 address (8 fields × 16 bits = 128 bits).

Example of IPv6 addresses:
ABCD:EF98:7654:3210:0123:4567:89AB:CDEF

Another rule when writing IPv6 addresses is not to write leading zeros in each of the fields (if there are any). Due to the longer length of IPv6 addresses, they may contain long strings of zero bits. With the goal of making easier the writing of zeros, the use of "::" is allowed to replace one or more groups of 16 bits (i.e., one or more neighboring fields in IPv6 address with all bits equal to zero). However, such compression of zeros in an IPv6 address can appear only once.

Example of IPv6 addresses with compression of zeros:
2000:0000:0000:0000:0000:000A:0BCD:1234 = 2000::A:BCD:1234
2000:0000:0000:0000:000A:0000:0BCD:1234 = 2000::A:0:BCD:1234

2.4.2 Allocation of IPv6 Address Space

The IPv6 address types are given in Table 2.6. Although global unicast IPv6 addresses include the whole address space, excluding multicast IPv6 addresses (i.e., FF00::/8) [31], currently IANA has assigned for global use only unicast range 2000::/3, which has been allocated to the five regional Internet registries (RIRs).

There are equivalent IPv4 addresses for most of the IPv6 addresses. For example, the loopback address is "::1/128" in IPv6, whereas it is "127.0.0.1" in IPv4 (see Table 2.6).

Additionally, similar to private address blocks in IPv6, there is a defined link local unicast IPv6 address block (i.e.., " FC00::/7") that is reserved for local use in home or enterprises and hence it is not part of the public IPv6 unicast address space [34]. These addresses might not be globally unique, so packets with these addresses in the source or destination fields are not intended to be routed over the public

Table 2.6 IPv6 Address Types

Address Type	Binary Prefix	IPv6 Representation	IPv4 Equivalent
Unspecified	00...0 (128 zeros)	::/128	0.0.0.0
Loopback	00...01 (128 bits)	::1/128	127.0.0.1
Multicast	11111111 (8 ones)	FF00::/8	224.0.0.0/4
Link-local unicast	1111111010 (10 bits)	FE80::/10	169.254.0.0/16
Unique-local unicast	1111110 (7 bits)	FC00::/7	10.0.0.0/8 172.16.0.0/12 192.168.0.0/16
Global unicast (includes all anycast)	001 (3 bits) is currently allocated by IANA	2000::/3 is currently allocated by IANA	Global unicast IPv4 addresses

Internet. However, they can be routed within "private" IPv6 networks in home or enterprise environments.

The unicast addresses are the most used in the public Internet. IPv6 currently has three types of unicast addresses: Global unicast, Link-local unicast, and Unique-local unicast. However, only the first two types are globally routable in IPv6 networks. Unicast addresses can have prefixes of arbitrary bit length similar to CIDR for IPv4 addresses.

The general format of IPv6 global unicast addresses is shown in Figure 2.7. It consists of three parts: a global routing prefix (with a length of n bits), subnet ID (m bits), and interface ID (128-n-m bits). However, except for IPv6 addresses that start with three zeros in binary notation (i.e., "000"), all Global unicast addresses have 64 bits dedicated to the interface ID. So, the global routing prefix and subnet ID also have a total length of 64 bits: $m + n = 64$. In contrast, IPv6 address types that start with three zeros (in binary form) are the Unspecified, Loopback, as well as IPv4-compatible and IPv4-mapped IPv6 addresses (the last two types are targeted for use in the IPv4-to-IPv6 transition, which is ongoing in the second decade of the 21st century). The IPv4-compatible IPv6 address has 96 zeros in binary form followed by an IPv4 address, while IPv4-mapped IPv6 addresses have the prefix "::FFFF/96" (i.e., binary 80 zeros followed by 16 ones) and the last 32 bits are used to embed IPv4 addresses in an IPv6 address (e.g., for use with a dual IPv4/IPv6 protocol stack in hosts [35]).

2.4.3 IPv6 Address Autoconfiguration

Internet hosts should be "plug-and-play" regarding their connection to the Internet. However, each network interface needs to have an IPv6 address that can communicate with other hosts on the Internet. IPv6 uses two main approaches for automatic address configuration:

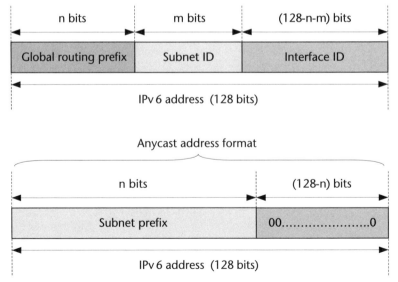

Figure 2.7 General format of IPv6 addresses.

- IPv6 stateless autoconfiguration: This is a process in which the router and the host create an IPv6 address and a default route.
- IPv6 stateful autoconfiguration: This process is based on use of DHCP version 6 (i.e., DHCPv6) similar to DHCP use in IPv4.

In stateless autoconfiguration a node automatically creates a Link-local IPv6 address (after booting) on each IPv6-capable network interface [36]. Such an address belongs to the Local-link unicast address block, which is FE80::/10 (see Table 2.6). The lower 64 bits of autoconfigured IPv6 addresses are constructed in so-called Modified EUI-64 format, in which the address is typically derived from the interface's physical address (i.e., the MAC address). Since the globally accepted standard for local access to the Internet is Ethernet (for wired access) and Wi-Fi (for wireless access), the MAC address used for creation of an IPv6 address has 48 bits (i.e., the Ethernet/Wi-Fi MAC address length).

Hence, let's look at an example for IPv6 address autoconfiguration based on an Ethernet MAC address. In such case stateless autoconfiguration uses the IEEE-defined 64-bit Extended Unique Identifier (EUI-64). The EUI-64 part of the address gives the interface ID of the unicast IPv6 address and has a length of 64 bits. The EUI-64 field is formed in the following manner: The first 3 bytes of the Ethernet MAC address become the first 3 bytes of the EUI-64; the next 2 bytes (4th and 5th of the EUI-64) are set to the hexadecimal value "FFFE"; and the last 3 bytes (of the EUI-64) are the last 3 bytes of the Ethernet address. The interface ID (the last 64 bits of the IPv6 address) is then formed from the EUI-64 by complementing its 7th bit in the first byte, which is called the universal/local (U/L) bit [37]. The U/L bit is used to distinguish between a universally administered IEEE 802 address or an EUI-64 (which always has "0" in the U/L bit position) and globally unique IPv6 Interface ID which is signified by value of "1" in the corresponding U/L position. For example, if Ethernet address in hexadecimal form is "AB-CD-EF-12-34-56," then the EUI-64 derived from that physical address is "AB-CD-EF-FF-FE-12-34-56". Each hexadecimal digit represents 4 bits in binary representation, hence the Interface ID derived from the this EUI-64 will be "A9-CD-EF-FF-FE-12-34-56" because "AB" in EUI-64 in binary representation is "1010 10_1_1" (the 7th bit is underlined) and after inversion of the 7th bit ("1" is inverted to "0"), we obtain "1010 1001," which is "A9" in hexadecimal representation. So, the IPv6 Link-local address for the Ethernet interface in this example will be "FE80::A9CD:EFFF:FE12:3456."

Stateless autoconfiguration uses Neighbor Discovery ICMPv6 messages (as given earlier in Table 2.5) [26], and it is illustrated in Figure 2.8. The stateful autoconfiguration is realized with DHCPv6.

2.4.4 Transition from IPv4 to IPv6

The transition from the IPv4 to the IPv6 world will be on a similar scale as the transition from analog to digital telephony in the past, or the transition from circuit-switching digital networks to packet-switching IP-based networks in the past decade. So, the long-term solution for Internet addressing is IPv6 due to the limited addressing space problem inherent to IPv4. The transition has already started and

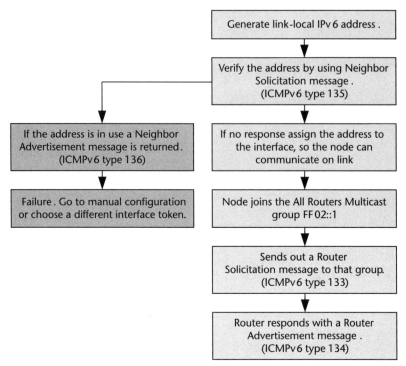

Figure 2.8 IPv6 stateless autoconfiguration.

has experienced exponential growth [14]. However, standardized approaches are needed for that process.

Generally, several transition strategies are possible:

- Edge-to-core: This strategy is convenient when services are important and IPv4 addresses are scarce. This strategy is user (i.e., customer) driven.
- Core-to-edge: This is a good strategy for operators (i.e., telecom operators, ISPs).
- By routing protocol area: This strategy can be used in scenarios where routing areas are small enough.
- By subnet: This strategy is less likely to be used because it may become too incremental.

In general, the several types of IPv4-to-IPv6 transition mechanisms can be classified as follows:

- Dual stacks: This represent the coexistence of the IPv4 protocol stack (IPv4 application/IPv4-based transport protocol/IPv4/link interface) and IPv6 protocol stack (IPv6 application/IPv6-based transport protocol/IPv6/link interface) on a single device or network node. However, in the transitional period it is typical for applications to be created that are able to use IPv4 and IPv6 network interfaces (such a case is referred to as a dual IP layer, also known as a dual stack) [38].

- Tunnels: In the first stage of the IPv4-to-IPv6 transition, tunnels are being used to encapsulate IPv6 packets into IPv4 packets for tunneling of IPv6 packets through IPv4 networks (at the beginning of the transition, the dominant address type is IPv4). When IPv6 finally prevails over IPv4 (estimated around the year 2020), then tunneling of IPv4 packets through IPv6 networks may become a more typical scenario.
- Translators: These are typically network nodes (e.g., routers) that translate IPv4 addresses to IPv6 and vice versa. Such nodes are typically placed on the boundaries between IPv4 networks and IPv6 networks.

2.5 Dynamic Host Configuration Protocol (DHCP)

DHCP is a standardized protocol [39] for dynamic provision of network configuration parameters to the Internet hosts in a given IP network. In general, it consists of two components: (1) a protocol for delivery of parameters specific to a given host and (2) a mechanism for dynamic allocation of IP addresses to hosts from a given "pool" (i.e., defined range) of IP addresses.

DHCP is built as a client–server model, where hosts have installed DHCP clients and the DHCP server is placed within the same IP network as hosts. In other words, DHCP messages do not pass through routers, except in cases when routers have configured a so-called DHCP relay mechanism, which can distribute an incoming DHCP message from a specified network interface of the router to one or more outgoing links.

There are two different versions: (1) DHCP for IPv4 [39] and (2) DHCP version 6 (DHCPv6) for IPv6 [40].

2.5.1 DHCP for IPv4

DHCP for IPv4 supports three mechanisms for IP address allocation: automatic, dynamic, and manual allocations. In the automatic allocation, the DHCP assigns a permanent IPv4 address to the client host. In dynamic allocation, which is most used in home and enterprise environments, the DHCP assigns a temporal IP address to the host (i.e., for a limited period of time). Finally, in the manual allocation the IPv4 address is assigned by the network administrator (i.e., the person who is responsible for maintenance of the given network) and the DHCP server is used in this mechanism to communicate with the DHCP client to set up the allocated address. However, it is possible for a single network to use more than one of the three DHCP allocation mechanisms, which is dependent on the network administrator.

The general purpose of DHCP is to provide dynamic allocation of IP addresses from a DHCP server when a client joins the network (e.g., booting a computer on the local Ethernet network or activating a wireless interface on laptop or smartphone). Historically, the DHCP was first standardized in 1993 (initially with RFC 1531) as an extension to its predecessor, the Bootstrap Protocol (BOOTP). However, the BOOTP had several drawbacks such as no ability to add configuration information to the clients and no ability to reclaim unused IP addresses. In contrast,

DHCP has the ability to provide an IP address to a client for a finite lease, and it can provide all IP configuration parameters (e.g., IP address, gateway address, DNS addresses) that are needed for establishment of a functional network interface.

Because DHCP is used for IP address allocations, as an application layer protocol it uses the UDP/IP protocol stack. DHCP clients use port 67 to send DHCP request messages, while DHCP servers use port 68 for responses sent to clients. In general, DHCP functioning is based on the exchange of the following messages (an example is given in Figure 2.9):

- DHCP discovery: This is the initial broadcast message sent by the host to the IPv4 broadcast address "255.255.255.255."
- DHCP offer: With this message a DHCP server replies to the "DHCP discovery" message sent by host in the network.
- DHCP request: This is sent by the host on receipt of the "DHCP offer" message to request an allocation of the offered IP address from the DHCP server.
- DHCP acknowledgment: This message is used by the DHCP server to acknowledge allocation of the offered IP address to the given host in the network.

However, a DHCP client may request more information than is offered by the server in the DHCP offer message, and in such cases it uses the "DHCP information" message to do so. Also, a client can send a "DHCP release" message to request the DHCP server to relinquish the IP address and cancel the remaining lease. If a network address is already in use, a DHCP client informs the DHCP server by means of a "DHCP release" message.

Additionally, DHCP has many configuration options (which were referred to as "vendor extensions" in BOOTP [39]) that can be used. Each option has an allocated code with a value in the range from 0 to 255 (the Code field in DHCP options has a length of 8 bits). Allocation of DHCP option codes is maintained by IANA [41]. Some important examples of options are as follows: subnet mask value (option code = 1), router addresses (option code = 3), DNS server addresses (option code = 6), IP address lease time (option code = 53), and DHCP server identification (option code = 54). Most of the option codes are already assigned by IANA, so their space is near exhaustion.

Overall, DHCP is one of the fundamental Internet technologies that provided a short- and mid-term solution to the problem of IPv4 address space exhaustion. It can be applied in all different pools of unicast IP addresses, including public and private ones.

2.5.2 DHCP Version 6 (DHCPv6)

DHCPv6 is used for stateful addressing in IPv6 by provision of IPv6 addresses, IPv6 prefixes, and other configuration data. Additionally, it can be also used in stateless IPv6 addressing when IPv6 hosts (that use stateless autoconfiguration) require information other than IPv6 address. For example, unlike DHCP for IPv4 (which typically is used to configure clients with IP addresses of DNS servers), the DHCPv6 is not necessarily used for configuring hosts with DNS addresses because in IPv6

2.5 Dynamic Host Configuration Protocol (DHCP)

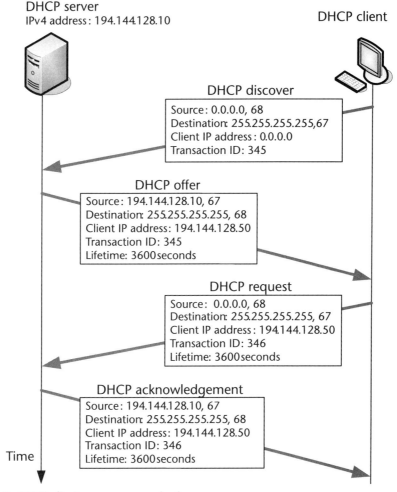

Figure 2.9 DHCP client–server communication.

networks such tasks can be accomplished by the Neighbor Discovery Protocol (in particular, DNS in IPv6 hosts can be configured with ICMPv6 router advertisement messages, type 134, as given earlier in Table 2.5).

In general, DHCPv6 is not completely based on DHCP for IPv4, because it uses new port numbers, has a new message format, and the options are restructured. Also, DHCPv6 and DHCP (for IPv4) are not compatible protocols.

Like DHCP for IPv4, DHCPv6 is also based on a client–server model [40]. Clients and servers use the UDP/IP protocol stack. DHCP clients typically use Link-local addresses for DHCP communication (i.e., sending and receiving messages from servers). The DHCP servers receive messages from a so-called link-scoped multicast address that is dedicated to that purpose. The DHCP clients transmit most of the messages to this multicast address. Hence, there is no need for a client configuration with addresses of DHCP servers. The DHCP uses the following reserved multicast addresses:

- All_DHCP_Relay_Agents_and_Servers (multicast address FF02::1:2): This is a link-scoped multicast IPv6 address used by a client to communicate with

neighboring relay agents or servers that, by default (when DHCPv6 is used), are members of this multicast group.
- All_DHCP_Servers (multicast address FF05::1:3): This is a site-scoped multicast address that is intended to be used by a relay agent communicating with DHCPv6 servers, which happens in two cases: (1) when a relay agent wants to send a message to all servers and (2) when a relay agent does not know the unicast addresses of the servers. By default, all DHCPv6 servers on a given site are members of this multicast group.

In all DHCPv6 communication between clients and servers, servers and relay agents listen for messages on UDP port 547 while clients listen on UDP port 546. There are two basic DHCPv6 client–server message exchanges:

- Four-message exchange: This approach is similar to DHCP client–server communication for IPv4, but with different messages. DHCPv6 communications start with a Solicit message sent by a client to an All_DHCP_Relay_Agents_and_Servers address (i.e., FF02::1:2) with the goal of locating a DHCPv6 server and request IPv6 address. Any server that can complete the request from the client responds with an Advertise message. In the case of multiple responses, the client chooses one of the DHCPv6 servers and sends it a Request message, asking that server to confirm assignment of the offered IPv6 address and accompanied configuration parameters. In the fourth step of this four-message exchange, the DHCPv6 server responds with a Reply message, thus finalizing the process. An illustration of this four-message exchange is given in Figure 2.10.
- Two-message exchange: This method is used when a client does not request assignment of an address, but contacts the servers for configuration information such as a list of available DNS servers or Network Time Protocol (NTP) servers through a single message exchanged with the DHCPv6 server. The client requests such information by sending an Information-request message to the multicast address "FF02::1:2." Any server that will receive that message responds with a Reply message, which contains the configuration information for the client.

DHCPv6 can have up to 255 message types (the Message type field has a length of 8 bits), from which 13 were standardized in the initial DHCPv6 recommendation [40], such as Solicit, Advertise, Request, Confirm, Renew, Rebind, Reply, Release, Decline, Reconfigure, Information-request, Relay-forward, and Relay-reply. The Message type field is followed by a 24-bit Transaction-ID field and a variable Options field. The options are stored serially (byte aligned) in the Options field without any padding bytes between different options. Examples of options are client identifier (i.e., client's IP address) and server identifier (i.e., server's IP address).

What are the differences between DHCPv6 and DHCP for IPv4? The DHCP (for IPv4) uses "0.0.0.0" as the IP address for DHCP clients that request an IPv4 address, whereas in IPv6 hosts always have a Link-local address (obtained with stateless IPv6 addressing, i.e., address autoconfiguration). Further, DHCPv6 uses reserved multicast addresses for servers and relay agents. DHCPv6 has a

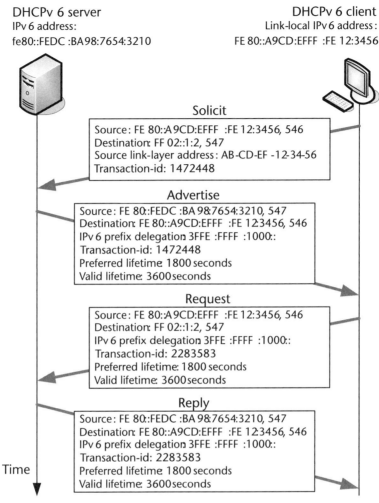

Figure 2.10 DHCPv6 client–server message exchange.

two-message exchange mechanism for host network configuration cases without IPv6 address assignment. However, a four-message exchange exists in both versions of DHCP, although with different message types for each version. Additionally DHCPv6 has several optional features, such as multiple unrelated requests to the DHCP server(s), a request for network settings reconfiguration of the client (via a Reconfigure message from the server), and so forth.

Overall, DHCPv6 and DHCP for IPv4 have similar functionalities, but they are not compatible. Rather, each protocol is accommodated to the IP version for which it is created, DHCPv6 for IPv6 and DHCP for IPv4.

2.6 Transport Protocols in the Internet

Transport protocols are placed on OSI layer 4, and they are located immediately above the IP. The Internet has two main transport protocols:

- User Datagram Protocol [42] and
- Transmission Control Protocol [43].

2.6.1 User Datagram Protocol (UDP)

The UDP is a connectionless transport protocol (on the fourth layer in the OSI reference model). It assumes that the IP is used as its underlying protocol in Internet hosts [42]. The UDP does not provide flow control, reliability, or any other control of data sent over the IP. The main function of UDP is to provide multiplexing of data from different applications (above the UDP) before sending it to the IP (below the UDP) by using an application identification called a port, and vice versa in the recipient host (to demultiplex incoming datagrams to different applications within the destination host, based on the port identification of applications). A port is a 16-bit number that acts as the application identification on the transport layer (it is defined in the same manner for UDP and TCP). Port numbers are application specific or process specific (e.g., a single application may use several different port numbers for different processes). The port numbers range from 0 to 65,535 (i.e., 2^{16} port numbers in total). All port numbers are assigned by IANA [44], which has divided ports in three ranges:

- System ports (i.e., well-known ports) range from 0 to 1,023 and are used for widely known types of Internet services and applications (e.g., HTTP, FTP, SMTP).
- User ports in the range from 1,024 to 49,151 are registered ports for registered services assigned by IANA (e.g., assigned to application layer protocols created by vendors).
- Dynamic ports are ports in the range from 49,152 to 65,535 (also referred to as private ports) that are never assigned by IANA, so applications may simply use any of these ports on a host. However, this also means these ports cannot be used as service identifiers because different applications may use the same dynamic ports for different purposes.

UDP is a transaction-oriented transport protocol that is suitable for applications and services for which reliability mechanisms are not necessary and in cases where higher-level protocols can provide flow control. In practice, it is initially used as a transport protocol for several well-known application protocols, such as NFS (Network File System), SNMP (Simple Network Management Protocol), TFTP (Trivial File Transfer Protocol), and DNS. Additionally, UDP is also a stateless protocol, which makes it suitable for applications with a very large number of clients (e.g., for IPTV). Also, due to its simplicity (no flow control, no delays due to retransmissions of lost datagrams) and connectionless approach (no need to establish a connection between the sender and the receiver prior to sending datagrams with user data), UDP is also used for real-time services such as VoIP and video streaming over the Internet.

Figure 2.11 shows the format of the UDP header, which includes the following fields:

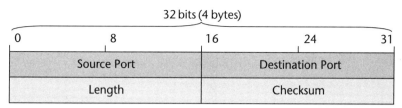

Figure 2.11 UDP header format.

- Source port (16 bits): This field indicates the port that is used by the sending process on the sender host. However, it is declared as optional and, if not used, all bits are set to zero. If it is then used, the response from the destination should be addressed to that port.
- Destination port (16 bits): This field contains the port that is used by the receiving process at the destination host.
- Length (16 bits): This field specifies the total length of the UDP datagram in number of bytes, including the header and the data. The minimum header length is 8 bytes.
- Checksum (16 bits): This field provides error control coding of the UDP header and data. In particular, it is the 16-bit one's complement of the one's complement sum of the so-called pseudo-header consisting of the IP header source and destination addresses (IPv4 or IPv6) and the Protocol field (for IPv4) or Next Header field (for IPv6), and UDP length. It is padded with all-zero octets (but not the last one) to ensure that it is a multiple of 32 bits. For UDP over IPv4, when the sending host does not care about the checksum (or it is not important), then it uses an all-zero checksum in the UDP header. Checksum is optional for UDP over IPv4 [42], but it is mandatory in the case of UDP over IPv6 because IPv6 nodes discard all datagrams with an all-zero checksum [3].

UDP and TCP are the fundamental transport layer protocols in the Internet. Together with the IP (both versions, IPv4 and IPv6) they define the nature of the Internet applications that typically run over the UDP/IP or TCP/IP protocol stack.

2.6.2 Transmission Control Protocol (TCP)

TCP is a connection-oriented transport protocol used in the Internet [43]. Most popular Internet services, such as WWW, FTP, and email, are based on TCP. Internet traffic measurements show that most of the traffic is based on TCP regarding the transport layer protocols (the reader may refer to Chapter 1 traffic statistics).

In its basic definition, the main purpose of TCP is to provide a reliable and secure logical connection service between two processes (typically running on two different hosts) [43]. The initial Internet best-effort philosophy (i.e., every connection and every packet is admitted from every host connected to the Internet) is in fact provided with TCP over IP as its underlying protocol. Applications that do not need the TCP functions use the UDP, which, like IP, does not provide functions for flow control. So, TCP has two interfaces within a given host, one toward the applications (it consists of set of calls described in Chapter 3 of this book, similar to calls provided by an operating system (OS) to application processes for manipulation with files), and the other toward the lower-level protocol, which is the IP. (This interface is in fact unspecified, and the two protocols, IP and TCP, can asynchronously pass information to each other.)

2.6.2.1 TCP Operations

The main goal of TCP is to provide reliable end-to-end communications over the Internet. Its operation is based on the following processes:

- Basic data transfer: This provides a continuous stream of octets (i.e., bytes) between two application processes (on two hosts) in both directions. If we label the two communicating hosts "A" and "B," then one TCP stream goes from host A to host B and the other one goes from host B to host A (i.e., two streams in parallel, which is known as full duplex in telecommunications terms). However, TCP does not recognize any boundaries in the transmitted data (it is left to the applications that use the TCP).
- Reliability: TCP was created to be able to recover from any data that have been delivered out of order, lost, or damaged (i.e., bit errors) on the way from the source to the destination. For the purpose of in-order delivery of all data, TCP assigns a sequence number to each octet that is sent from the sender. To provide lossless communications between the sender and the receiver, the TCP requires a positive acknowledgment (ACK) to be sent by the receiver to the sender. Finally, errors in the data (i.e., damaged packets) are managed by error control coding in the form of a checksum (similar to the UDP approach).
- Flow control: The TCP uses a mechanism called a sliding window to control the amount of data that can be sent by the sender to the receiver. The window defines the amount of data (i.e., number of octets) that the sender may transmit before further change of the window (i.e., further permission to transmit).
- Multiplexing: The multiplexing of data to/from different application layer protocols is performed by using the ports (used for application and process identification, similar to UDP) that are connected with IP addresses via an artificial form called sockets. Each host can have multiple IP addresses. Each host independently binds ports to processes (i.e., applications) and creates sockets. A pair of sockets on each end of the TCP communication between two hosts uniquely identifies that connection.
- Connections: Unlike UDP, which is stateless, TCP maintains certain status information for each data stream of that host. The combination of such information, together with the window sizes (used for flow control), sequence numbers, and sockets, is called a connection [43].

Overall, regarding its operations, the TCP provides stream-based connection-oriented full-duplex completely reliable end-to-end data delivery. It is implemented as a transport protocol (together with UDP) in OSs in all hosts (e.g., personal computers, smartphones, servers) and network nodes (e.g., routers and gateways).

2.6.2.2 TCP Segment Format

The individual unit that TCP uses for data transmission is called a segment. Each segment begins with a TCP header followed by a TCP payload that is used to store the application layer protocol unit, which is referred to as data from TCP's point of view. The minimum length of the header is 20 bytes (without options), and it can be extended up to 60 bytes (with options included at the end of the header).

The format of the TCP header is shown in Figure 2.12. It consists of the following fields:

- Source port (16 bits): This is the port number of the sending host, which identifies the application program that is sending the segment.
- Destination port (16 bits): This identifies the port number of the destination host that is receiving the segment.
- Sequence Number (32 bits): This field stores the sequence number of the first data byte (i.e., in TCP payload) in the given segment (as discussed in TCP operations, the TCP numbers each byte of a data stream with a sequence number). An exception to this case is the initial segment for TCP connection establishment, which carries a randomly generated initial sequence number (ISN) in the range from 0 to $(2^{32} - 1)$.

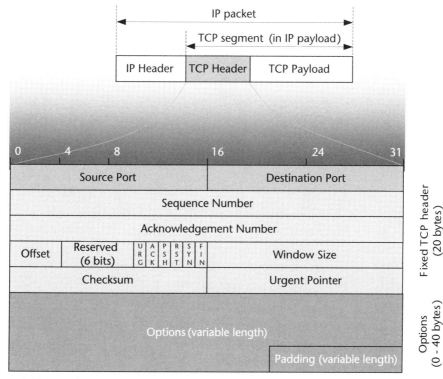

Figure 2.12 TCP header format.

- Acknowledgment Number (32 bits): This field contains the next sequence number that the TCP receiver is expecting to receive from the TCP sender. To provide higher efficiency for the transmission and considering the full-duplex nature of the TCP, the acknowledgments sent from host B to host A for the data sent from host A to host B can be piggybacked with the data sent from host B to host A (i.e., in the opposite direction), and vice versa. Segments that carry an acknowledgment number in this field have an ACK bit (in the TCP header) set to 1.
- Data Offset (4 bits): This field specifies the length of the TCP header in the number of 32-bit words, which is necessary to indicate to the receiving host where the data begins in the segment.
- Reserved (6 bits): These bits are reserved for future use and they must be set to zero.
- Control bits (6 bits):
  - URG (1 bit): When set to 1, this bit indicates that the Urgent pointer filed in the TCP header contains significant value.
  - ACK (1 bit): When set to 1, this bit indicates that the Acknowledgment field in the TCP header contains an acknowledgment number.
  - PSH (1 bit): When set to 1, this bit means that receiving hosts should immediately push the data toward the application. Normally, the data on the receiving side is buffered first and then sent from the TCP to the application.
  - RST (1 bit): When set to 1, this bit triggers a reset of the connections (TCP connection establishment and termination is explained later in this chapter).
  - SYN (1 bit): This bit is used to synchronize the sequence numbers (i.e., it is a synchronization bit), and it should have a value of 1 only in the first segment that is sent from a given source to a given destination.
  - FIN (1 bit): This field is set to 1 to indicate finishing of a given TCP connection, meaning that the sending host (which has set the FIN bit to 1) has no more data to send.
- Window (16 bits): This field indicates the size of the receive window for the sending host of a given TCP stream (in bytes), which is the number of bytes that the receiving host (which is the sender of the segment carrying the given Window field) is willing to accept. The maximum receive window size is $2^{16} - 1 = 65,535$ bytes.
- Checksum (16 bits): This field contains the checksum, which is computed in exactly the same manner as for UDP, by using the same pseudo headers for IPv4 and IPv6. However, the Protocol field of an IPv4 header is set to 6, which is the protocol number for TCP (from IP's point of view).
- Urgent Pointer (16 bits): This field carries value that is an offset from the sequence number in the given segment (i.e., from the number of the first byte of the TCP payload).
- Options (variable length from 0 to 40 bytes): Different options may be used in the Options field, but when it is used, it is included in the header checksum.

Each option is a multiple of 8 bits (i.e., integer number of bytes) and can start on any octet boundary within the TCP header. The options end with End-of-Option option, which is followed by header padding with zeros up to the 32-bit boundary (i.e., the TCP header length must be an integer number of words, where a word has a length of 32 bits).

Each TCP segment can have a length of up to 65,535 − 20 = 65,515 bytes (due to a minimum IPv4 header length of 20 bytes). One of the TCP options is maximum segment size (MSS), which is specified as the largest amount of data (in bytes) that TCP is willing to receive in a single segment. However, two independent MSS values are possible for a given TCP connection, one MSS per direction. This is done because the host capabilities on each of the two ends of a single TCP connection might be different (e.g., one host might have a limited amount of memory). However, with the goal of achieving the best TCP flow performance in each direction, the MSS should be set to a value that will be small enough to avoid unnecessary IP fragmentation. Since most LANs connected to the Internet nowadays are Ethernet or Wi-Fi, which have a data-link layer MTU length of 1,500 bytes, the MSS for TCP should be less than that value decreased by the IP header length (because the TCP segment is encapsulated in an IP packet payload before it is relayed to the network interface, i.e., the link layer). If the MSS option is not used, the default value for TCP segment size is 536 bytes, which is defined as the maximum IP packet size (called effective MTU to receive, i.e., EMTU_R [45]) that all hosts on the Internet are required to accept or reassemble. It is set to 576 bytes for IPv4 [45, 46], minus 40 bytes (to accommodate the IP header). However, the EMTU_R is set by the IP layer and it must be at least 576 bytes and up to 65,535 bytes. But, due to local network access to the Internet, the EMTU_R value is between 576 and 1,500 bytes. Then, the TCP MSS is calculated as follows: EMTU_R − fixed IP header − fixed TCP header (the fixed IPv4 header is 20 bytes, the fixed IPv6 header is 40 bytes, and the fixed TCP header is 20 bytes, where in all cases fixed refers to a header without options).

2.6.2.3 Establishment and Termination of a TCP Connection

TCP is a connection-oriented transport protocol and therefore requires connection establishment before the data transfer and connection termination at the end. Overall, every TCP connection has three phases: (1) connection establishment, (2) data transfer, and (3) connection termination.

A TCP connection is always initiated by a client toward a server (TCP is a client–server transport protocol). However, TCP connections are generally full duplex, so each machine (on both ends of the connection) must initialize communication in one direction. A TCP connection progresses through a series of states during its duration, including the following: LISTEN, SYN-SENT, SYN-RECEIVED, ESTABLISHED, FIN-WAIT-1, FIN-WAIT-2, CLOSE-WAIT, CLOSING, LAST-ACK, TIME-WAIT, and the last state CLOSED (which is fictional, because it is a state when the TCP connection does not exist because it is closed) [43].

Before any client connects to a given TCP server, the server must be open for accepting the incoming connection; that is, the server must bind to a given port to listen for an incoming connection targeted to a given application identified with that port number. This is called passive-open for the server's side (the LISTEN state in the TCP state diagram). A TCP client can initiate an active-open to TCP servers that are listening for incoming connections. For connection establishment TCP uses a so-called "three-way handshake" (Figure 2.13), which consists of the following three steps:

- SYN (client to server; client goes into SYN-SENT state): A TCP connection is initiated by the SYN segment (a segment with control bit SYN set to 1) sent by the client to the server. The sequence number in the TCP header in this segment is a random number X, set by the client.
- SYN+ACK (server to client; server goes into SYN-RECEIVED state): The server responds to the SYN segment (from the client) by sending a segment with two control bits, SYN and ACK, set to a value of 1. The ACK number in this segment is set to X + 1. On the other side, the sequence number in the segment sent from the server is another random number Y, set by the server.
- ACK (client to server; after this step client and server are both in the ESTABLISHED state): In the third step the client sends an ACK segment to the server, with the ACK number set to Y + 1 (to acknowledge receipt of the segment sent by the server in the step 2) and the sequence number set to X + 1.

After a successful three-way handshake, the TCP connection is established. It then follows with the data transfer over the established connection. In this case (when connection is established), the TCP is in the SEND state for sending segments

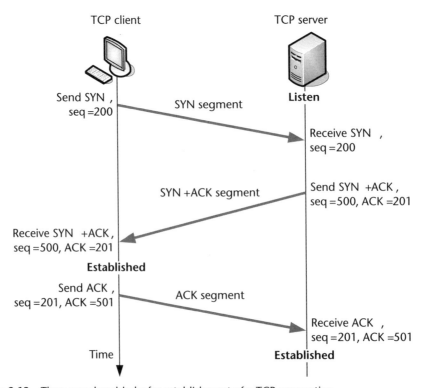

Figure 2.13 Three-way handshake for establishment of a TCP connection.

2.6 Transport Protocols in the Internet

and in the RECEIVE state for receiving segments from the remote end point. (The TCP connection state diagram is shown in Figure 2.14.)

A TCP connection is full duplex, but at termination it is referred to as two-way simplex connections. When a given TCP end point wants to terminate its half of the connection (i.e., the connection established to transfer data sent from that host to the other end point), it sends a FIN segment, which has its FIN control bit set to 1. The receiving end point replies with an ACK (i.e., ACK control bit set to 1) in the reverse direction. The other half of the connection (in the opposite direction) is similarly terminated by using a pair of FIN and ACK segments. Hence, the connection termination process uses a four-way handshake between the two end points (e.g., the hosts and routers). In general, there are three possible cases for closing a TCP connection [43]:

- Local user-initiated close: In this case a FIN segment is created and placed in the outgoing TCP queue, so no new outgoing segments can be created for that direction (i.e., no further sends issue from the user, such as the local

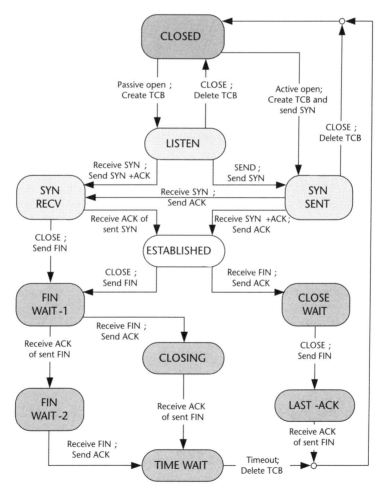

TCB – Transmission Control Block (record of variables that define a TCP connection)

Figure 2.14 TCP connection state diagram.

application). This state in the TCP state diagram is noted as FIN-WAIT-1. TCP flow control will retransmit all unsuccessfully transmitted segments prior to and including the FIN segment. However, the remote TCP that will receive the FIN will send an ACK (for that FIN segment), but it may continue to use the other half of the connection (i.e., the remote host is not obliged to send its own FIN segment until its user closes its TCP connection).

- Remote user-initiated close: In this case the TCP receives a FIN segment from the remote host (i.e., remote TCP end point), which can be responded to by an ACK to tell the remote end that the connection is closing. After sending any remaining data in the outgoing direction, the TCP can send a FIN segment and then delete the connection after receiving an ACK. However, if the ACK does not appear (e.g., it might get lost in the network) after a certain time period, called a time-out, the TCP connection is terminated.

- Both users simultaneously initiated close: In this case both hosts send FIN segments (i.e., segment with FIN control bit set to 1) to terminate the connection. Upon receiving an ACK for the FIN segment (of course, after acknowledgment for successful receipt of all segments prior to the FIN segment), each TCP terminates the connection. However, when both sides simultaneously request termination of the connection, the client before reaching the CLOSED state passes through the FIN-WAIT-1, CLOSING, and TIME WAIT states.

2.6.2.4 TCP Windows

TCP uses a sliding window mechanism for end-to-end flow control. There is one send and one receive window in every direction, so there are a total of four windows in a single TCP connection. In each direction the receive window size is returned with an ACK segment (which also can carry data in the opposite direction). With the receive window the receiver specifies the number of bytes (i.e., amount of data) that the TCP sender may transmit before receiving further permission by the receiver.

When a TCP sender sends a segment with a given sequence number, it also starts a timer. After the segment reaches the destination, the receiving TCP entity responds with an ACK segment regardless of whether it has useful information to send or not. This ACK segment provides an acknowledgment number equal to the sequence number of the next byte the receiver expects to receive. If the timer expires before receipt of an ACK for a sent segment, the TCP sender retransmits the same segment (for which the timer has expired). To avoid the process being stalled by segment losses, the TCP sender uses a sliding window mechanism, which is illustrated in Figure 2.15.

The sliding window mechanism uses three pointers as given below:

- Left boundary pointer: This pointer marks the left boundary of the sliding window, which is the last octet of already acknowledged data from the TCP receiver.

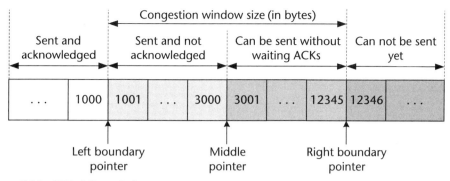

Figure 2.15 TCP sliding window.

- Middle pointer: This pointer denotes the already sent data within the TCP sliding window that are not acknowledged (i.e., the outstanding data in the network).
- Right boundary pointer: This pointer marks the right boundary of the sliding window, which is the octet with the highest sequence number that can be sent without awaiting ACKs for already sent data.

Positively acknowledged octets by the receiver indicate movement of the boundaries of the sliding window to the right (to higher sequence numbers), whereas newly sent data result in movement of the middle pointer to the right.

Additionally, there is also the probability that bit errors could occur in TCP segments during the transport from end to end via various links in different networks based on different media (e.g., copper, fiber, wireless). Bit errors are detected by using the checksum in the TCP header, and erroneous segments (errors occur with higher probability in wireless networks than in wired ones) are treated in the same manner as lost segments.

2.7 TCP Mechanisms and Versions

The best-effort design concept of the Internet (to accept every connection and every packet in the network) creates network congestion. In general, congestion refers to overloading the capacities of outgoing links from hosts and routers with IP packets. Packets that cannot be sent are buffered. Buffers, however, are limited in memory, so after a certain congestion period, losses occur. The Internet does not imply congestion control in routers, but it is supposed to be done by the end points of the Internet connection. The transport protocol that is designed to provide end-to-end congestion control is TCP. In order for TCP to deal with congestion problems, its initial standard was further enhanced with congestion control mechanisms. However, several congestion control mechanisms have been standardized by IETF, and many proprietary ones exist as well. Each TCP implementation must implement the standardized congestion control mechanisms, but it is also possible to implement several other additional proprietary mechanisms (implemented in the OSs by certain OS vendors). In general, different congestion control mechanisms define different TCP versions.

2.7.1 TCP Congestion Control

The best-effort Internet is based on TCP end-to-end control of congestion, without direct network assistance. TCP congestion control is standardized with four algorithms: Slow Start, Congestion Avoidance, Fast Retransmit, and Fast Recovery (as shown in Figure 2.16). Initially these algorithms were standardized for use in TCP in 1989 with RFC 1122 [7] when the Internet expanded outside the academic environment and a significant increase in traffic volume occurred that resulted in network congestion. However, the current versions of congestion control algorithms are specified in RFC 5681 [47].

Initially, TCP had no congestion control mechanisms, so the source would start by sending a whole window of data. However, after the introduction of congestion control in TCP, each TCP module must use Slow Start and Congestion Control to control the amount of data injected into the network. Two window variables are added to TCP to control the sending of data: (1) a sender's congestion window (cwnd) and a receiver's advertised window (rwnd). However, cwnd = min(cwnd, rwnd) is the window size that is used to govern the data transmission. Another variable called the slow start threshold (ssthresh) is used to specify the window size when Slow Start transits to Congestion Avoidance. When cwnd < ssthresh, the Slow Start mechanism is used, whereas Congestion Avoidance is in force when cwnd > ssthresh. (If cwnd = ssthresh, then the sender may use either of the two mechanisms.)

The Slow Start has a cold start with cwnd equal to an initial window (IW) size, which initially was set to one segment (in RFC 2001 from 1997), then changed to up to two segments (in RFC 2581 from 1999). However, currently the IW size can be up to four segments (in RFC 5681 from 2009), specified by the following rules:

If SMSS > 2,190 bytes, then IW = 2 × SMSS bytes (up to 2 segments)
else if SMSS > 1,095 bytes then IW = 3 × SMSS bytes (up to 3 segments)
   else IW = 4 × SMSS bytes (up to 4 segments)

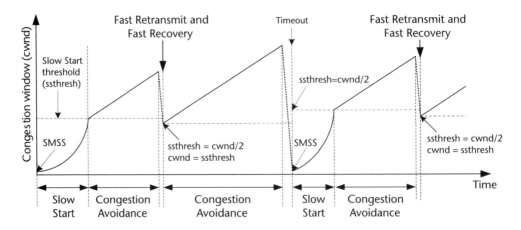

ACK: Acknowledgement
SMSS: Sender Maximum Segment Size

Figure 2.16 TCP congestion control mechanisms.

where SMSS is sender maximum segment size (without TCP/IP headers), which is the largest segment that a given TCP sender can transmit, based on maximum transfer units of underlying physical network (e.g., less than 1,500 bytes for Ethernet) and other factors.

Such an increase in IW in the Slow Start phase during the past two decades is driven by spreading of broadband access to the Internet (e.g., with several tens of megabits per second) while the MTU of the underlying technologies (e.g., Ethernet, Wi-Fi) on the link layer have remained unchanged.

During the Slow Start the TCP increases cwnd by SMSS bytes with each received ACK that cumulatively acknowledges new data. That results in an exponential increase in the cwnd over the time, as shown in Table 2.7, where the round-trip time (RTT) is the time needed for the transmitted segment to reach the receiver and then the ACK for that segment to come back to the sender.

The bit rate of the TCP connection can be calculated as cwnd/RTT. For example, if RTT = 250 ms, SMSS = 500 bytes, and cwnd = 16 SMSS, then the initial bit rate of the TCP connection will be 500 bytes/250 ms = 16 Kbps (note that 1 byte = 8 bits); after a time period of $4 \times RTT = 1,000$ ms = 1 sec, the bit rate will become 16×500 bytes/250 ms = 256 Kbps, and so on. Generally, if there are no losses and there is no Slow Start threshold, the bit rate will reach a value of $2^N \times$ SMSS/RTT after a time period of $N \times RTT$ seconds. However, bit rates for every connection are limited by the capacity of the channel (i.e., maximum bit/sec over the link). Also, a given link may be shared by multiple TCP (and non-TCP) connections. Hence, the bit rate is always limited and that limit may be constant or dynamic. For example, available capacity may change during movement of a mobile user due to changes in the radio propagation environment or access technology, or it may change due to traffic conditions. For example, new connections are established and other are closed on the same links, where a given TCP connection may share the link capacities of different hops through the Internet with different TCP connections and UDP traffic at the same time. So, an exponential increase in cwnd cannot continue on a longer time scale.

When ssthresh > cwnd, the TCP goes from the Slow Start to the Congestion Avoidance phase, with the goal of delaying the loss of segment, that is, congestion. In the Congestion Avoidance phase, the congestion window (cwnd) increments linearly with the time (assuming that RTT is constant between the two TCP end points) by increasing the cwnd for min(Bytes acknowledges with the ACK, SMSS) for each single positive ACK that arrives at the sender in RTT intervals.

Table 2.7 Congestion Window Increase in Slow Start

Time	ACKs Received	Congestion Window	Segments Sent
0	0	$1 = 2^0$ SMSS	1
RTT	1	$2 = 2^1$ SMSS	2
$2 \times$ RTT	2	$4 = 2^2$ SMSS	4
$3 \times$ RTT	4	$8 = 2^3$ SMSS	8
$4 \times$ RTT	8	$16 = 2^4$ SMSS	16

However, because congestion windows increase either exponentially (in Slow Start) or linearly (in Congestion Avoidance), the losses must occur after a certain time because the capacity of each individual link on the Internet is limited. (In general, each TCP segment travels over multiple hops, i.e., links, on its way from the source to the destination.)

Two mechanisms are available for recovery from losses in TCP connections: Fast Retransmit and Fast Recovery. When a segment arrives at a receiver out of order, the TCP receiver must immediately send an ACK, which will acknowledge again the last in-order received segment, so it will appear as a duplicate ACK to the sender. The duplicate ACK is an indication to the sender that possible loss might occur. However, there is the possibility for reordered receipt of segments without them being lost. Also, a duplicate ACK can be caused by duplication of the ACK by the network. In that manner, the receiving TCP must send immediate an ACK when an incoming segment fills all gaps in previously received segments, with the goal of ensuring the sender receives information in a timely manner. The sender can detect losses in two ways:

- Retransmission time-out (RTO): When sending a segment, the sender starts an RTO timer for that segment. When a time-out occurs, the ssthresh = max (FlightSize/2, 2 × SMSS) where FlightSize is the amount of outstanding data in the network, the cwnd is set to the loss window size (which equals one full-sized segment), and the sender enters the Slow Start phase. The RTO calculation is based on measurements of RTT and calculation of a smoothed RTT (SRTT) based on the following equations:

$$\text{RTT} = (\alpha \times \text{SRTT}) + [(1 - \alpha) \times \text{RTT}] \tag{2.1}$$

$$\text{RTO} = \min\{U, \max[L, (\beta \times \text{SRTT})]\} \tag{2.2}$$

where U is the time-out upper bound (e.g., 1 min), L is its lower bound (e.g., 1 sec), α is the so-called smoothing factor (e.g., 0.8 to 0.9), and β is a delay variance factor (e.g., 1.3 to 2).

- Duplicate ACK: After receipt of three duplicate ACKs (an indication that loss has occurred with high probability), the sender retransmits the segment that appears to be lost without waiting for the retransmission timer to expire. This is the Fast Retransmit mechanism. After the Fast Retransmit, the Fast Recovery mechanism governs recovery of the TCP connection after the loss (by avoiding Slow Start, because duplicate ACKs show that segments are continuing to arrive at the receiver) until a nonduplicate ACK arrives.

So, the Fast Retransmit and Fast Recovery mechanisms are used together to recover TCP connections from losses [47]. However, on the first and second duplicate ACK, the sender should send a segment with previously unsent data (if it is allowed by rwnd), while the FlightSize should be up to cwnd + 2 × SMSS (because there are up to two unacknowledged sent segments in the network). When third duplicate ACK arrives, the sender must set ssthresh = max(FlightSize/2, 2 × SMSS). Then, the lost segment must be retransmitted and cwnd = ssthresh + 3 × SMSS, because three sent segments have left the network and have been buffered at the receiver, which resulted in three duplicate ACKs. In the same manner for additional duplicate ACKs (fourth, fifth, and so on), the ssthresh increases for one SMSS per duplicate ACK. While waiting for the ACK for the lost segment, the sender is allowed to transmit SMSS bytes of previously unsent data (if, of course, rwnd allows it). Finally, when the next ACK arrives, which acknowledges all previously acknowledged segments, then TCP sets cwnd = ssthresh, where ssthresh is already set to max(FlightSize/2, 2 × SMSS). This ends the Fast Recovery phase, and TCP continues with the Congestion Avoidance phase.

Overall, the main TCP mechanisms for flow and congestion control are created to provide fairness among multiple flows that share the same link. So, when N TCP sessions share the same bottleneck link with total used bit rate R (which must be less than the link capacity in bits per second, because TCP cannot provide 100% utilization of network resources due to its continuously changing window size), then each session should achieve an average rate of R/n.

2.7.2 TCP Versions

TCP is available in many different implementations, based on variations of its initial definition. However, all Internet software must implement standardized flow control, error control, and control congestion mechanisms, as described in the previous section. Considering that, the most known TCP versions are TCP Tahoe, TCP Reno, TCP New-Reno, and TCP SACK.

The TCP Tahoe version implements the Slow Start, Congestion Avoidance, and Fast Retransmit phases. This version was the first one implemented in 4.3 BSD (Berkeley Software Distribution) Unix in 1988. TCP Tahoe recovers from single losses in a single congestion window (or more accurately, in a given flight), but starts with a Slow Start after the loss, which significantly degrades the bit rate at losses.

Adding a Fast Recovery mechanism to TCP Tahoe resulted in another version called TCP Reno, which improves performance for single losses in a TCP flight, but it cannot recover efficiently from multiple packet losses (i.e., segment losses) and it enters the Slow Start phase in such a case (due to a retransmission timer that expires). The reason for that is the lack of a recursive run of the Fast Retransmit mechanism for lost segments when TCP Reno is in the Fast Recovery phase (it enters this phase after the first lost segment).

Several TCP versions can handle multiple losses in a TCP flight, including TCP NewReno and TCP SACK.

TCP NewReno [48] is the same as TCP Reno, but it implements a modification to the sender's algorithm during Fast Recovery. To avoid a retransmission time-out, as occurs with TCP Reno, in TCP NewReno the sender remembers the sequence number of the last segment that was sent before entering the Fast Retransmit phase. So, when TCP NewReno enters the Fast Retransmit phase, it remains in that phase even when partial ACKs are received (which acknowledges some of the lost segments, but not all of them). Some variants of NewReno reset the retransmission timer with each partial ACK received. Finally, the sender leaves the Fast Retransmit phase when it receives an ACK for the segment with the highest sequence number sent by the source before entering the Fast Retransmit phase. In this way TCP NewReno allows recovery from X lost segments for X RTT intervals.

A TCP version that is capable of efficiently recovering from multiple losses is TCP SACK (TCP with Selective Acknowledgments) [49]. For single lost segments TCP SACK reacts in the same way as TCP Reno. TCP SACK works by appending to a duplicate ACK a TCP option containing a range of noncontiguous data blocks received. Support for SACK is negotiated at the beginning of a TCP connection (because it is optional), and may only be used if both hosts support it. However, TCP SACK can also go into Slow Start from the recovery phase, which happens when a retransmitted segment is lost and the retransmit timer expires. Overall, TCP SACK can restore multiple segment losses in a given flight (i.e., TCP outstanding packets) in a single RTT period, which is better congestion control performance than Tahoe, Reno, and NewReno [50]. The SACK is optional for TCP implementations, but it is mandatory in the Stream Control Transmission Protocol (SCTP), a protocol initially created for transfer of signaling over all-IP networks (SCTP is covered in Chapter 4).

2.8 Internet Governance

From its beginning until 1995, the Internet backbone was primarily funded from U.S. federal funds. From 1995 onward, no single person or single organization or government has run the Internet. However, the Internet has two naming spaces, network addresses (IPv4 and IPv6), and domain names, which must be managed on a global scale with the goal of having a functional global Internet as it currently is. Additionally, the Internet's initial concept for network neutrality (regarding applications and services) still survives in best-effort Internet usage, but it is changing with the transition of telephony and television (as legacy telecommunication services) to all-IP due to different requirements for QoS as well as other addressing schemes (e.g., telephone numbers). In legacy telecommunications each country is responsible for governance of the telecommunications and telecommunication infrastructure on its territory, including the numbering and addressing schemes as well as content. Nowadays the Internet is completely replacing legacy telecommunications and further extending it with many application ecosystems (run in a best-effort manner, which is nowadays regarded as OTT); hence, each country is becoming responsible for governance of the Internet infrastructure on its own territory. The global harmonization for legacy telecommunications (telephony and television) regarding the telephony numbering (and signaling) and frequency bands used for telephony and television is under the umbrella of ITU, which is the largest agency for ICT in the world and part of the United Nations. Such global harmonization is also needed for the Internet for the two name spaces (Internet addresses and domain names) as well as for AS numbers (because the current Internet is organized globally in a flat architecture consisting of interconnected ASs). However, the Internet has kept its own governance bodies, which have roots in ARPANET and later in NSFNET.

Several organizations play major roles in Internet governance:

- Internet Corporation for Assigned names and Numbers (ICANN): ICANN is a nonprofit U.S.-based organization, created in 1998, that has taken over most of the technical governance of the Internet, with its focus being the governance of the two name spaces: IP addresses (IPv4 and IPv6) and top-level domain (TLD) names such as .edu, .com, .int, and .org. Regarding the numbering, ICANN realizes assignment of IP addresses and port numbers through its Internet Assigned Numbers Authority (IANA) department. Additionally, IANA distributes the AS numbers, which are 16- or 32-bit numbers used to uniquely identify the autonomous systems.

- Internet Society (ISOC): ISOC is an international nonprofit organization that includes the Internet Engineering Task Force (IETF) as its major standardization organization for Internet technologies, and the Internet Research Task Force (IRTF). The management of ISOC is through the Internet Architecture Board (IAB), a committee initially created by DARPA.
- WWW Consortium (W3C): W3C is a global standardization organization for the World Wide Web (WWW), formed in 1994 by Web inventor Tim Berners-Lee. Its main task is to standardize key parts of what makes the WWW work.
- Regional Internet registries (RIRs): There are five RIRs in the world, and each of them manages allocation of IP addresses and AS numbers (allocated by IANA to each RIR) within a particular region in the world. RIRs further delegate parts of regional allocations of network addresses and AS numbers to local Internet registries (LIRs) and other customers (ISPs such as telecom operators, different organizations such as universities, administrations, and so on). The five RIRs, part of the Internet Numbers Registry System [51], are as follows: African Network Information Centre (AfriNIC), American Registry for Internet Numbers (ARIN), Asia-Pacific Network Information Centre (APNIC) , Latin America and Caribbean Network Information Centre (LACNIC), and Réseaux IP Européens Network Coordination Centre (RIPE NCC).

Regarding the regulation of contents, the global Internet is open to all contents and applications. However, the contents are regulated by national (local) laws for electronic communications and/or content (e.g., web content, video content). In that respect, some contents may not be available in certain countries or regions, which is typically dependent on local (national) legislation as well as the business strategies of telecom operators.

The current model of Internet governance is likely to evolve further in the near future. From its current model it is expected to transit to a multistakeholder model, which should allow the private sector to take leadership for DNS management (e.g., take over from IANA functions). Such an approach is relevant for the best-effort Internet. On the other hand, PSTNs are transiting to carrier-grade VoIP, which is logically separated from the best-effort traffic, but it is fully carried over the same Internet infrastructure using the same Internet technologies standardized by IETF. Numbering for telephony is governed by the ITU. Hence, one may conclude that Internet governance has already started to change during the 2010s and it is developing toward multistakeholder Internet governance on a global scale.

References

[1] J. Postel, "Internet Protocol," RFC 791, September 1981.
[2] S. Deering, R. Hinden, "Internet Protocol, Version 6 (IPv6) Specification," December 1995.
[3] S. Deering, R. Hinden, "Internet Protocol, Version 6 (IPv6) Specification," RFC 2460, December 1998.
[4] J. Touch, "Updated Specification of the IPv4 ID Field," RFC 6864, February 2013.
[5] K. Egevang, P. Francis, "The IP Network Address Translator (NAT)," May 1994.
[6] J. Postel, "Internet Control Message Protocol," September 1981.
[7] R. Braden, "Requirements for Internet Hosts—Communication Layers," RFC 1122, October 1989.
[8] D. C. Plummer, "An Ethernet Address Resolution Protocol," RFC 826, November 1982.
[9] R. Finlayson, et al., "A Reverse Address Resolution Protocol," RFC 903, June 1984.
[10] Y. Rekhter, T. Li, "An Architecture for IP Address Allocation with CIDR," RFC 1518, September 1993.
[11] Y. Rekhter et al., "Address Allocation for Private Internets," February 1996.
[12] M. Cotton et al., "Special-Purpose IP Address Registries," RFC 6890, April 2013.
[13] J. Weil et al., "IANA-Reserved IPv4 Prefix for Shared Address Space," RFC 6598, April 2012.
[14] Google, "IPv6 Statistics," www.google.com/intl/en/ipv6/statistics.html, accessed September 2014.
[15] J. Reynolds, J. Postel, "Assigned Numbers," RFC 1700, October 1994.
[16] S. Amante et al., "IPv6 Flow Label Specification," RFC 6437, November 2011.
[17] R. Braden et al., " Resource ReSerVation Protocol (RSVP)—Version 1 Functional Specification," RFC 2205, September 1997.
[18] H. Schulzrinne, R. Hancock, "GIST: General Internet Signalling Transport," RFC 5971, October 2010.
[19] S. Kent, R. Atkinson, "IP Authentication Header," RFC 2402, November 1998.
[20] S. Kent, R. Atkinson, "IP Encapsulating Security Payload (ESP)," RFC 2406, November 1998.
[21] S. Kent, R. Atkinson, "Security Architecture for the Internet Protocol," November 1998.
[22] S. Krishnan et al., "A Uniform Format for IPv6 Extension Headers," RFC 6564, April 2012.
[23] Internet Assigned Numbers Authority, "Protocol Numbers," www.iana.org/assignments/protocol-numbers/protocol-numbers.xml, accessed September 2014.
[24] B. Carpenter, S. Jiang, "Transmission and Processing of IPv6 Extension Headers," RFC 7045, December 2013.
[25] A. Conta, S. Deering, M. Gupta, "Internet Control Message Protocol (ICMPv6) for the Internet Protocol Version 6 (IPv6) Specification," RFC 4443, March 2006.
[26] T. Narten et al., "Neighbor Discovery for IP Version 6 (IPv6)," RFC 4861, September 2007.
[27] S. Deering, W. Fenner, "Multicast Listener Discovery (MLD) for IPv6," RFC 2710, October 1999.
[28] B. Haberman, J. Martin, "Multicast Router Discovery," RFC 4286, December 2005.
[29] C. Perkins, D. Johnson, J. Arkko, "Mobility Support in IPv6," RFC 6275, July 2011.
[30] IANA, "Internet Control Message Protocol Version 6 (ICMPv6) Parameters," www.iana.org/assignments/icmpv6-parameters/icmpv6-parameters.xhtml, accessed September 2014.
[31] R. Hinden, S. Deering, "IP Version 6 Addressing Architecture," RFC 4291, February 2006.
[32] A. Conta, "Extensions to IPv6 Neighbor Discovery for Inverse Discovery Specification," RFC 3122, June 2001.

[33] J. Arkko et al., "SEcure Neighbor Discovery (SEND)," RFC 3971, March 2005.

[34] R. Hinden, B. Haberman, "Unique Local IPv6 Unicast Addresses," RFC 4193, October 2005.

[35] Y.-G. Hong et al., "Application Aspects of IPv6 Transition," RFC 4038, March 2005.

[36] S. Thomson, T. Narten, T. Jinmei, "IPv6 Stateless Address Autoconfiguration," RFC 4862, September 2007.

[37] M. Crawford, "Transmission of IPv6 Packets over Ethernet Networks," RFC 2464, December 1998.

[38] E. Nordmark, R. Gilligan, "Basic Transition Mechanisms for IPv6 Hosts and Routers," RFC 4213, October 2005.

[39] R. Droms, "Dynamic Host Configuration Protocol," RFC 2131, March 1997.

[40] R. Droms et al., "Dynamic Host Configuration Protocol for IPv6 (DHCPv6)," RFC 3315, July 2003.

[41] Internet Assigned Numbers Authority, "Dynamic Host Configuration Protocol (DHCP) and Bootstrap Protocol (BOOTP) Parameters," www.iana.org/assignments/bootp-dhcp-parameters/bootp-dhcp-parameters.xhtml, accessed September 2014.

[42] J. Postel, "User Datagram Protocol," RFC 768, August 1980.

[43] J. Postel, "Transmission Control Protocol," RFC 793, September 1981.

[44] Internet Assigned Numbers Authority, "Service Name and Transport Protocol Port Number Registry," www.iana.org/assignments/service-names-port-numbers/service-names-port-numbers.xhtml, accessed September 2014.

[45] D. Borman, "TCP Options and Maximum Segment Size (MSS)," July 2012.

[46] J. Postel, "The TCP Maximum Segment Size and Related Topics," RFC 879, November 1983.

[47] M. Allman, V. Paxson, E. Blanton, "TCP Congestion Control," RFC 5681, September 2009.

[48] S. Floyd, T. Henderson, A. Gurtov, "The NewReno Modification to TCP's Fast Recovery Algorithm," RFC 3782, April 2004.

[49] M. Mathis et al., "TCP Selective Acknowledgment Options," RFC 2018, October 1996.

[50] K. Fall, S. Floyd, "Simulation-Based Comparisons of Tahoe, Reno and SACK TCP," *Computer Communication Review*, July 1996.

[51] R. Housley et al., "The Internet Numbers Registry System," RFC 7020, August 2013.

CHAPTER 3

Internet Networking

3.1 What Is a Socket?

A socket is an end point of a given Internet connection. However, the next question that arises is about the connection to the Internet and its definition. The connection is uniquely identified by the 5-tuple (source and destination IP addresses, source and destination ports, and protocol) for IPv4 communication, and by the 3-tuple (source and destination IPv6 addresses, and flow label) for IPv6 connections (although it is possible to use 5-tuple with IPv6 also). Then, the socket is a virtual end point of a given Internet connection that binds the IP address, port, and protocol with a given application protocol (above the transport protocol layer) installed on a given host. Internet communication is based on information exchange between two processes running on different hosts across the Internet (although communication through the local Loopback address is an exemption of this principle). Currently all Internet communication is realized via sockets.

The initial definition of a "socket" was created at the beginning of 1970s [1]. The socket was defined to be the unique identification to or from which information is transmitted in the network. However, at that time it was defined as a 32-bit number for use by the Network Control Protocol (NCP), which later split into the existing two protocols, IP and TCP (standardized in 1981, and implemented on all Internet hosts on January 1, 1983, i.e., the Flag Day). Since its first definition for the Internet, the socket has evolved with the introduction of the TCP/IP protocol stack into its current form, as a combination of an IP address, port number, and transport protocol, on each end of the given connection. The protocol for a given socket is a transport layer protocol (e.g., TCP, UDP) in end hosts, or network layer protocol (i.e., the IP) in network nodes (e.g., routers). Generally, the socket identifies the application process or thread in a similar manner as a telephone number in the PSTN identifies the socket in the wall that connects a telephone device. In that manner, for two hosts to communicate over the Internet, each one must have an open socket. So, every end-to-end connection in the Internet is established between a pair of sockets. Hence, sockets can be created to establish a connection and then terminated after the connection had ended. However, certain hosts (called servers), have open sockets for listening to incoming connections from other hosts (called clients).

How is a socket implemented in machines connected to the Internet? A socket is implemented via a so-called socket interface. In practice, that is a set of instructions that provides capability to a given application program running on a given machine with a certain OS to communicate over the Internet with another application program running on another machine. However, different machines may have different OSs (e.g., UNIX, Linux, Windows, Macintosh, or Android), and therefore it is not adequate to standardize the interface between the applications (i.e., application processes) and the Internet protocol suite (e.g., TCP/IP, UDP/IP) implemented on the side of machine's OS. So, the main reason why the socket interface has not been standardized is because of the dependence of such an interface on the particular OS used by the Internet host. The implementation of the socket interface is made via the Application Programming Interface (API), which is a programming interface (based on library routines and calls implemented by the OS) that can be used by applications to interact with the Internet protocols on the transport and network layers. Most known APIs are Berkley sockets and System V transport layer interface (TLIs) [2, 3].

Another alternative to the socket interface is the STREAM-based TLI in UNIX System V. However, the most used implementation in the past decades has been the Berkley sockets, which were introduced with the 4.2 BSD (Berkeley Software Distribution) Unix operating system, released in 1983 and made publicly available in 1989 with the 4.3 BSD UNIX version (without licensing constraints). The interface implementation of Berkeley sockets is the original API used in the Internet hosts. Although API is not standardized, almost all OS vendors have based the socket interface implementation on Berkeley sockets.

The UNIX operating system was created in the late 1960s and early 1970s. It was created as single processor system by dividing the central processing unit (CPU) time between applications that are being executed simultaneously (i.e., a timesharing scheme). The basic element in UNIX is the process. Application programs are executed as processes. The communication of a given application with the operating system is performed through so-called system calls, which behave like certain procedures from the standpoint of a programmer. System calls specify certain arguments as inputs (which are specific to the given system call), and after its execution they provide a value or values as a result.

In UNIX operating systems, the input/output characteristics are based on the philosophy of open–read–write–close. In fact, in UNIX all I/O operations are seen as read-and-write operations to files, including regular files, devices, or network communications. A user process begins I/O operations by using a system call "open" that, when executed, returns an integer response called a descriptor (e.g., file descriptor) that is associated with the called file or device. When the target object (e.g., a file or device) is open for I/O operations, one can read from it or write into it using calls "read" and "write," respectively. (Both calls use three main input arguments: file descriptor, the address of the buffer, and the number of bytes to be transmitted.) When the I/O operations are finished, the process closes communication with the object by calling for a system call "close." (However, if the process that initiated the "open" terminates, then the OS automatically closes all I/O operations associated with that process.)

At the beginning of the implementation of TCP/IP in the early 1980s, network communication in the UNIX OS used the same approach as for I/O operations on files. The original TCP/IP implementation of BSD UNIX was developed by Bolt, Beranek, and Newman in DARPA in 1981. However, because network protocols are more complex than locally connected I/O devices, the communication between user processes and network protocols appears to be more complex than communication between user processes and common I/O operations in the local file system. Therefore, 4.x BSD UNIX has introduced several new system calls to the OS (defining the socket interface), with new library routines. At that time UNIX needed to support simultaneous existence of different communication protocols (e.g., UNIX, Internet, and Xenox domains), which increased the complexity of the network programming (i.e., creation and use of API). Here are some features that were taken into consideration in network programming, which are different than I/O operations with files:

- Client–server communication is not symmetric. The process needs to know whether it is a client or server. However, it is possible for a given application to have implemented both types of socket interfaces.
- Network communication can be connection oriented (based on TCP/IP) or connectionless (based on UDP/IP).
- Names in network I/O operations are more important than names used in I/O operations with local files. Namely, the application must know the name of the remote process (on the remote host) to be able to communicate with it.
- Every network communication must specify five parameters (in I/O operations with files there are just three parameters): local address, local process, external address, external process, and protocol (i.e., the 5-tuple).
- Some communication protocols have restrictions on the size of transmitted messages. The Unix-enabled I/O system is stream oriented and it is not based on messages (i.e., there are no defined boundaries on data transmitted between the operating system and the application).

These characteristics of the UNIX OSs have influenced the creation of Berkeley sockets as a socket interface, which further was accepted conceptually in all operating systems for Internet hosts that appeared later (Windows, Macintosh, Android, and so forth).

Whenever possible, Berkeley sockets use the same API-like UNIX files or devices. So, a socket appears as a generalization to I/O UNIX systems, to files hosted on other machines (not only to local files in the given host). Similar to local file access (for read or write operations), the application program (used for communication with a remote host) requests the operating system to create a socket when it needs to send/receive data to/from a remote machine (i.e., host). After the creation of the socket, the operating system returns a small integer that is further used as socket descriptor. So, the application program uses the socket descriptor to send or receive data from the created socket. However, a socket descriptor can be created without being bound to a specific destination address, which is different from file descriptors that are always bound to a certain local file or device [4].

In all hosts the socket interface is placed between the operating system and the application programs (Figure 3.1). Additionally, in the Internet network nodes such as routers the sockets are also placed between the program (e.g., a routing

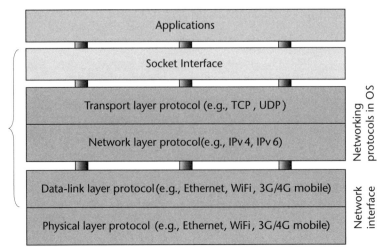

Figure 3.1 Socket interface in the Internet hosts and network nodes.

protocol) and the IP layer (Figure 3.1), using IP packets directly without a transport layer protocol (such as TCP or UDP).

Based on the location of the sockets in the protocol layering and the type of the protocol used for socket creation, there are three types of sockets:

- Datagram sockets (UDP/IP sockets): These are based on the UDP on the transport layer, and hence used for connectionless communication.
- Stream sockets (TCP/IP sockets): These are based on TCP or SCTP, and hence used for stream-based connection-oriented communication.
- Raw sockets (raw IP sockets): These are typically used in network nodes such as routers, bypassing the transport layer and making IP packet headers directly accessible to the application (e.g., routing protocols, security applications).

Further classification of sockets can be done based on the type of IP addresses (IPv4 and IPv6), for which there can be two types of sockets:

- IPv4 sockets: These bind an IPv4 address (of the network interface) with the port number and protocol.
- IPv6 sockets: These bind an IPv6 address (of the network interface) with the port number and protocol. (Port numbers have same length and allocation regardless of the IP address type.)

If we want to be more general, note that there are other types of sockets (or similar interfaces) in packet-based networks that are not IP based. Such examples are ITU's X.25 sockets. However, due to convergence of all telecommunication networks and services toward Internet technologies, socket types that are not related to Internet communication are considered not important in the present time (and for the near future).

3.2 TCP Sockets (Stream Sockets)

Stream sockets provide two-way communication streams with reliable transmission of packets. They are based on TCP on the transport layer and IP on the network layer. Their characteristics can be summarized as follows:

- Transmission is stream based, and large amounts of data can be transmitted between two application programs (i.e., user processes), where the data are divided into 8-bit octets (i.e., bytes).
- The stream socket interfaces on both ends of the connection create a type of virtual circle (because it is two-way communication), allowing data transfer similar to the streams of voice data (analog or digital) over an established telephone connection in the PSTN. Actually, one machine generates the call to establish a connection, and the other one accepts it. For the purpose of connection establishment, both application programs (in both machines) must inform their local operating system to create sockets for streaming data transmission. The connection is a virtual one because application programs see the sockets as assigned hardware, but in fact it is an illusion provided by the service programs aimed to control the transmission stream.
- The transfer is buffered. Application programs stream data in virtual time, where the size of the data stream is determined by the application program. (It is always an integer number of bytes, so the smallest stream size can be 1 byte.) However, lower-layer protocols can repack the data to provide flow control for efficient utilization of network resources (e.g., creation of TCP segments on transport layer, and IP packets on network layer). The packets must arrive at the receiving stream socket in the same order in which they were sent by the sending stream socket. Buffers are used on both sides of the connection for storing bytes to be transmitted or bytes that are received.
- Before the start of transmission, both sides agree on the format of the stream of data that will be used in the transmission.
- Communication over stream sockets is full duplex, which mean that simultaneous communication is enabled in both directions. Communication in one direction can be completed without disturbing the other direction. Two-way transfer reduces the network traffic load, since it provides feedback control information to be transmitted within the datagrams transmitted in the opposite direction.

The characteristics of stream sockets are compatible with the characteristics of TCP; hence, they are also referred to as TCP sockets. This socket type is only used over TCP as the transport protocol and IP as the network protocol. However, at the time of initial creation of the socket interface with Berkeley sockets, IPv4 was the only available version of the Internet Protocol. With the introduction of IPv6, TCP sockets for IPv6 network interfaces were created.

TCP sockets are based on the client–server model, because TCP is based on the client–server model. (There is always a TCP server listening to incoming connections from TCP clients.) In general, the client–server communication model in the Internet involves two sides with different functions:

- Server: Provides service through a well-defined server interface.
- Client: Requests service from the server through a well-defined client interface.

TCP/IP communication is based on the client–server model (Figure 3.2). The server is basically a program that listens on a particular TCP port on the server host. The client is a program that initiates a connection to the application program on a server (across the Internet). Initially, all Internet applications were based on the client–server model. One of the first applications based on the client–server model was Telnet, which typically provides access to a Telnet client on one machine to a command-line interface on a remote machine that has a Telnet server. The Telnet server listens on TCP port 23 for incoming connections from Telnet clients. Typically, server programs such as Telnet server run as daemon processes, which is a term that is usually used for computer programs that run as background processes without direct interactive control by the user (i.e., without a user interface). A machine that has server programs running on it is typically referred to as a server machine or, simply, a server. However, a single machine can act as a server for certain application programs, but can then act as a client machine for other application programs. In the existing Internet architectures, servers typically have statically assigned IP addresses, so they can listen to incoming connections for longer time periods, while a client machine or, simply, a client needs to have an IP address (on a given network interface connected to the Internet) only when it initiates and then maintains a connection to a given server on the Internet (or in the local IP network). Other client–server examples that use TCP sockets are FTP, HTTP, and SMTP.

In client–server communication, binary data (bits) are transmitted from one machine to another one across the Internet. The data sent from one host must be "understandable" on the receiving side, which is possible when the order of the transmitted data bits is structured in a given predefined way.

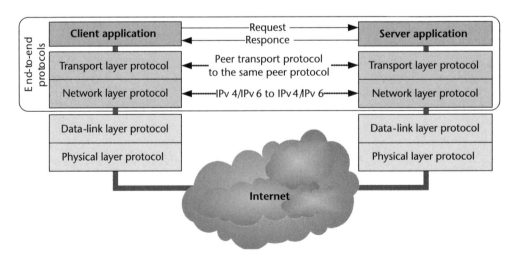

Figure 3.2 Client–server communication model.

3.2.1 Interpretation of Binary Data in Internet Networking

Different hosts and network nodes have different operating systems. Each OS has its own method of interpretation of data stored in its local memory. In all hosts and network nodes, the memory typically stores binary data organized into 8-bit units called bytes (or octets). Each byte in the memory has its own memory address (internal to given host or node), so it can be accessed by local programs that are running on that machine. When a certain application program reads data from memory (which is organized in bytes) or writes into it, then the order of the bytes stored in the memory defines the interpretation of the binary data. Overall, two types of operating systems are used in machines:

- Big-endian: This type of system stores the most significant byte of a word in the smallest memory address, and the least significant byte in the highest address.
- Little-endian: This type of system stores the most significant byte in the highest address, and the least significant byte in the smallest memory address.

Examples of big-endian and little-endian data interpretations are shown in Figure 3.3. The two types of endian systems create potential problems for different interpretations of exchanged data to occur between machines that use different ways to interpret bytes in their OSs.

To avoid the misinterpretation of transmitted data, the order of transmission of bytes across the Internet, called the network byte order, has been standardized. The convention for the Internet is to use the big-endian type of data interpretation for communication, as stated in RFC 1700 [5]. Although the network byte order in the Internet is big-endian, different OSs can still use different orders, such as little-endian, locally. However, any data transmitted across the Internet must be converted from local byte order (in the machine) to network byte order (i.e., to big-endian) before the transmission, and vice versa, from network byte order to local byte order after reception of the data.

3.2.2 TCP Socket Interface

For TCP/IP (as well as UDP/IP) communication, the ideal API would be one that can "understand" IP addresses and port numbers. Because the library of system calls for sockets in the operating system typically is used for multiple networking protocols on the network and transport layers, addresses are placed in more

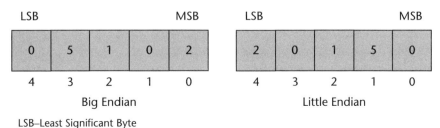

Figure 3.3 Big-endian and little-endian data interpretations.

general data structures that can be used by most of the protocols currently in use. Note, however, that the two standardized types of Internet Protocol (IPv4 and IPv6) require separate data structures for their sockets (regardless of the transport layer protocol), at least because of the different sizes of IPv4 and IPv6 addresses. In general, the socket interface API consists of the following components [6]:

- Main socket functions: These are designed to be transported independent of the layer and are used to set up and tear down TCP connections, as well as to send and receive UDP packets.
- Address data structures: These structures are protocol specific and hence defined for each protocol that is supported by the socket functions (i.e., different structures are required for IPv4 and IPv6).
- Name-to-address translation functions: These functions provide translation of a host name to an IP address and vice versa.
- Address conversion functions: These are specifically created to convert the IPv4 addresses (32 bits) between binary and decimal form, but functions also exist that convert both IPv4 and IPv6 addresses.

When the socket structures are defined, they are used as arguments in system calls to the socket interface. However, system calls are different for the client and server side, as shown in Figure 3.4 for TCP sockets (i.e., stream sockets).

In the following section we illustrate system calls used for the creation, operation, and closing of TCP (i.e., stream) sockets, as shown in Figure 3.4. Some of them are also used in UDP (i.e., datagram) sockets, such as the "socket," "bind," and "close" system calls. The standard socket interface, based on 4.3BSD UNIX implementation, is written in the C programming language. That version has been adapted for use in almost all operating systems used by Internet hosts today.

The creation of a TCP socket is realized through the following functions:

- Socket creation: This is the first step in the creation of every Internet connection, the opening of a socket in order for a process to perform network I/O operations. Socket creation is executed by a call to a function "socket," which specifies the protocol to be used for communication (e.g., SOCK_STREAM for TCP, SOCK_STREAM UDP, and SOCK_RAW for raw sockets). This system call creates a communication end point for the application program. After execution it returns an integer value that is used as a socket descriptor for the created socket, which is further used as an argument to almost all other system calls used for that connection.
- Binding a socket to a network interface: Before the application can start sending and receiving data through the created socket, it must bind the created socket to a local port and IP address assigned to an existing network interface on the local machine. The mapping of the socket to a given port (e.g., TCP port, UDP port) and IP address is called binding, which is realized with the system call "bind." In TCP-based Internet communication, the "bind" function is typically used on the server side to bind a socket with an IP address and a port to which the server listens to incoming connection requests

3.2 TCP Sockets (Stream Sockets)

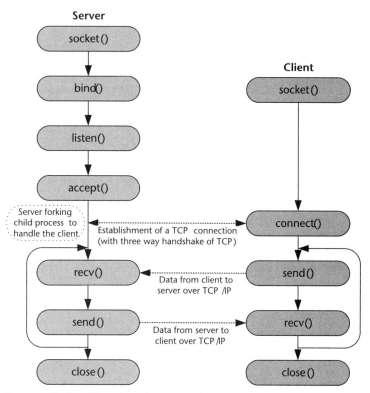

Figure 3.4 System calls for TCP sockets (stream sockets).

from clients. In UDP-based communication, the "bind" function is used by both sides, client and server, before the data transfer.

- Connecting to a remote host: Initially each socket is not connected to any remote host, and it is not related to any destination. Then, the "connect" function is used by clients to connect to a particular server in the case of connection-oriented communication, such as TCP-based communication. UDP sockets, however, do not need to connect before use, but they also may use the "connect" function to provide data transmission without specifying the destination address for each datagram that is sent from the client to the server.
- Sending and receiving data through a socket: When an application program has created a socket, bound it to a local address (i.e., local IP address and port), and finally connected the socket to a given destination address (i.e., destination IP address and port), then it can start to send and receive data through that socket. To transfer data through a TCP stream socket or through a UDP datagram socket connected to the destination address, the application uses the "send" function to send data, and the "recv" function to receive data. To send and receive datagrams without connecting to the destination address, the application program uses the functions "recvfrom" and "sendto," respectively. Functions "send" and "recv" are used only in cases when a previously established connection exists between the client and the server (by calling the "connect" function). Both functions for sending or

receiving data through a socket return a value that is the number of bytes that are read or written in the buffer memory.

- Server listening on incoming connections from clients: In TCP-based connection-oriented communication, the server listens for incoming connections from clients through a given socket, which has been previously created and bound to a local address (IP address and port) of the server. This is accomplished with the "listen" function.
- Accepting connections by the server: After the established socket on the server side starts to listens to incoming connections from clients (by using the "listen" function), the server waits for a connection. The server uses the "accept" function to inform TCP (locally) that it waits for connections from clients, thus blocking itself until the connection request from a client arrives. For each new connection request, a new socket is created. It can be handled iteratively or concurrently [4]. In the iterative case, the server closes the new socket created with the "accept" function upon receiving a connection request from a client, and handles the request by itself. Afterward it calls the "accept" function for the next connection request. In the concurrent approach, the server process creates a duplicate process for each connection request (in UNIX terminology, the server parent process forks a child process for each connection request from a client), which communicates with the client process via the newly created socket (as a return of the "accept" function call). The concurrent processes on the server side are possible because each connection in the Internet is identified by the IP address and port on each end of the connection and the protocol in use (TCP, UDP, and so forth). So, when different clients from different destinations connect to the same network interface of a given server, then there is one different end point in each connection (the end point on the client side). After the parent process on a server forks a child process (within the same server), it returns to listen to the next connection request, while the CPU gives processing time to each of the processes alternatively. When an established TCP connection between a client and the server is closed, the child process (i.e., the slave process at the server) closes the socket and terminates.
- Termination of communication through a socket: When data transfer through a socket is completed, it can be simply closed by calling the "close" function. For UDP datagram sockets, this system call releases the port that was previously bound to the socket. In the case of TCP stream sockets, the connection is first closed in both directions (client to server, and vice versa) and then the port is released. However, TCP sockets can also be closed in one direction of the connection by using the "shutdown" function. If the goal, however, is to finally close the socket, the application process must call the "close" function.

Besides functions for handling the sockets, the socket API uses a set of so-called library routines. While system calls are part of the OS of the host, the library routines are similar to other procedures that programmers use in their programs. An application may use system calls in a computer's OS, such as "socket," "bind," "connect," "accept," and so forth, and library routines, which belong to the application space. In addition, library routines may call other library routines or system calls. Many library routines are used in network programming; the most used ones are as follows:

- Byte ordering functions: Information in machines (computers, smartphones, and so on) is in host byte order (hbo), which can be little-endian or big-endian. However, network byte order (nbo) is big-endian. To provide for the possibility that each machine uses arbitrary byte ordering yet still allow for data exchange with a host, each host before sending the data through a socket calls routines for changing hbo to nbo. Similarly, byte ordering functions (from nbo to hbo) are used at the receiving end.
- Address conversion functions: These were defined first for IPv4 (AF_INET address family) and later for IPv6 (AF_INET6 address family) [6]. These functions convert an IP address given in its standard text presentation form (which is different for IPv4 and IPv6) into its numeric binary form, and vice versa.
- Memory management functions: These functions manage values that are stored in the memory, such as storing bytes (received or ready for transmission) on a given memory address, copying bytes from one to another memory address, and so forth.

3.3 UDP Datagram Sockets

UDP-based connectionless communication does not require explicit identification of who is a server and who is a client. When initiating a call the client application program must specify its own IP address and the UDP port number on the local machine through which communication will be realized.

The characteristics of the UDP protocol are also characteristics of the UDP datagram sockets. UDP actually uses IP as an underlying network protocol and it is providing the same service (datagram transfer) as IP (transfers packets from a given source to a given destination), but with one important difference: the ability of the UDP to differentiate between different processes on the same machine by using the ports as application addresses on the transport layer.

System calls used for UDP-based communication (i.e., UDP sockets) are shown in Figure 3.5. Note that UDP communication, like TCP, also starts with creation and binding of a socket with system calls "socket" and "bind," respectively. In contrast, however, the "connect" function is optional for datagram sockets, because UDP clients do not need to connect to the destination, as is required for TCP-based communication. However, the "connect" function can be used for UDP sockets, but it is not used for establishing a connection to the remote socket; it is instead used for storing the remote socket address locally (at the sending host) for the following datagrams that have to be sent to the same destination address. When UDP communication is not using the "connect" function (on the client side) before the data transfer, the application program typically uses functions "sendto" and "recvfrom" for sending and receiving datagrams, respectively.

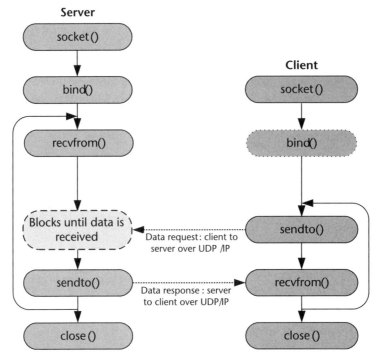

Figure 3.5 System calls for UDP sockets (datagram sockets).

Similar to the "send" and "recv" functions used for TCP sockets, the "sendto" and "recvfrom" functions return the number of bytes that are read (i.e., sent) or written (i.e., received) in the buffer memory. For each open datagram socket that is bound to a given port and IP address on the local machine, the application can use functions "sendto" and "recvfrom" as many times as necessary.

When there are no datagrams to be sent or received, the UDP socket is closed by calling the "close" function, in the same manner as it is used for the TCP sockets.

Overall, the datagram sockets are used for connectionless communication, which is convenient for delay-sensitive applications such as VoIP or live video streaming (e.g., IPTV), as well as specific applications or systems that require higher reliability that is managed by the application layer processes (e.g., DNS).

An illustration of the Internet network programming regarding client and server programs is given in Figure 3.6.

3.4 Client–Server Architectures

The Internet as a global network architecture for all telecommunication services, including Internet native services (e.g., e-mail, FTP, WWW) and traditional telecommunication services (e.g., telephony, radio, and television), is based on the client–server principle in which both the clients and the servers use the same Internet protocol stack on the network layer (e.g., IPv4, IPv6) and transport layer (e.g., TCP, UDP) implemented within the operating systems in both end hosts of a given communication session. Each communication application uses a defined API in the given operating system to send and receive data across the Internet through the

3.4 Client–Server Architectures

Code	Description

`#include <stdio.h>`
`#include <sys/socket.h>`
`#include <netinet/in.h>`
`#include <string.h>`
	Header files that include all system calls and library routines needed for the given program.

`int main(){`
  `int ServSocket, ChildSocket;`
  `char buffer[1024];`
  `struct sockaddr_in ServerAddress ;`
  `struct sockaddr_storage servStorage ;`
  `socklen_t addr_size ;`
	Definition and initialization of variables and structures.

`ServSocket = socket(PF_INET, SOCK_STREAM , 0);`	Creation of stream socket in Internet domain

`ServerAddress .sin_family = AF_INET;`
`ServerAddress .sin_port = htons(7777);`
`ServerAddress .sin_addr.s_addr = inet_addr("127.0.0.1");`
`memset(ServerAddress .sin_zero, '\0', sizeof(ServerAddress .sin_zero));`
	Setup of the server address, with IP address, port, address family and padding bits set to zero.

`bind(ServSocket , (struct sockaddr *) &ServerAddress , sizeof(ServerAddress ));`	Binding ListeningSocket to the server address.
`if(listen(ServSocket ,5)==0) printf("Server is listening \n");` `else printf("Error!\n");`	Listen on the socket for incoming connections
`addr_size = sizeof(servStorage );` `ChildSocket = accept(ServSocket , (struct sockaddr *) &servStorage , &addr_size);`	Accept of client's request.
`strcpy(buffer,"Hello Internet\n");` `send(ChildSocket,buffer,16,0);`	Sending data through the stream socket toward the client.
`return 0;` `}`	

a) Simple server code (in c programming language)

Code	Description

`#include <stdio.h>`
`#include <sys/socket.h>`
`#include <netinet/in.h>`
`#include <string.h>`
	Header files that include all system calls and library routing needed for the given program.

`int main(){`
  `int clientSocket;`
  `char buffer[1024];`
  `struct sockaddr_in serverAddr;`
  `socklen_t addr_size ;`
	Definition and initialization of variables and structures.

`clientSocket = socket(PF_INET, SOCK_STREAM , 0);`	Creation of stream socket in Internet domain

`serverAddr.sin_family = AF_INET ;`
`serverAddr.sin_port = htons(7777);`
`serverAddr.sin_addr.s_addr = inet_addr("127.0.0.1");`
`memset(serverAddr.sin_zero, '\0', sizeof serverAddr.sin_zero);`
	Setup of the server address, with IP address, port, address family and padding bits set to zero.

`addr_size = sizeof serverAddr ;` `connect(clientSocket, (struct sockaddr *) &serverAddr, addr_size );`	Listen on the socket for incoming connections
`recv(clientSocket, buffer, 1024, 0);`	Receiving data through the socket from the server
`printf("Data received: %s",buffer);`	
`return 0;` `}`	

b) Simple client code (in c programming language)

Figure 3.6 Illustration of client and server programs.

socket interfaces (e.g., TCP sockets, UDP sockets). TCP connection-oriented and UDP connectionless communication are both based on the client-server model.

In client-server network architectures typically the server provides services through well-defined interfaces, by listening for requests from clients with open sockets on predefined or well-known ports. The client is the host that requests a certain service through the given interface on the server. The server responds to a client's requests with a certain response (e.g., the server accepts the request and provides the service to the client, or rejects it). An Internet architecture that consists of hosts that are either clients or servers, interconnected by switches and routers, is called a client-server network architecture (Figure 3.7).

In practice, servers are more powerful computers than clients, because usually one server services many clients (depending on the service type provisioned by the server). In contrast, clients are typically end-user machines such as PCs, laptop computers, smartphones, and smaller devices. However, in certain cases a single machine may act as a server to clients, and as a server to another machine (for another application process), something that is typical for machines (e.g., machines that host application servers, databases, and so on) in the core network of network providers and services providers (e.g., centralized network controllers and gateway nodes). Introduction of the client-server architecture by the Internet was a crucial development from previous network architectures (e.g., PSTN) that were based on centralized mainframe computers (on the network side) and simple "dumb" devices (on the user side).

Finally, the client-server architecture is inseparably defined with the Internet protocol architecture and its transport layer protocols. Note that both end points (i.e., hosts) in a communication over a client–server network architecture must use the same transport protocol (e.g., TCP, UDP) and the same type of sockets on both sides (e.g., stream sockets, datagram sockets). The peer application on both end hosts in each communication must use the same type of applications (Web clients

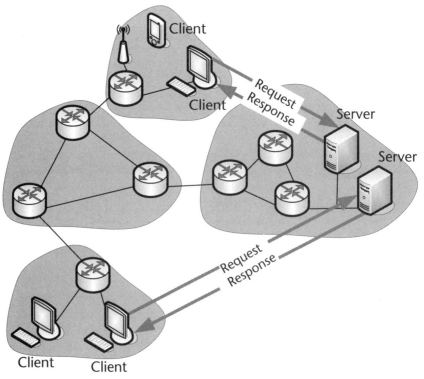

Figure 3.7 Client-server network architecture.

3.5 Peer-to-Peer Architectures

communicate with Web servers, FTP clients communicate with FTP servers, and so on).

3.5 Peer-to-Peer Architectures

P2P architectures are network architectures in which each host (e.g., computer, smartphone) or network node (e.g., router) has similar capabilities and functionalities (Figure 3.8). Such an approach is different than client-server architectures in which certain hosts or nodes are dedicated to serving others. Note, however, that a single host (e.g., a PC) may be simultaneously a part of a client-server network architecture (e.g., for Web browsing or email) and a P2P network architecture (e.g., for Skype or BitTorrent). The same statement can be made for network hosts and network nodes. So, the application design (used on a given host or node) defines the current type of network architectures for that application. However, typically the owner of each host or node on a P2P network is supposed to set up a certain amount of resources to be shared with other hosts or nodes (e.g., processing power, maximum bit rates, maximum memory that can be used on the local hard disk).

There are different definitions of P2P systems. According to the IETF definition [7], a system is P2P if the elements that form the system (e.g., hosts, network nodes) share their resources in order to provide the service the system has been

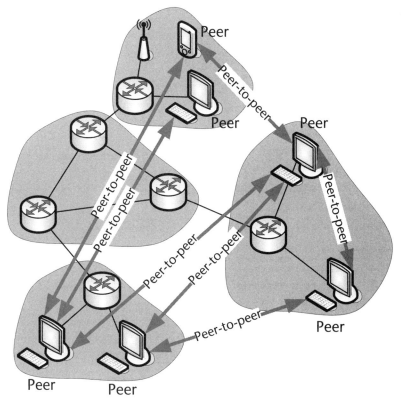

Figure 3.8 Peer-to-peer network architecture.

designed to provide. At the same time, such elements in the P2P system both provide services to other elements and request services from other elements. However, some exceptions are allowed. For example, certain P2P systems (i.e., P2P network architectures) may use centralized servers (e.g., for management of user data) and still be considered P2P.

In general, a P2P network is based on establishment of an overlay network in the Internet consisting of peers (nodes in the P2P networks) that act as clients or servers to other nodes in the P2P network, allowing shared access to different resources such as files, video/audio streams, various devices (e.g., cameras, sensors), and so forth.

In terms of P2P network architectures [8], two main types are used:

- *Structured P2P network:* In this type of network architecture, the peers are organized according to defined algorithms, policies, and specific topologies. In structured P2P architectures, peers join the P2P system by connecting themselves to well-defined peer nodes. Such P2P networks are typically used for specific large-scale service implementations that demand higher availability or reliability.
- *Unstructured P2P networks:* In these P2P networks peers join the system by connecting themselves to any other existing peers, so there is no defined, specific structure of an overlay P2P network. Unstructured P2P also shows a certain level of structure such as special nodes with more functionality (e.g., centralized node for indexing functions of the peers). In that manner, one may distinguish among the following types of unstructured P2P network architectures:
  - *Pure P2P:* In this type of architecture, there are no centralized nodes and all peers are equal. Communication between peers is typically established by using the IP addresses of the peers, which are stored locally in the peers.
  - *Centralized P2P:* This architecture has centralized nodes in the network for indexing (of the peers) and bootstrap functions (whom to contact when a P2P application is started on a given peer host).
  - *Hybrid P2P:* This architecture has both types of nodes: dedicated infrastructure nodes in the P2P network with defined functionalities (e.g., super nodes) and pure peer nodes.

The P2P network architectures are designed to be used by so-called P2P applications. Note that popular P2P applications for residential users had appeared by the end of the 1990s with Napster being the first globally widespread P2P application for sharing of music files. It was followed by other file-sharing systems (e.g., Gnutella, Kazaa, eDonkey, BitTorrent), P2P VoIP (e.g., Skype), and so forth. All such P2P applications are based on best-effort (i.e., network-neutral) P2P communication, without any QoS support. Generally, the existing P2P applications can be classified into the following types:

- *Content distribution:* This is a P2P application used to provide legitimate distribution of different contents, such as prerecorded TV programs, live TV, free software, and game updates. Such use of P2P systems provides higher scalability.

- *Distributed computing:* In this case each task is divided into separate subtasks that can be completed simultaneously by different peers and eventually delivered to the peer that initiated the task. In this way ordinary end-user machines (e.g., PCs) can complete tasks using the performance capabilities of more powerful computers.
- *Collaboration:* This type includes real-time P2P applications for communications between humans, such as VoIP and Instant Messaging (IM). In this group also belong standardized P2P communication, such as communication based on Session Initiation Protocol (SIP) signaling (e.g., used for VoIP by telecom operators). Additionally, P2P collaboration can be useful for creation of ad hoc communication infrastructures (e.g., in disaster areas).
- *Platforms:* This type of P2P system is typically used to build applications on top of them. Hence, these P2P platforms provide a convenient environment for application developers, which can benefit from the good scalability characteristics of P2P network architectures.

3.6 Internet Access Networks

Internet access networks are used for connecting end-user hosts to the Internet, including residential and business users. In the past access to the Internet was based on dial-up modems (e.g., 56-kbps modems) for residential and different types of access network for business users (e.g., token ring, FDDI, and Ethernet). Currently, access has converged toward unified access in local networks, based on Ethernets for wired local access (i.e., IEEE 802.3 standard and its amendments) and Wi-Fi (i.e., IEEE 802.11 standard and its amendments) and its wireless extension in homes, offices, and public places (i.e., hotspots).

Nowadays the range of hosts can be wide, such as PCs, laptops, smartphones, smart TV devices, smart home devices, smart vehicles, and different "things" (sensors, cameras, and so on) with Internet connectivity.

3.6.1 Ethernet and Wi-Fi as Unified Local Access

The Ethernet was standardized by the IEEE at the same time as the main Internet protocols (IP, TCP, UDP, and so on) at the beginning of the 1980s. From its first standard [9], the Ethernet was primarily targeted to the Internet protocol model on the network layer (i.e., to IPv4 at that time). So, Ethernet standards include standardization only on the lowest two protocol layers, the physical layer (OSI-1 layer) and the data-link layer (OSI-2 layer), while on the network layer is IP (standardized by IETF). The Ethernet standards were typically based on asynchronous access to a network by the attached hosts (e.g., computers), where the typical access scheme was Carrier Sense Multiple Access with Collision Detection (CSMA-CD), which was based on contention among the hosts attached to the Ethernet for access to the medium.

The different Ethernet standards use different media such as copper and fiber. Bit rates up to 1 Gbps are supported by copper-based Ethernet, while bit rates of 1 Gbps and above are provided by using fiber as a medium. The first version of Ethernet was IEEE 802.3 (i.e., 1Base5) with a bit rate of 1 Mbps over twisted pair (a copper medium) with CSMA-CD as the medium access technique. However, the successor versions of Ethernet, such as 10BASE-T, 100BASE-TX, and 1000BASE-T, provided bit rates of 10, 100, and 1,000 Mbps (i.e., 1 Gbps), respectively. These three versions have also been used over twisted pairs and traced the Ethernet as the main LAN technology globally. The first standard for Ethernet over fiber was Gigabit Ethernet (IEEE 802.3z; commonly referred to as 1000BASE-X), which was finished in 1998 [9]. Four bit rates are currently supported by fiber-based Ethernet: 1, 10, 40, and 100 Gbps [10]. However, the Ethernet fiber cable can be single-mode and multimode (single-mode fiber is better than multimode in terms of the supported distances, but it is also more expensive due to its much smaller diameter). Table 3.1 gives an overview of the existing versions of the Ethernet. Future developments are aimed at 400-Gbps and 1-Tbps Ethernet standards.

The wireless equivalent of the Ethernet is Wi-Fi (see the IEEE 802.3 group of standards), which is referred to as wireless LAN (WLAN). The development of Wi-Fi started in the 1990s. It is defined by and operates in the so-called unlicensed bands on 2.4 and 5 GHz, which are globally harmonized to be an unlicensed frequency spectrum. The main IEEE 802.11 standard provided for data rates up to 2 Mbps (operating on 2.4 GHz). The Wi-Fi standard specifies the lowest two protocol layers, the physical and data-link layers, similar to Ethernet and most of the IEEE standards. Amendments to Wi-Fi at the physical layer include IEEE 802.11b, IEEE 802.11a/g, IEEE 802.11n, IEEE 802.11ac, and emerging IEEE 802.11ad (see Table 3.2). Also, several amendments have been made to the Wi-Fi standard on OSI-2 layer for introduction of QoS support in Wi-Fi (IEEE 802,11e), security (IEEE 802.11i), and others.

The WLAN architecture can be in two modes: infrastructure mode and ad hoc mode. In the infrastructure mode, Wi-Fi access points (APs) are connected to the Internet via a fixed network architecture based on Ethernet. In contrast, the ad hoc mode in WLAN is used for direct communication between end-user wireless hosts using IEEE 802.11 standards.

Table 3.1 Ethernet Standards

Ethernet Name	Bit Rate	Cabling	Standard
10BASE-T	10 Mbps	Twisted pair (Category 3)	802.3i
100BASE-TX	100 Mbps	Twisted pair (Category 5)	802.3u
1000BASE-SX, 1000BASE-LX/EX	1 Gbps	Multimode fiber (MMF) or single-mode fiber (SMF)	802.3z
1000BASE-T	1 Gbps	Twisted pair (Category 5 or higher)	802.3ab
10GBASE-SR, 10GBASE-LR/ER	10 Gbps	Laser optimized MMF or SMF	802.3ae
10GBASE-T	10 Gbps	Twisted pair (Category 6A)	802.3an
40GBASE-SR4/LR4	40 Gbps	Laser optimized MMF or SMF	802.3ba
100GBASE-SR4/LR4	100 Gbps	Laser optimized MMF or SMF	802.3ba

3.6 Internet Access Networks

Table 3.2 Wi-Fi Standards

Wi-Fi Standard	Bit Rate	Frequency Bands
IEEE 802.11	Up to 2 Mbps	2.4 GHz
IEEE 802.11b	Up to 11 Mbps	2.4 GHz
IEEE 802.11a	Up to 54 Mbps	5 GHz
IEEE 802.11g	Up to 54 Mbps	2.4 GHz
IEEE 802.11n	Up to 600 Mbps	2.4 and 5 GHz
IEEE 802.11ac	Up to 6.93 Gbps	5 GHz

The Ethernet LAN architectures are based on several different types of network nodes, such as repeaters, hubs, bridges, and switches, that operate on the physical and/or data-link layer. Although all of them existed in the past, current Ethernet LANs are based on interconnected switches that connect with other switches (and routers toward the global Internet) on one side, and to end hosts on the other side, as shown in Figure 3.9. Typical hosts include PCs and laptops in homes and offices, or servers in the data centers of network operators and service providers.

All Internet hosts are connected to the Internet via a given LAN, MAN, or WAN. Mobile networks typically have a MAN or WAN architecture, whereas a fixed access network is based on a LAN (Ethernet or Wi-Fi). Each LAN is connected to one or more routers, and has assigned IP addresses with the same network ID to all network interfaces attached to it. Generally, network elements that can exist in an Ethernet-based IP network are as follows:

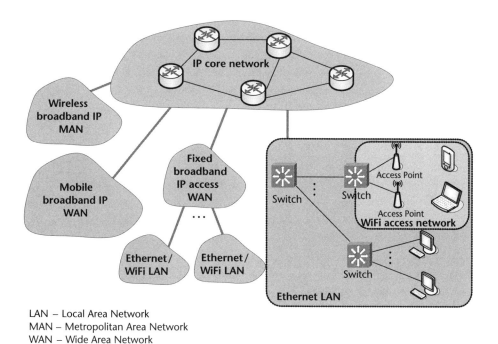

LAN – Local Area Network
MAN – Metropolitan Area Network
WAN – Wide Area Network

Figure 3.9 LANs for access to the Internet (Ethernet and Wi-Fi).

- Host: This is an end-user machine that is attached to the Internet through a network interface, which contains the Internet protocol stack and application layer communication protocols.
- Repeater: This is a network device used to regenerate digital signals; therefore, it operates mainly at the OSI-1 physical layer.
- Hub: This is a network device for connecting multiple devices to act as a single network segment (e.g., Ethernet segment). Hubs operate at the physical layer.
- Bridge: This is a network device used to aggregate networks or network segments in the Internet at the OSI-2 layer. The bridge does forwarding of data-link layer data frames by means of MAC addresses.
- Switch: This is a network device used for bridging multiple LAN segments (usually Ethernet network segments) by means of data forwarding based on MAC addresses.
- Router: This is a network node (equivalent to a computer with multiple network interfaces) that forwards IP packets between different IP networks, such as Ethernet LANs, connected to its network interfaces. This operation is called routing, and it is based on the source and/or destination IP addresses that are placed in the IP packet header. Because IP is a network layer protocol, note that routers work at the OSI-3 network layer.

3.6.2 Fixed Broadband Access Networks

Broadband access to the Internet is offered to individual (i.e., residential) users and to business (i.e., enterprise) users. Fixed broadband access is provided via copper lines (e.g., twisted pairs and coaxial cables) and fiber-optic cables. Then, the main fixed broadband access technologies over copper lines are DSL technologies as well as cable access networks.

Typically, ADSL uses the upper frequency band, which is not used for telephony calls, by means of splitting of the frequency bandwidth of the twisted-pair line using a so-called splitter (or DSL filter).

The initial standard for ADSL was created by ANSI [11], but the most commonly used standard for ADSL on a global scale is the ITU-T's ADSL standard [12], also known as G.DMT, because it is uses discrete multitone (DMT) modulation. For longer lengths of the subscriber's local loop, there is also the ITU-T standard [13] known as G.Lite for its lower data rates of up to 1.5 Mbps in the downlink, which represents a trade-off for the longer lengths of the subscriber's line.

The successor to ADSL is ADSL2, defined in G.992.3 [14] and G.992.4 [8]. The next standard, G.992.5 (i.e., ADSL2+), provides data rates that are 2 times higher than those of ADSL2, as shown in Table 3.3.

The integrated network with cable access can be divided into several segments; they are, going from the user's side toward the core network of the cable operator, the home IP network, DOCSIS, PacketCable, and core IP network, which constitute a functional environment that provides transparent transfer of various types of data from different services, as well as implementation of new services in the future over the same access network.

Table 3.3 ADSL Standards and Bit Rates

ADSL Name	Maximum Downstream Bit Rate (Mbps)	Maximum Upstream Bit Rate (Mbps)	Standard
ADSL	8	1	ANSI T1.413-1998
ADSL G.DMT	12	1.3	ITU G.992.1
ADSL G.Lite (splitterless)	1.5	0.5	ITU G.992.2
ADSL2	12	1.3	ITU G.992.3
Splitterless ADSL2	1.5	0.5	ITU G.992.4
ADSL2+	24	3.3	ITU G.992.5

The bit rates of different versions of DOCSIS standards are summarized in Table 3.4. In the downstream direction one can distinguish between DOCSIS (6-MHz bandwidth per TV channel) and Euro DOCSIS (8-MHz bandwidth per TV channel). The given bit rates are aggregated on a given coaxial cable, so they are shared between users, which are using a single coaxial cable in the DOCSIS access network (e.g., in a residential building).

Table 3.5 summarizes the main passive optical network (PON) standards in terms of access bit rates, wavelengths used in the downstream and upstream, and transmission technology. All PON technologies use different wavelengths in different directions; downstream, wavelengths of 1,490 and 1,550 nm are used, whereas upstream the 1,310-nm is used in all PON standards. The highest bit rates are provided with GPON, considering legacy PON technologies.

Future standardization of PONs is also carried by ITU and IEEE. The ITU continues with further standardization toward a 10-Gbps-capable PON called XG-PON (also referred to as Next Generation PON 1, or NG-PON1), the G.987 standard [15], and the next-generation 40-Gbps-capable PON (referred to as NG-PON2), the G.989 standard [16].

3.6.3 Mobile Broadband Access Networks

Mobile technologies have grown in parallel with the Internet. While the first mobile technologies, such as analog first-generation (1G) and digital second-generation (2G) mobile systems, were based on circuit switching, the success of both Internet and mobile technologies (especially the success of the GSM in the 1990s, which continues into this century) provided ground for their convergence. So, the two

Table 3.4 DOCSIS Bit rates Including Overhead

	Downstream		Upstream
Version	DOCSIS (Mbps)	Euro DOCSIS (Mbps)	Both
DOCSIS 1.0 and 1.1	42.8	55.62	10.24
DOCSIS 2.0	42.8	55.62	30.72
DOCSIS 3.0 (4 channels)	171.52	222.48	122.88
DOCSIS 3.0 (8 channels)	343.04	444.96	122.88

Table 3.5 Standardized Passive Optical Networks for Broadband Access to the Internet

Standard Organization	Passive Optical Network (PON)	Maximum Downstream Bit Rates	Maximum Upstream Bit Rates	Standard
ITU-T	BPON (Broadband PON)	1.244 Gbps	622 Mbps	ITU-T G.983
	GPON (Gigabit PON)	2.5 Gbps	2.5 Gbps	ITU-T G.984
	XG-PON (NG-PON1)	10 Gbps	2.5 Gbps	ITU-T G.987
	XLG-PON (NG-PON2)	40 Gbps	10 Gbps	ITU-T G.989
IEEE	EPON (Ethernet PON)	1.25 Gbps	1.25 Gbps	IEEE 802.3ah
	10G-EPON	10 Gbps	10 Gbps	IEEE 802.3av

winning technologies by the end of the 1990s, GSM in mobile world and the Internet in the packet-switching world, converged. This convergence was accomplished by the development of GPRS for Internet access via the GSM radio access network (although with smaller data rates, in the range of several tens of kilobits per second).

The GPRS added two additional centralized routers in the GSM core network architecture to handle the Internet traffic. One, called the serving GPRS support node (SGSN), acts as a centralized node and the other, the gateway GPRS support node (GGSN) acts as an edge node in the network. Introduction of Internet connectivity to GSM, called GSM/GPRS, is also referred to as 2.5G. Note that systems similar to GSM/GPRS were being used in other parts of the world, but they were discontinued in the following generations. GSM/GPRS was standardized by the 3G Partnership Project (3GPP) [8] and has evolved toward third-generation (3G) mobile systems as the Universal Mobile Telecommunication System/High-Speed Packet Access (UMTS/HSPA) with available bit rates in the radio access network of several megabits per second to several tens of megabits per second, a speed that is referred to as mobile broadband (Table 3.6). The beginning of the 2010s saw the start of 4G mobile broadband implementation, where 4G mobile systems are represented by two technologies: LTE/LTE-Advanced from 3GPP [17] and Mobile WiMAX 2.0 [18,19]. The 4G mobile broadband technologies can offer up to several gigabits per second of aggregate throughput in the radio access networks, which gives up to several hundreds of megabits per second per individual mobile user.

Note that the data rates given in Table 3.6 are theoretical maximums for given 3GPP releases (i.e., UMTS/HSPA and LTE/LTE-Advanced) or Mobile WiMAX versions. In practice, the bit rates available to end mobile users (for access to the Internet) depend on several factors, including mobility (higher mobility results in lower bit rates using the same spectrum), distance between the mobile terminal and base station (e.g., longer distances result in a worse signal-to-noise ratio in the radio link and hence in lower bit rates due to the different modulation and coding schemes used in such cases), capabilities of the mobile terminals (some may have no support for certain mobile access technologies), as well as the number of mobile participants at a given location that are simultaneously using the same mobile network to access the Internet. (The bit rates listed in Table 3.6 are aggregate bit rates on a given frequency bandwidth in a given cell of a given radio access network, and hence they are shared among active users.)

Mobile communications are personal, because mobile users typically carry their mobile handheld terminals with them. With mobile users connected to the

3.6 Internet Access Networks

Table 3.6 Mobile Broadband Internet Access

Standard Organization	Mobile Network	Maximum Downstream Bit Rate	Maximum Upstream Bit Rate	Frequency Bandwidth (FDD)	Standard Development Organization	Generation
3GPP	UMTS/HSPA+ (Release 8)	42 Mbps	11.5 Mbps	2 × 5 MHz	3GPP	3G
	UMTS/HSPA+ (Release 10)	168 Mbps	46 Mbps	2 × 20 MHz (each 4 × 5 MHz)	3GPP	3G
	LTE (Release 8)	300 Mbps	75 Mbps	2 × 20 MHz	3GPP	3.9G/4G
	LTE-Advanced	3 Gbps	1.5 Gbps	2 × 100 MHz	3GPP	4G
IEEE	Mobile WiMAX 1.5 (IEEE 802.16e)	141 Mbps	138 Mbps	2 × 20 MHz	IEEE, WiMAX Forum	3G
	Mobile WiMAX 2.x (IEEE 802.16m)	365 Mbps	376 Mbps	2 × 20 MHz	IEEE, WiMAX Forum	4G
	Mobile WiMAX 2.x (IEEE 802.16m)	>1 Gbps	>100 Mbps	2 × 100 MHz	IEEE, WiMAX Forum	4G

Internet via mobile broadband access (using smartphones, lap-tops, and so on), mobile traffic (i.e., the volume of transferred data to/from mobile users) becomes a significant share of the total Internet traffic (like never before) and it is continuously increasing. Nowadays, huge application ecosystems exist in the Internet (which provide OTT services by using network neutrality of the Internet) that are targeted to mobile users for provision of user-centric Internet services (e.g., social networking, cloud services, gaming, messaging, video on demand, data sharing).

Due to the importance of mobile broadband access to the Internet, new spectrum is continuously being added for mobile broadband technologies (listed in Table 3.6). Such spectrum is referred to as the International Mobile Telecommunications (IMT) spectrum and includes IMT-2000 (IMT-2000 is an umbrella requirement for 3G, set by ITU-R) and IMT-Advanced mobile technologies (IMT-Advanced is an umbrella requirement for 4G, set by ITU-R). It is globally harmonized by ITU-R, and regulated in each country by so-called national regulators. The usable bandwidth for mobile networks is below 5 GHz (due to radio propagation characteristics and attenuation of radio signals). Note, however, that there is a continuous demand for higher bit rates in mobile networks, which is accomplished by using enhanced modulation and coding schemes, multiple antennas, and allocation of new spectrum for mobile broadband (which is lately defined as technology neutral regarding the IMT mobile technologies, which are listed in Table 3.6).

Whereas 3G mobile networks from 3GPP (i.e., UMTS/HSPA) had a circuit-switched (CS) part for mobile telephony (similar to GSM) and a packet-switched (PS) part for Internet-based services, all 4G mobile technologies are all-IP based, which means that they are using the standardized Internet technologies (by IETF) from the network layer above to the application layer (e.g., IPv4 and IPv6, transport protocols such as TCP and UDP, routing protocols, signaling protocols such as SIP, various application layer protocols such as HTTP, SMTP, and so on). However, certain Internet technologies are standardized in a given context for use in mobile networks (for certain services), for which typical examples are the IP multimedia subsystem (IMS), standardized by 3GPP, and next-generation networks (NGNs), standardized by ITU.

Regardless of the type of access networks to the Internet (fixed or mobile one), the core and transport networks are based on the Internet architecture. Hence, they are using Internet routing between interconnected network nodes.

3.7 Internet Routing

The Internet was initially developed as a network of interconnected nodes called routers that route IP packets from a given source (host or node) to a given destination (host or node). The routing is primarily based on source and destination addresses that are specified in the header of every IP packet. However, the IP packet typically travels through multiple network nodes called routers. Each router is usually connected to several IP networks, where an IP network is defined with the network part in IP addresses in all hosts attached to it. In that manner, a link that directly connects two routers (i.e., has only two network interfaces attached to it) is also considered to be a separate IP network. Each packet is routed from one router to another until it reaches the destination router, which has the destination IP network attached to a LAN that is directly connected to one of its network interfaces.

In contrast, when the network part of the source and destination IP addresses of a given IP packet are the same, the datagram is sent directly to the destination interface without having to pass any router (because in such case both hosts are located in the same IP network).

Definition: Routing in the Internet is a process of selection of a route (path) through which the packet will be sent, based on defined routing algorithms. The network nodes in the Internet that make the choice about the path for the packets are called routers.

According to the IETF definitions [20], an IP router can be distinguished from other sorts of PS devices in that a router examines the IP protocol header as part of the switching process. So, generally it removes the OSI-2 data-link layer header that a packet was received with, modifies the IP header (e.g., recalculates the Header Checksum in IPv4 header due to a change in the time-to-live field), and adds a new data-link layer header for its transmission through the selected outgoing link of the router (which corresponds to the link technology used on the outgoing link of the router for the given packet).

Routing is performed using routing tables (table-driven routing), which are databases, referred to as a routing information base (RIB), that contain lists of routes to particular network destinations in the Internet. However, the traditional method of routing in the Internet is hop-by-hop routing (i.e., next-hop routing, where a hop is a logical link between two adjacent routers in the Internet), in which each router routes the packet to the next router. This process is used because it requires only partial routing information and reduces the size of the routing tables. The routing tables may contain two types of routes:

- Static routes: These routes are created by manual entries or, in other words, by nonautomatic means.
- Dynamic routes: These routes are automatically learned by the router using dynamic routing protocols.

A set of rules that specify how the information in routing tables is being used and updated via routing information exchange between routers is called a routing algorithm. A general model for routing algorithms is given in Figure 3.10.

Routing tables typically contain information only for the destination IP networks (of destination IP prefixes), not the destination host addresses, since the number of networks is many times smaller than the number of hosts on the Internet. Otherwise, routing tables would be too large. However, where necessary it is also possible to set up routes to individual IP addresses in specific cases (for security reasons, network administration, and so on).

IP addresses can be either unicast (for communication between two Internet nodes) or multicast (one node to many nodes communication). Each network interface in every host or router must have a unique unicast IP address (globally unique for public IP addresses, and locally unique for private IP addresses). In such a case, if a given host has multiple network interfaces (Wi-Fi interface, Ethernet interface, and so on), then it should have a different unicast IP address for each network interface. However, each network interface can also have an assigned multicast IP address (for multicast communication) besides the unicast address. Hence, routing algorithms can also be grouped in unicast routing algorithms (and respective routing protocols that implement them) and multicast routing algorithms (implemented via multicast routing protocols).

```
Routing-of-Datagram(Datagram Routing Table)
Extract Destination IP Address (DA) from the datagram and calculate Network Prefix (NP);
    if {NP equals network prefix of directly connected IP network to this router}
        then {send the datagram to through that network interface}
    else if {routing table contains host-specific route for DA}
        then {send the datagram to next hop specified for DA in the routing table}
    else if {routing table contains a route for the network NP}
        then {send the datagram to the next hop specified for NP in the routing table}
    else if {routing table contains a default route/gateway}
        then {send the datagram through the default route given in the routing table}
    else {send a "routing error" message};
```

Figure 3.10 General model for routing algorithms.

3.7.1 Unicast Routing

The purpose of propagation mechanisms for distribution of routing information in a given network is not only to find a set of routes, but to update continually the routing information. Network administrators cannot respond to such changes on a regular and timely basis, so such tasks in larger IP networks are accomplished with routing algorithms and protocols. To summarize, changes in routing tables may be performed in two ways:

- Manually: Network administrators manually define routes through the user interface of the router.
- Automatically: Routers exchange information with each other at certain time intervals or after certain changes using routing protocols that are based on certain routing algorithms.

To be able to understand the essence of the routing algorithms, a graph abstraction of a network that performs routing is shown in Figure 3.11. Namely, each of the outgoing links from a given router to other routers with which it is connected has a cost (or metric) associated with it. Generally, different links (i.e., hops) may have different costs. The default cost of a link is 1. Of course, the cost may be influenced by other parameters such as the traffic load on the link (a link with a higher traffic load will have a higher cost and vice versa) or link delays (e.g., the cost of a satellite link between two routers is typically higher than the cost of terrestrial links, due to the bigger packet delay that results when satellite links are used).

The path from a given source to a destination is composed of a sequence of links. (An exception is the case in which both hosts belong to the same network, and hence routing is not needed.) The cost of the path from a given source to a given destination is the sum of the costs of all links in the path:

$$\text{Path\_cost}(x_1, x_2, x_3, \ldots, x_p) = \text{cost}(x_1, x_2) + \text{cost}(x_2, x_3) + \ldots + \text{cost}(x_{p-1}, x_p) \quad (3.1)$$

where $\text{cost}(x_i, x_j)$ is the cost of the link between routers i and j. The purpose of the routing algorithm is to find the path with the lowest cost for a given pair of source and destination addresses.

Definition: A routing algorithm is an algorithm that finds the path with the lowest cost, that is, the best path in a network.

In general, routing algorithms can be divided into the following groups:

- Distance-vector routing: This type of algorithm is based on cost numbers assigned to each link between each pair of nodes in the network, and calculation of the path with the least cost.
- Link-state routing: This type of algorithm is based on a graph-map of the network to which the given router belongs, and each router independently determines the least-cost path from itself to all other routers in the given network.

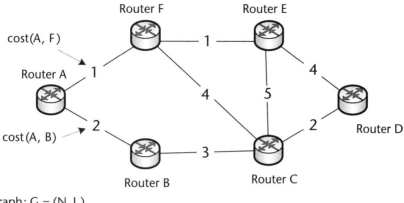

Graph: G = (N, L)
N = set of router = {A, B, C, D, E, F}
L = set of links = {(A, B), (A, F), (B, C), (C, D), (C, E), (C, F), (D, E), (E, F)}

Example of path cost from Router A to Router D
Path 1: cost(A, B) + cost(B, C) + cost(C, D) = 7
Path 2: cost(A, B) + cost(B, C) + cost(C, E) + cost(E, D) = 14
Path 3: cost(A, B) + cost(B, C) + cost(C, F) + cost(F, E) + cost(E, D) = 14
Path 4: cost(A, B) + cost(B, C) + cost(C, F) + cost(F, E) + cost(E, C) + cost(C, D) = 17
Path 5: cost(A, F) + cost(F, C) + cost(C, D) = 7
Path 6: cost(A, F) + cost(F, C) + cost(C, E) + cost(E, D) = 14
Path 7: cost(A, F) + cost(F, E) + cost(E, D) = 6
Path 8: cost(A, F) + cost(F, E) + cost(E, C) + cost(C, D) = 9

Shortest path (i.e., least cost path) from Router A to Router D is Path 7 (in this example).

Figure 3.11 Graph abstraction of a network performing a routing function.

- Hierarchical routing: This type of algorithm is based on different routing algorithms used for intra-AS routing and inter-AS routing. The default protocol for inter-AS routing in the current Internet is BGP-4, which is also considered to be a path vector protocol.

3.7.1.1 Distance-Vector Routing

In distance-vector routing algorithms, each router informs all neighboring routers of topology changes in the network. A given router has a routing table that contains one entry per destination network (present in the table). Each entry in the table identifies only the direction to the destination network (i.e., the next router to which the packet should be forwarded) and gives the distance to that network, usually measured in hops (where a hop is a link between two logically neighboring routers). Periodically each router sends a copy of its routing table (or part of it) to every neighboring router. The term distance vector comes from information sent in periodic updates to other routers. Each update message contains a list of pairs (V, D), where V denotes the destination (called a vector), and D is the distance to the destination. A typical representative of the distance-vector routing algorithms is the Bellman-Ford algorithm. According to this algorithm packets are routed by the route with the lowest cost, where the least-cost path between two routers x and y is calculated as follows:

$$D_x(y) = \min\{\text{cost}(x,v) + D_v(y)\} \quad (3.2)$$

where $D_x(y)$ is the least-cost path from router x to y, cost(x, v) is the path cost from router x to router v, and $D_v(y)$ is the least-cost path from router v to router y.

The distance-vector algorithm has some disadvantages. One of them is the slower convergence of the routing information in all routers in the network, due to the propagation of route changes on a hop-by-hop basis, which may lead to inaccurate routing information in some routers for a certain time period.

The oldest and most known distance-vector routing protocol is the Routing Information Protocol (RIP) [21]. It is an intra-AS routing protocol that uses hop count as a cost metric, which is the number of links that have to be passed to reach the destination. To avoid the appearance of routing loops, in which a packet returns to the same router, the RIP limits the maximum number of hops to 15 (so 16 is considered as infinite distance). In RIP regular update messages are typically sent at time intervals of 30 sec. However, the protocol is limited to smaller IP networks in which the longest path (i.e., the network's diameter) is 15 hops. In addition, it uses only fixed metrics. RIP uses UDP as the transport protocol on port number 520 [21]. Also, at the time of RIP standardization the Internet was using classful IPv4 addressing. RIP version 2 (RIPv2) [22] was standardized in 1998, with the goal of providing support to RIP for CIDR addressing (by adding a field for a subnet mask). It has since become the main IPv4 addressing scheme in the Internet. Additionally, support for IPv6 addressing was included in RIP next generation (RIPng) [23].

3.7.1.2 Link-State Routing

The link-state routing algorithm requires each router to have complete topological information for the network. The easiest way to understand the topological information is to imagine that each router has a map that shows all other routers and links that connect them.

A link-state algorithm performs two tasks. First, each router actively tests the status of all neighboring routers. In terms of graph abstraction of the network, two routers are neighbors if they share a link. Also, each router periodically propagates information about the status of its links to all other routers. Second, each router independently calculates the shortest path from itself to all other nodes in the network.

The most well-known example of a link-state routing algorithm is the Dijkstra algorithm. Calculations with the Dijkstra algorithm for a change in the status of a router's links result in corresponding changes in the path costs using the following formula:

$$D(v) = \min\{D(v), D(w) + \text{cost}(w,v)\} \quad (3.3)$$

where $D(v)$ is the current cost of the path from the router to the destination v, $D(w)$ is the current cost of the path from the router to the destination w, and cost(w, v) is the current cost of the path from w to v.

Each node floods the network with information about other routers that it can connect to, so each router independently calculates the least-cost path from itself to every other node in the network (e.g., by using the Dijkstra algorithm). The final result is a tree graph rooted at the given router, and paths through such a tree from the root (i.e., itself) to any other node are used to construct the routing table.

However, a link-state routing algorithm has its drawbacks. For example, it may give an incorrect cost for a given link and rooting loops may appear. A routing loop can appear when each of two neighboring routers thinks that the other is the best path to a given destination.

The most known standardized link-state routing protocol is Open Shortest Path First (OSPF) [24], which is used in intra-AS routing (similar to RIP). It includes explicit support for subnetting in IPv4 (i.e., for CIDR), routing based on the ToS values in IPv4 headers (for QoS support), authentication of routing updates, and use of IP multicast for sending or receiving the updates. Overall, OSPF routes IP packets based on the destination IP address and ToS information in each IP header. Each router in a given AS contains a database describing the AS topology. OSPF can detect changes in the topology (e.g., link failures), and converges on a new loop-free routing tree within seconds. However, unlike other routing protocols (e.g., RIP, BGP), the OSPF does not use transport protocols (such as TCP and UDP), but runs directly over IP (encapsulating OSPF messages directly in IP packets), using IP protocol type 89 [24]. Additionally, it allows grouping of a set of IP networks in so-called OSPF areas (as shown in Figure 3.12). Modifications to OSPF with the goal of supporting IPv6 were added later in a new version of the protocol, OSPF for IPv6 [25], which is also referred to as version 3 (OSPFv3).

Another standardized intra-AS (i.e., interior) link-state routing protocol is Intermediate System to Intermediate System (IS-IS) [26]. However, there are also vendor-specific implementations of intra-AS routing protocols such as the Enhanced Interior Gateway Routing Protocol (EIGRP).

3.7.1.3 Hierarchical Internet Routing

An Internet network is built on the principle of a network of many networks that are interconnected through routers. Certainly, there must be routers between different IP networks. However, several smaller or larger networks could be administered by a company, such as an ISP. In such a case we actually have independently operated networks. For example, we can have a network of an ISP that is composed of multiple IP networks, we can have a network of a mobile operator providing Internet service (e.g., 2.5G, 3G, or 4G mobile networks), and so forth. Each of these networks of networks (where a network is identified by a unique routing prefix or network ID) actually forms an autonomous system. So, an AS is a collection of routers under the control of one authority (a company, an operator, and so on). These autonomous systems are connected to each other by means of gateway routers to the Internet (i.e., to other ASs).

Each AS is uniquely identified by its AS number. AS numbers are delegated by the IANA as a global authority. AS numbers were defined until 2007 as 16-bit

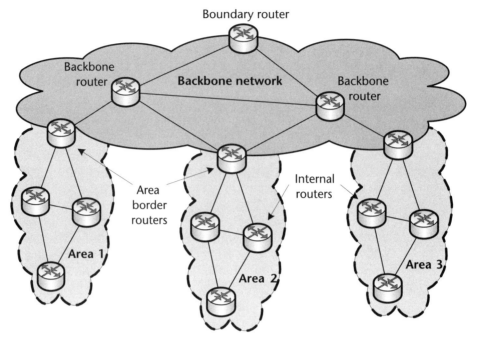

Figure 3.12 OSPF areas in an autonomous system.

numbers, which allowed for the existence of a maximum of 65,535 autonomous systems. However, because of the exponential increase in the Internet in terms of the number of hosts and networks, IANA introduced 32-bit AS numbers [27], which are currently being allocated.

With autonomous systems as the basic architectural approach in the current Internet, we actually have two-tier routing, which is referred to here as hierarchical routing. Namely, it is a concept, not a specific algorithm. Thus, it is possible for different algorithms to be used in each of the two tiers in the hierarchy (Figure 3.13):

- Intra-AS routing: This is routing between routers belonging to a single autonomous system. Examples of intra-AS routing protocols are RIP and OSPF.
- Inter-AS routing: This is routing between routers belonging to different autonomous systems. Currently, the globally accepted protocol for inter-AS routing is BGP-4 [28]; however, its operation is based on several parameters, such as path and routing policies and rules.

Hierarchical routing is practiced today in the Internet. Additionally, the owner of an AS (e.g., ISP, company, or institution) can independently choose and implement the desired routing protocols within its own autonomous system.

3.7.2 Multicast Routing

The transmission of a packet from one source to multiple destinations is more complex to implement in the Internet than unicast transmission (i.e., point-to-point). Routing of a single packet to multiple destinations on the Internet is called multicast routing. Three general multicast routing approaches are used:

3.7 Internet Routing

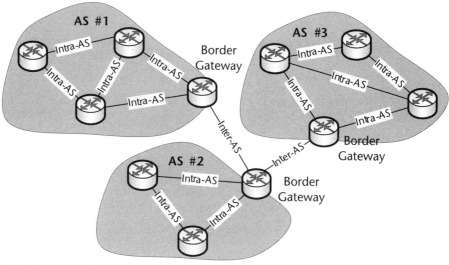

AS – Autonomous System
Inter-AS – Routing between Autonomous Systems
Intra-AS – Routing within an Autonomous System

Figure 3.13 Hierarchical Internet routing: intra-AS and inter-AS.

- One-to-many (e.g., live video, digital TV, distance learning, corporate communications),
- Many-to-one (e.g., auctions, voting, data collection), and
- Many-to-many (e.g., distance learning, collaboration, multimedia conferencing, online gaming with multiple players, chat groups).

3.7.2.1 Internet Group Management Protocol (IGMP)

Multicast is based on so-called multicast groups and use of multicast routing protocols. Creation of multicast groups in IPv4 networks is done with IGMP [29]. With IGMP the routing of multicast IP packets is accomplished by multicast routing protocols. In that manner, the IGMP is not a routing protocol per se, but a protocol that manages group memberships. In particular, IGMP can be used for one-to-many applications. For example, if one host (e.g., a server) sends a video stream that requires 3 Mbps on average (e.g., a standard definition IPTV stream) to 1,000 users, then the backbone networks for unicast distribution will need a capacity of 1000 × 3 Mbps = 3 Gbps on each link that carries the aggregated traffic (in the backbone network); with multicast one-to-many distribution of the same stream, however, the network will need a capacity of only 3 Mbps (i.e., the actual bit rate of the stream in this example). Hence, IGMP provides highly efficient resource utilization in the case of multicast applications from a single source. There are three versions of IGMP:

- IGMPv1 [29]: This is the main standard that defined IGMP.
- IGMPv2 [30]: This version adds to IGMPv1 functionality for a client host to signal to a local IGMP router the intention to leave the multicast group.

- IGMPv3 [31]: This version of IGMP added support for source filtering, that is, the ability for a client host to report interest (to the IGMP local router) in receiving packets only from specific source addresses (or from all but specific source addresses), sent to a particular multicast address.

A client host can join or leave a multicast group through IGMP communication with a local multicast router. IGMP as a protocol operates on the network layer and its messages are directly encapsulated into IP packets (similar to ICMP for IPv4). It is part of the Internet network protocol suite, together with IPv4 and ICMP. Each host that joins the multicast group gets a multicast address (besides its unicast IP address). IPv4 addressing has a dedicated class of multicast IP addresses (refer to the discussion about IPv4 addressing in Chapter 2).

3.7.2.2 Multicast Listener Discovery (MLD)

The multicast component of the IPv6 is called multicast listener discovery (MLD). Like IGMP for IPv4, MLD is used by IPv6 hosts for discovery of an IPv6 router that are multicast listeners on a directly attached link. It has two versions:

- MLDv1 [32]: This version is derived from IGMPv2 for IPv4, and therefore has the same functionalities. But, unlike IGMP, which uses IGMP message types, the MLD uses ICMPv6 messages for communication with MLD listeners (for ICMPv6 refer to Chapter 2).
- MLDv2 [33]: This version is compatible with MLDv1, but additionally provides the ability for a node to report interest in listening to packets sent to a multicast address only from specific source addresses (or from all sources except for specific source addresses), which is similar to IGMPv3 capabilities.

MLD is implemented in all hosts and routers that have an OS that uses the IPv6 protocol stack. Both MLD (for IPv6) and IGMP (for IPv4) are multicast group membership discovery protocols, and hence they are not multicast routing protocols.

3.7.2.3 Multicast Routing Protocols

Multicast routing protocols are used for communication between multicast routers with the goal of creating multicast distribution trees. A multicast tree has a route to every recipient of the multicast packets that has joined to a multicast group by using IGMP (for IPv4) or MLD (for IPv6). It typically has a root node that is the source of the information transmitted by using multicast.

A multicast router uses a multicast routing table that contains more information than the standard (unicast) routing table. However, if a given router on the path does not support multicast routing, then it may encapsulate the multicast packet into common unicast packets and tunnel them to the next multicast router that will decapsulate the unicast packets (as the end point of the unicast tunnel) and then route the multicast packets. Encapsulation of an IP packet is done by placing the whole packet as the payload of a new IP packet, that is, by adding a new IP header with source address set to the IP address of the entry point of the tunnel and destination address set to the IP address of the end point of the tunnel.

The main initial goal for multicast is to create a multicast tree consisting of multicast-capable routers. Two main types of multicast trees are as follows:

- Source tree: In this case the root is the source of the multicast tree and its branches form a spanning tree through the Internet to the receivers (members to the multicast group). Examples of source trees include the following types:
  - Shortest path tree (SPT): This is a tree composed of the shortest paths from the source to group members.
  - Reverse path forwarding (RPF): This tree is based on the information stored in the multicast router about the shortest path from the source to itself.
- Shared tree: In this case the same tree is used by all members of the multicast group, with a common root placed at some chosen node in the network (regardless of the multicast traffic source).

An example of an SPT is shown in Figure 3.14. The tree is created from the root by adding branches that are the shortest paths to the destination networks in each router along the way. (A destination network is an IP network that has members of the multicast group attached to it.)

Different standardized Internet multicast routing architectures are in use [34]. The most used multicast routing protocols in practice include the following:

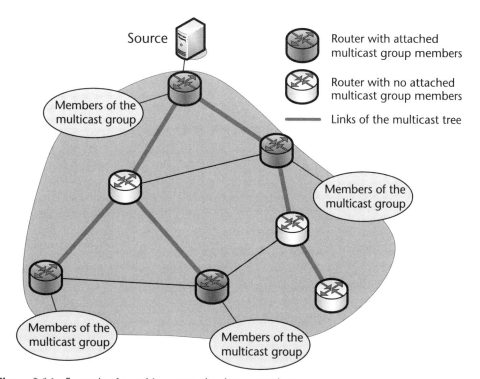

Figure 3.14 Example of a multicast tree, the shortest path tree.

- Protocol Independent Multicast (PIM): PIM is mainly implemented in the following two modes:
    - PIM Sparse Mode (PIM-SM) [35]: This is the most common multicast routing protocol worldwide [34]. In this mode it is assumed that a small number of designated routers (DRs) are included in the multicast tree. The DR is responsible for PIM messaging and source registration on behalf of the end hosts. The PIM-SM builds a unidirectional (group-specific) distribution tree, and uses a pull model to deliver the multicast traffic; that is, it delivers only to networks with active receivers that have requested the data (avoiding unnecessary flooding of the networks). This is the only active interdomain (and intradomain) multicast routing protocol.
    - PIM Dense Mode (PIM-DM) [36]: In this mode it is assumed that almost all routers will be included in the multicast tree, and every subnet has at least one active receiver of the multicast traffic. It uses a push model to flood multicast traffic to every network attached to the multicast tree. PIM-DM is no longer in widespread use, because it has been replaced by PIM-SM.
- Distance-Vector Multicast Routing Protocol (DVMRP): This was the first multicast routing protocol based on the distance-vector routing algorithm. It also included tunneling capabilities as part of its multicast architecture. Currently, it is rarely used and has been replaced by PIM-SM.
- Multicast extensions to OSPF (MOSPF): This is based on intradomain OSPF, but it is not adapted for interdomain multicast routing. Hence, it has been rarely used because network operators prefer multicast protocols that provide both intradomain and interdomain routing capabilities.

Besides those listed above, several other multicast routing protocols were standardized, but never deployed [34]. Overall, multicast routing in the Internet has not been widely deployed, although multicast mechanisms have existed for several decades and a multicast backbone was created for the Internet (as an overlay network) at the beginning of the 1990s (mainly for research purposes). However, inter-AS (or interdomain) multicast is currently a failure due to the absence of "killer" applications as well as different policies in various ASs. In contrast, multicast has been used in the past years for delivery of QoS-enabled video streaming (e.g., IPTV) within a given ISP network. The multicast routing protocol used in such an implementation (intra-AS implementations) is PIM-SM, and therefore it is the most widely used. An example of PIM-SM multicast routing is shown in Figure 3.15.

PIM-SM multicast routing makes use of a rendezvous router (in the example of Figure 3.15, it is the router R6). All multicast-capable routers (that have members of the multicast group in the networks directly attached to them) that want to join the multicast tree send join messages to the rendezvous router. The rendezvous router distributes information about the newly joined members to the other routers in the given tree.

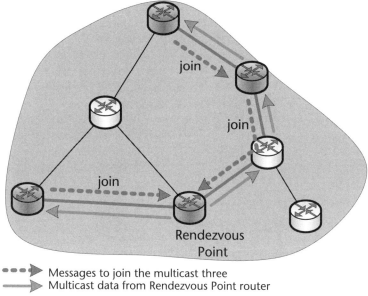

Figure 3.15 PIM-SM multicast routing.

3.8 Border Gateway Protocol (BGP)

Internet architecture is based on interconnected autonomous systems, thus it has a flat architecture. While the choice of intra-AS routing protocol (RIP, OSPF, and so on) is left to the AS itself (i.e., to its network administrators), an inter-AS routing protocol has been unified as the Border Gateway Protocol (BGP). The first version of BGP appeared in 1989; the current version of BGP is version 4 (BGP-4), standardized initially with RFC 1654 in 1994 [37] and later updated several times up to the current version that is defined with RFC 4271 [28].

In general, the main task of BGP is to act as the so-called speaking system for exchange of network reachability information with other BGP speaking systems located in other ASs, which includes list of ASs that such reachability information traverses. BGP assumes that the Internet is an arbitrarily interconnected set of ASs.

The ASs carries all Internet traffic (web, email, video, voice, and so on). Similar to the approach in legacy telecommunication systems (e.g., telephone networks), the Internet traffic in a given AS can be classified into the following two types:

- Local traffic starts or ends within the given AS.
- Transit traffic passes through the given AS.

Autonomous systems vary with regard to their capabilities (architecture, capacities, and so on):

- Stub AS: This type of AS has a single connection to another AS and carries local traffic only. This AS type can be used by large organizations or enterprises.

- Multihomed AS: This type has connections to two or more ASs, but it does not carry transit traffic. This is the typical type used by ISPs, which connect to other network providers.
- Transit AS: This type has connections to more than one AS, and carries both transit and local traffic. These ASs are used to transfer traffic between ASs.

In particular, BGP-4 acts as a so-called speaker for the whole AS. Typically, a single AS may have several border routers (depending on the AS type), where one of them acts as the BGP speaker. The BGP speaker establishes BGP sessions with other BGP speakers and advertises local network names and path information including path weights and withdrawn routes. In the case of transit ASs, it also announces other reachable AS networks. The main target of BGP is to find loop-free paths. For that purpose it has a built-in mechanism that allows BGP routers to avoid the import of any routes that contain themselves in the AS path.

Figure 3.16 illustrates the BGP's advertisements mechanism. Because BGP advertises the reachability of ASs, the number of active ASs directly influences the size of the RIB in BGP routers. There are more than 70,000 allocated AS numbers, from which about 64,000 ASs have allocated 16-bit AS numbers, while the remaining ASs have 32-bit AS numbers (note that every new AS in the future will have a 32-bit number, because 16-bit AS numbering space is drained out) [38]. However, the number of active ASs is in fact the number of advertised ASs by the BGP (there are nearly 50,000 active ASs [38]), which is less than the total number of assigned AS numbers, as one may conclude. The number of active ASs has seen a linearly increasing curve in the past 15 years, from around 5,000 ASs in 1999 up to nearly 50,000 in 2014. But, the total number of BGP RIB entries (i.e., entries in BGP routing tables) has seen a similar increase during the same time period, from over 50,000 entries in 1999 up to over 500,000 RIB entries in 2014 [39]. The difference

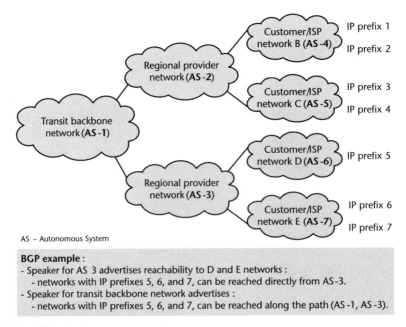

Figure 3.16 BGP advertisement example.

in number of active ASs and the total number of BGP RIB entries is due to BGP mechanisms that include support for advertising a set of destinations as an IP prefix (thus eliminating the concept of IP addressing classes within BGP) [28]. However, larger Internet size (in terms of prefixes and ASs) means larger tables in BGP routers, which may cause scalability problems. However, BGP provides for the possibility of route aggregation (based on classless addressing, i.e., CIDR). Note, however, that the processing power of machines increases two times each 1.5 to 2 years per Moore's law, which can compensate for the increased processing burden in BGP routers.

In practice, not all ASs have the same significance. Only a small portion of the total number of active ASs are considered "tier 1" ASs, which are in fact the Internet backbone.

3.8.1 BGP Operations

BGP as an inter-AS routing protocol is based on the experience gained with its predecessor protocol, the Exterior Gateway Protocol (EGP) [40].

One of the most important innovations in BGP is the path vector. The BGP is not a link-state routing protocol, nor a distance vector routing protocol. It is more correctly referred to as a distance-path routing protocol. It makes the routing decisions based on paths and additional network policies set up by network administrators.

In BGP neighbors are called peers. Communication among peers is based on manual configuration of the TCP sessions between routers on TCP port 179. Use of TCP as the transport protocol eliminates the need to implement explicit update fragmentation, sequencing, acknowledgment, and retransmissions to provide the required reliability for information exchange between peers (i.e., peering).

When two peers belong to the same AS, the protocol is called Internal BGP (IBGP). When two peers belong to different ASs, it is called External BGP (EBGP), which is interchangeably used with BGP (in this chapter). In the latter case, routers from the ASs that exchange information by using BGP are called border (or edge) routers. The edge routers from different ASs that have established BGP sessions are connected directly (on the network layer). However, during connection establishment, peers may negotiate certain optional capabilities, such as multiprotocol extensions [41], recovery modes, and so forth. Initially, BGP (or BGP-4, which became the actual worldwide version) was created for exchange of IPv4 routing information. Afterward, multiprotocol extensions to BGP-4 [41] were standardized with the goal of extending its capabilities to carry routing information for multiple network layer protocols, such as IPv4 (as the default one), IPv6, and IPv4 and IPv6 VPN. Hence, BGP can also be used to carry routing information over secured connections such as VPNs, which are not part of the public Internet.

The routing information that is exchanged via BGP supports only the destination-based forwarding, which assumes that a router forwards the packets based only on the destination addresses in the IP headers of the packets. Such an approach directly influences the type of policies that can and cannot be implemented with BGP. Additionally, BGP-4 also provides mechanisms for aggregation of routes, including aggregation of AS paths.

For its operations BGP uses a finite state machine model (FSM), as shown in Figure 3.17. It consists of six states: Idle, Connect, Active, OpenSent, OpenConfirm, and Established. Each BGP implementation has, at most, one FSM for each configured peering, plus one FSM for each incoming TCP connection for which the peer has not yet been identified. That is due to the fact that each FSM corresponds to one TCP connection. So, one FSM is maintained per BGP connection. Then, a BGP router sends Keep-Alive messages every 30 sec to maintain the connection with the other peer for each established BGP connection.

Regarding the FSM, the first state that BGP enters is the Idle state. In this state the BGP refuses all incoming connections for the peer, initializes all BGP resources for the peer connection (which has to established with another BGP peer on the other end), and initiates a TCP connection to the other BGP peer (on port 179) while listening for a connection that may be initiated by the remote peer. Further, in the Connect state the BGP FSM is waiting for completion of the initiated TCP connection, and after success it transits to the OpenSent state (otherwise, it returns to the Connect state). In the Active state the BGP FSM attempts to obtain a connection with the remote peer by listening to the incoming TCP connection from it. When the FSM of the BGP connection is in the OpenSent state, the router sends an open message and then transits to the OpenConfirm state to wait for an acknowledgment message from the remote peer. After successful receipt of the acknowledgment, the BGP FSM transits to the Established state (Figure 3.17).

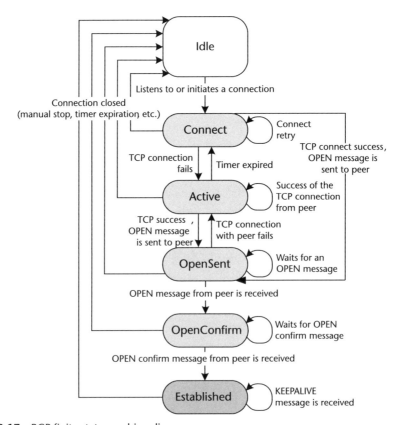

Figure 3.17 BGP finite state machine diagram.

To provide its functionalities, the BGP uses four types of messages:

- Open message: This is the first message sent by each peer after the connection is established. Each peer announces its AS by specifying its AS number in this message. To accept the Open message and neighborhood relation, each peer responds to it by sending a Keep-Alive message. This message also contains the Hold Time interval, which is reset with each new message received. If it expires, then it means that the other peer is no longer reachable.
- Update message: This is used to transfer routing information between BGP peers, which includes advertisements of feasible routes and withdrawing of unfeasible routes. Based on exchanged routing information, each peer will be able to construct a graph containing the relationships of the various ASs.
- Keep-Alive message: This is used to maintain an established connection between BGP peers. Its task is to keep the Hold Time interval from expiring. This message contains only the BGP message header and has 19 bytes.
- Notification message: This is sent when an error is detected. After this message the BGP connection between peers is closed.

The information exchange between BGP peers is controlled by the export policy in the AS [28]. In general, BGP does not require refreshing of the table. However, to provide for the possibility of changes in the local policies (e.g., in the AS) without the need to reset any of the established BGP connections, the BGP can either retain the current versions of all routes or use the Route Refresh extension [42], which is also a type of BGP message (besides messages listed above).

3.8.2 BGP Discussion

Because the architecture of the Internet network is constantly changing (e.g., new ASs and IP prefixes are added to the Internet), many BGP updates are flying over the global network. How long does it take to route a change via BGP updates? Because BGP updates are typically sent in intervals of 30 sec, and they have to propagate across several concatenated BGP peerings (between different pairs of BGP routers), the convergence time for a changed route ranges from 30 sec to several minutes. However, only a small portion of BGP routes are updated in shorter time periods (e.g., per second). For example, globally there are around 10 prefix updates per second, with peaks in the range from 1,000 to 10,000 updates per second [43].

Regarding BGP operation, the goal is to have higher BGP stability, which means fast route convergence, minimal updates, and path redundancy (for higher reliability). The convergence is delayed for up to several minutes, because it is slowed by the timers and by the BGP tendency to explore many alternate paths before giving up or setting a new path. Additionally, there is no global knowledge regarding the vector protocols (as BGP is using), because each AS can set it own policies and change them after a certain time.

However, several mechanisms are available for squashing updates. One of the most efficient is route flap dampening [44]. With route flap dampening the routes are penalized for changing. If the penalty exceeds the so-called suppress limit, the route is dampened. On the other hand, when the route is not changing, then its penalty decays exponentially. If the penalty goes below a defined reuse limit, then it is announced again. For example, if the suppress limit = 800, reuse limit = 2000, and penalty = 600, then after several penalties in a shorter time interval (e.g., within an hour), the route will exceed the suppress limit and will be dampened until it again goes below the reuse limit. (The penalty decreases exponentially over the time, so it will decrease if there are no new penalties.)

Why do BGP routers use updates all the time? That happens because IP networks come and go (i.e., new networks are implemented and other are dumped), router failures happen (hardware failures, software failures, and so on), congestion may occur and cause the BGP connections to reset, route flap dampening, vendor-specific implementations in certain cases, and so forth. However, such behavior is normal and expected for dynamic routing protocols.

Considering the importance of BGP for the functioning of the global Internet, it is necessary to analyze also its vulnerabilities and security threats [45]. BGP is vulnerable to attacks (as are other routing protocols) from various malicious sources that try to disrupt the Internet by injecting bogus routing information into the BGP distributed routing database. This is done by modifying, forging, or replaying BGP messages during their transport among peers. To prevent such attacks, the current BGP standard requires a support of authentication mechanism [45], but its use is optional. Another way to attack BGP communication (among many of them) is through TCP port 179 (used by BGP), by using SYN flooding, that is a denial–of–service (DoS) attack. Therefore, network providers that administer their BGP routers have to use some form of filtering technique to prevent attacks from outside the ISP network.

In the short term, BGP in its current version (BGP-4) is irreplaceable for interconnection of the ASs as the main building blocks of the existing Internet; in the long term, it will certainly have a successor. It is used in the Internet much like Signaling System 7 (SS7) is used for signaling between different legacy telecommunication networks (e.g., PSTN, PLMN).

As a short summary, routing is done in two tiers:

- Intra-AS routing: All choices are independent from one AS to another AS.
- Inter-AS routing: BGP-4 is the de facto standard for inter-AS "talk."

A high-level view of a BGP router is shown in Figure 3.18.

Note that the main routing protocols, standardized by IETF many years ago, are not going to change drastically overnight because there are too many autonomous systems and routers on the ground and faster changes are possible only within ASs. Besides that, with convergence of traditional telecommunication real-time services (telephony and television) to the Internet, the standardization and implementation of end-to-end QoS is becoming mandatory for Internet routers on their way toward the next-generation networks and future networks, which will be all-IP by default.

3.8 Border Gateway Protocol (BGP)

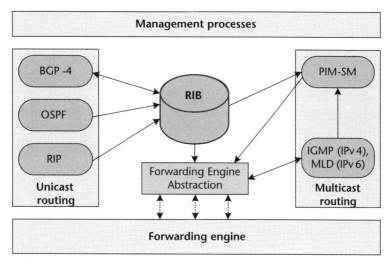

BGP-4 – Border Gateway Protocol version 4
IGMP – Internet Group Management Protocol
MLD – Multicast Listener Discovery
OSPF – Open Shortest Path First
PIM-SM – Protocol Independent Multicast- Sparse Mode
RIB – Routing Information Base
RIP – Routing Information Protocol

Figure 3.18 High-level view of routing protocols inside a router.

References

[1] J. M. Winett, "The Definition of a Socket," RFC 147, May 7, 1971.
[2] W. R. Stevens, *Unix Network Programming*, Upper Saddle River, NJ: Prentice Hall, 1990.
[3] G. R. Wright, W. Richard Stevens, *TCP/IP Illustrated, Volume 2: The Implementation*, Reading, MA: Addison Wesley, 1995.
[4] D. E. Comer, J. A. Jones, *Internetworking with TCP/IP—Principles, Protocols, and Architecture*, 4th ed., Upper Saddle River, NJ: Prentice Hall, 2000.
[5] J. Reynolds, J. Postel, "Assigned Numbers," RFC 1700, October 1994.
[6] R. Gilligan et al., "Basic Socket Interface Extensions for IPv6," RFC 2553, March 1999.
[7] G. Camarillo, "Peer-to-Peer (P2P) Architecture: Definition, Taxonomies, Examples, and Applicability," RFC 5694, November 2009.
[8] T. Janevski, *NGN Architectures, Protocols and Services*, New York: John Wiley & Sons, April 2014.
[9] IEEE Std 802.3z-1998, "Gigabit Ethernet," 1998.
[10] IEEE Std 802.3ba-2010, "40 Gb/s and 100 Gb/s Ethernet," 2010.
[11] ANSI T1.413, Issue 2, "Network and Customer Installation Interfaces—Asymmetric Digital Subscriber Line (ADSL) Metallic Interface," 1998.
[12] ITU-T Recommendation G.992.1, "Asymmetric Digital Subscriber Line (ADSL) Transceivers," June 1999.
[13] ITU-T Recommendation G.992.2, "Splitterless Asymmetric Digital Subscriber Line (ADSL) Transceivers," June 1999.
[14] ITU-T Recommendation G.992.3, "Asymmetric Digital Subscriber Line Transceivers 2 (ADSL2)," April 2009.
[15] ITU-T Recommendation G.987.1, "10-Gigabit-Capable Passive Optical Networks (XG-PON): General Requirements," January 2010.

[16] ITU-T Recommendation G.989.1, "40-Gigabit-Capable Passive Optical Networks (NG-PON2): General Requirements," March 2013.
[17] H. Holma, A. Toskala, *LTE for UMTS: Evolution to LTE-Advanced*, 2nd ed., New York: John Wiley & Sons, March 2011.
[18] WiMAX Forum, "WiMAX Forum Network Architecture: Detailed Protocols and Procedures—Base Specification," April 17, 2012.
[19] IEEE 802.16m-2011, " Part 16: Air Interface for Broadband Wireless Access Systems, Amendment 3: Advanced Air Interface," May 2011.
[20] F. Baker, "Requirements for IP Version 4 Routers," RFC 1812, June 1995.
[21] C. Hedrick, "Routing Information Protocol," RFC 1058, June 1988.
[22] G. Malkin, "RIP Version 2," RFC 2453, November 1998.
[23] G. Malkin, R. Minnear, "RIPng for IPv6," RFC 2080, January 1997.
[24] J. Moy, "OSPF Version 2," RFC 1247, July 1991.
[25] R. Coltun et al., "OSPF for IPv6," RFC 5340, July 2008.
[26] R. Callon, "Use of OSI IS-IS for Routing in TCP/IP and Dual Environments," RFC 1195, December 1990.
[27] J. Haas, J. Mitchell, "Reservation of Last Autonomous System (AS) Numbers," RFC 7300, July 2014.
[28] Y. Rekhter, T. Li, S. Hares, "A Border Gateway Protocol 4 (BGP-4)," RFC 4271, January 2006.
[29] S. Deering, "Host Extensions for IP Multicasting," RFC 1112, August 1989.
[30] W. Fenner, "Internet Group Management Protocol, Version 2," RFC 2236, November 1997.
[31] B. Cain et al., "Internet Group Management Protocol, Version 3," RFC 3376, October 2002.
[32] S. Deering, W. Fenner, B. Haberman, "Multicast Listener Discovery (MLD) for IPv6," October 1999.
[33] R. Vida, L. Costa, "Multicast Listener Discovery Version 2 (MLDv2) for IPv6," June 2004.
[34] P. Savola, "Overview of the Internet Multicast Routing Architecture," RFC 5110, January 2008.
[35] B. Fenner et al., "Protocol Independent Multicast—Sparse Mode (PIM-SM): Protocol Specification (Revised)," RFC 4601, August 2006.
[36] A. Adams, J. Nicholas, W. Siadak, "Protocol Independent Multicast—Dense Mode (PIM-DM): Protocol Specification (Revised)," RFC 3973, January 2005.
[37] Y. Rekhter, "A Border Gateway Protocol 4 (BGP-4)," RFC 1654, July 1994.
[38] "The 32-Bit AS Number Report," www.potaroo.net, accessed November 2014.
[39] "Growth of BGP Table—1994 to Present," www.potaroo.net, accessed November 2014.
[40] D. L. Mills, "Exterior Gateway Protocol Formal Specification," RFC 904, April 1984.
[41] T. Bates et al., "Multiprotocol Extensions for BGP-4," January 2007.
[42] E. Chen, "Route Refresh Capability for BGP-4," RFC 2918, September 2000.
[43] "The BGP Instability Report," http://bgpupdates.potaroo.net, accessed November 2014.
[44] C. Villamizar, R. Chandra, R. Govindan, "BGP Route Flap Damping," RFC 2439, November 1998.
[45] S. Murphy, "BGP Security Vulnerabilities Analysis," RFC 4272, January 2006.

CHAPTER 4
Fundamental Internet Technologies

4.1 Domain Name System (DNS)

The Internet has two name spaces: (1) Internet addresses, including both IPv4 and IPv6, and (2) domain names. Currently there is no third one. So, we can assume that connection between the two names spaces in the Internet is crucial for its functionalities. Such connections, that is, mapping between the name spaces, is accomplished with the Domain Name System (DNS) as one of the fundamental Internet technologies in the past and in near future.

Why are domain names needed? Well, communication between different machines connected to the Internet (e.g., hosts, routers) is based on IP addresses as identifications of their network interfaces. In many cases, however, humans are using machines to communicate over the Internet, to access certain contents, or to configure Internet hosts and routers. However, for humans words are more suitable than numbers such as IPv4 or IPv6 addresses (either they are presented in binary form or decimal-dot for IPv4 and hexadecimal for IPv6).

The mapping between IP addresses and domain names has been used in Internet hosts from the beginning (i.e., from early days of ARPANET). When the number of hosts in the Internet was small, the "hosts.txt" file was used for mapping domain names to numerical addresses (similar to the manual telephone exchange after the invention of telephony in 1876). However, the growth of the Internet raised the need for an automated system for maintaining the mappings between names and IP addresses. That led to creation of the DNS in 1983, with the first DNS software written by Paul Mockapetris. DNS was standardized with RFC 882 and RFC 883, which later were updated with the actual DNS standards, RFC 1034 [1] and RFC 1035 [2].

DNS can be compared to a phonebook. Similar to finding telephone number of a person with a given name and using the phone number to contact that person, the DNS provides an IP address (and optionally additional information) for a given domain name; the IP address obtained is then used to contact the host that has that address assigned to one of its network interfaces.

DNS is characterized by two independent conceptual aspects. The first is an abstract one that defines the syntax of the names and the rules for the delegation of authority for the names. The second specifies the implementation of a distributed computing system that efficiently maps names to IP addresses.

DNS can be thought of as an application layer protocol and as a distributed database. As an application layer protocol the DNS is based on the client–server principle and typically uses UDP as its transport protocol on well-known port 53 (e.g., DNS servers on this port receive requests from DNS clients). The size of UDP messages in DNS is limited to 512 bytes [2]. However, DNS can also use TCP on port 53. TCP is used for DNS when messages exceed 512 bytes or in specific cases (e.g., DNS zone transfer, i.e., replication of DNS databases across a set of DNS servers).

Overall, all DNS functionalities are provided by three major DNS components [1]:

- Domain name space and resource records (RRs): This is a tree-structured name space with information associated with the domain names stored in the resource records.
- Name servers: These store information about the domain tree's structure. In general, a name server stores (or caches) only a subset of the domain space called a zone (for which it is called an "authority"). Also, a name server has pointers to other name servers that can be used to provide information about any other part of the domain name tree.
- Resolvers: These are programs that extract information from name servers as a response to client requests. These are typically routines which are called by the application programs on a given host in the network. For that purpose, each resolver must know at least one name server [2]. For reliability purposes, there should be two domain name servers (primary and secondary), specified in the hosts either manually or automatically (via DHCP).

4.1.1 Domain Name Space

Domain name space is defined as a hierarchical tree structure, with the root on the top, as shown in Figure 4.1. Each domain name consists of a sequence of so-called labels, separated by dots. The domain name is written (or read) starting from the label of a node in the tree and going up to the root (which is a null label and therefore is not written in the domain name); that is, an inverted-tree structure. For example, domain name www.example.com consists of three labels divided by dots. The label on the left side is always lower in the name hierarchy, and hence defines a subdomain of the domain to the right. The right-most labels are called top-level domains (TLDs). So, in the given example (www.example.com) the top-level domain is the "com" domain. However, the size of the domain names is limited to 255 octets and each label may have a size of 63 octets or less (i.e., the maximum label size is 63 characters).

In terms of TLDs, two main types are used today:

- Generic TLD (gTLD): These include generic domains such as "com" (for commercial organizations), "edu" (for educational institutions), "gov" (for

4.1 Domain Name System (DNS)

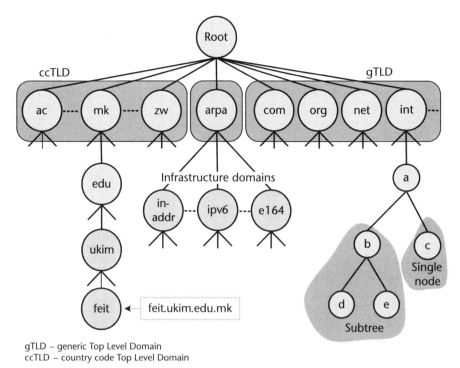

Figure 4.1 Domain name space structure.

governmental institutions), "info" (for information service providers), "int" (for international organizations), "mil" (for military groups), "net" (for network support centers), "org" (for nonprofit organizations), and so forth. Some of the gTLDs are limited to specific organizations/institutions (e.g., "edu," "gov," "mil"), while other gTLDs are not limited to the type of entity that owns the domain (e.g., "com," "net," "org," "info").

- Country code TLD (ccTLD): These TLDs were introduced later than the generic TLDs. Their purpose is to distinguish between the different domain spaces of different countries (similar to the network country codes used in traditional telephony).

However, the first Internet domain name is "arpa," a name that originated during the ARPANET era. Although originally conceived to exist only temporarily to implement migration of the names of the hosts in the ARPANET to the DNS system, it has remained in use as a so-called infrastructure highest domain that is used exclusively for infrastructure purposes in the Internet such as inverse DNS. Although conventional DNS resolution (lookup) is used to determine the IP address of a host by a domain name of that host, inverse DNS is used to determine a domain name based on the known IP address.

When a new domain needs to be added to a DNS, it is done via a "registrar," which is a commercial entity accredited by the ICANN, which verifies whether the given domain name is globally unique, and after that it enters the DNS database.

4.1.2 Name Servers

Name servers are DNS server programs that know the parts of the domain tree for which they have the complete information. Such name servers are referred to as the authorities for that part of the name space. So, the hierarchical name space tree is subdivided into DNS zones. Each zone may consist of one or multiple domains and subdomains for which the administrative authority has been delegated to a single managing entity. On the other hand, the zones can be automatically distributed to the name servers. So, name servers see the domain name system as sets of local information called zones. A name server has locally stored copies of some of the zones, which are used for processing queries that arrive from resolvers. However, a name server and resolver may be running on the same machine.

A given zone is typically available through several name servers so that its availability is ensured (e.g., in the case of a hardware, software, or connection failure of a given name server that serves a given zone). Although a name server has authoritative information for one or several zones, it may also have cached nonauthoritative data from other parts of the domain tree (i.e., from other zones).

Globally there is no single server that has all domain names to IP address mappings; instead, they are distributed on many name servers, which can be classified into the following main groups:

- Local name servers: Each ISP, either fixed or mobile/wireless, has a local name server that is set as a default for hosts in that network (typically that is done by DHCP). So, DNS queries from hosts first go to the local name servers.

- Authoritative name servers: This type of server is the name server for a given Internet host (e.g., web server, email server) that stores that host's IP address and name mappings. An authoritative name server can perform name-to-address translation for a given host name.

- Root servers: These servers are contacted by local name servers that cannot resolve a name. In such a case the contacted root name server contacts the authoritative name server if the name mapping is not known, obtains the requested mapping, and returns it to the local name server that has requested the mapping. There is a limited number of root servers in the world (currently there are 13 root server domain names), and this list is maintained by IANA [3]. The root server names are "a.root-server.net," "b.root-server.net," and following the same pattern (differentiated only by the left-most label, which is a single letter) up to "m.root-server.net." However, each root server may have one or more sites, which are typically distributed at different locations throughout the world so that they require less time to resolve names (when root servers are queried by local name servers).

Also, a given DNS name server may delegate the authority of its subdomains to other organizations. A simple DNS example is shown in Figure 4.2.

An authoritative name server can be a master server (i.e., primary DNS server) or slave server (i.e., secondary DNS server). For a given zone or zones, a master server is a DNS server that stores original (i.e., master) copies of all zone records.

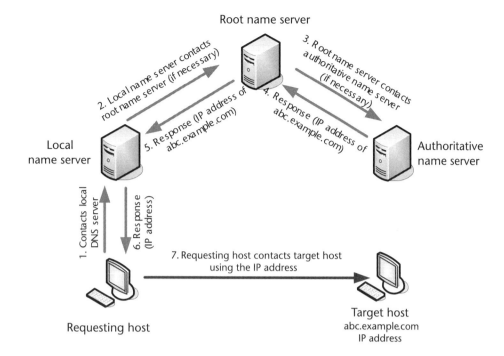

Figure 4.2 DNS simple example.

The slave DNS server typically uses an automatic updating scheme to synchronize with the data stored by the master DNS server.

When a name server learns a given mapping (name to address), it caches that mapping. The cache mappings expire after a certain time period called time-to-live (TTL), which is associated with every record in a DNS database. The minimum TTL value is typically set to 1 day (3 or more days are typically recommended), but its value typically should not exceed 2 weeks [4].

4.1.3 Resolvers and Resolution

From a user's point of view, the DNS is accessed through a simple call of the OS to a local resolver, which is the DNS client. The DNS client gives an abstraction to the end user that the domain space consists of a single tree and every part of that tree can be accessed. However, from the DNS client's (i.e., resolver's) point of view, the domain system is composed of an unknown number of name servers (e.g., local, authoritative, root name servers), where each name server has one or more parts of the whole domain's tree.

The resolvers initiate the resolving processes by sending a query to the local name server. Typically, the names of two (e.g., Windows OS) or three (e.g., Unix OS, stored in file "/etc/resolv.conf") local name servers are stored in each host for higher reliability (i.e., to create a higher probability of connecting to an accessible and working name server).

In most of the cases, the resolver (local DNS client in the Internet host, either a fixed or mobile one) gives the domain name to the name server (via a DNS query) and asks for a corresponding IP address. The query is sent to the local name server (i.e., local DNS server), which is typically placed in the telecom operator's (i.e., ISP's) core network. However, if the local server cannot resolve a given query, then it uses other DNS servers to help resolve the query. In such a case, two main approaches are used for resolution of domain names (Figure 4.3):

- Recursive query (recursive resolution): The contacted local name server provides a resolution by finding the authoritative name server for the given domain name.
- Iterative query (iterative resolution): In this case, the contacted name server replies to the query from the DNS client with name of the name server to contact (in a case in which the contacted name server cannot resolve the query).

For the resolution the DNS protocol uses request and response DNS messages, both in the same format. Each DNS message consists of a 12-byte header and a payload. A request and a response to that request are connected by using the same 16-bit (2-byte) identification number that is stored at the beginning of the header. The payload of each DNS message has four types of fields: questions (i.e., queries from the resolver's request), answers (i.e., responses from DNS servers), records

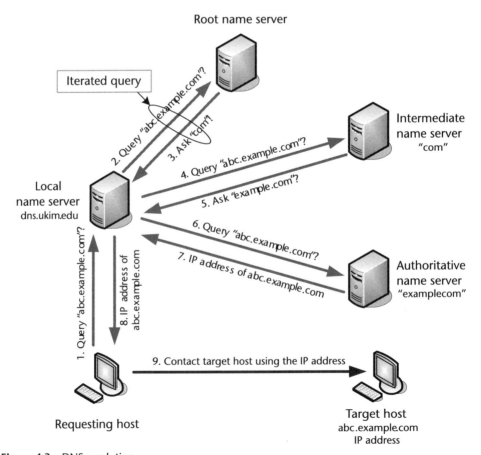

Figure 4.3 DNS resolution.

for authoritative servers, and additional information (variable number of resource records).

The DNS data are stored in distributed DNS databases in the form of resource records (RRs). Several types of RRs are used, some of which are illustrated in the following example.

DNS Resource Record Format Example

(name, value, type, ttl)
If "type" = "A," then "name" is hostname and "value" is IP address.
If "type" = "NS," then "name" is domain (e.g., example.com) and "value" is IP address of authoritative name server for that domain.
If "type" = "CNAME," then "name" is an alias name for some canonical (i.e., real) name.
If "type" = "MX," then "name" is a name of an email server, and "value" is the hostname of the mail server associated with the "name."

So, many resource records are available in a given DNS database. The distributed character of the DNS provides good scalability, which allows it to cope with the exponential growth of the Internet (regarding the number of hosts as well as traffic volume), and therefore growth in both name spaces, IP addresses (IPv4 and IPv6), and domain names.

4.1.4 DNS Discussion

As discussed, the Internet has only two name and addressing spaces, the IP address space and the domain name space. The DNS provides mappings between them. Without the DNS, the current operations in the Internet would not be possible, at least in a form that most Internet users are used to (web, email, and so forth). Hence, we can easily conclude that the DNS is one of the most fundamental Internet technologies in today's Internet.

However, at the time of DNS standardization, no one could really predict the unprecedented growth of the Internet that would follow. Due to the large number of hosts with a given IP address and domain names, manually making changes in the DNS master records is not efficient (adding a new host, removing a host, changing the IP address in a given mapping, and so forth). For that purpose, the DNS master records are dynamically updated by using Dynamic DNS (DDNS). In particular, when a given binding (between a name and an address) is determined by the DHCP, it sends such information to the primary DNS server, which updates the zone data.

Finally, due to its importance to the Internet and its operation on a global scale, DNS is a possible target for attackers, which may read DNS messages (if not protected) to find the user's profile, intercept DNS responses and change them, or flood DNS servers. With the goal of protecting DNS communication, the IETF has standardized Domain Name System Security Extensions (DNSSEC) [5], which add data origin authentication and data integrity to the DNS.

Overall, the DNS has a very important position in the Internet world and currently it is irreplaceable. Note, however, that the DNS specifications will be continuously updated and changed/extended in parallel with the process of convergence of all telecommunication services onto the Internet and with the development of new Internet services in the future.

4.2 File Transfer Protocol (FTP)

FTP is an application for copying files from one machine (i.e., host) to another. It was one of the first important application layer protocols in the Internet from its inception. The first file transfer mechanisms were proposed in 1971 with RFC 114. FTP is a client–server protocol that uses the TCP/IP protocol stack.

FTP requires two connections [6]: a data connection and a control connection, as shown in Figure 4.4. The control connection is established prior to the data connection(s), by exchanging commands (from the FTP client) and replies (from the FTP server). Each FTP server listens on port 21 for incoming control connections from FTP clients. A single control connection remains open during the entire FTP session, and it may be used for control of many file transfers (over data connections) such as server-to-client file transfer, client-to-server file transfer, and transfer of a list of directories. However, before any file transfer the client must log in to the FTP server (i.e., perform authentication). FTP login utilizes a username and password scheme, which is based on the commands USER (for the username) and PASS (for the password). FTP also allows the server to provide anonymous access to given files by using the username "anonymous" during the login process.

After establishing the control connection with the server (it includes authentication as well as definition of file type, data structure, and transmission mode), the FTP client establishes a new TCP connection for each data transfer to be accomplished. The data connections are established on TCP port 20 on the server's side. Each data connection is initiated by the FTP client on a random TCP port, and the client issues commands for transferring files. However, two FTP modes ares used for the data transfer:

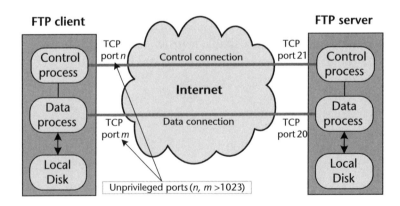

Figure 4.4 FTP processes and connections.

- Active mode: A client initiates a control connection from a random unprivileged TCP port n ($n > 1,023$) to the FTP server's control TCP port (i.e., port 21), and sends the FTP command "PORT $n + 1$" to tell the server to connect to the next-higher-than-n unprivileged port. The FTP client listens for server responses on port n + 1. So, the server connects from its data port (i.e., TCP port 20) to the client's data port, which is set to n + 1.

- Passive mode: In this mode the client opens two random unprivileged ports n and n + 1 (n > 1,023) and then connects port n to the server side FTP control port (port 21) and issues a PASV command over the control connection. After receiving the PASV command from the client, the server indicates which port should be used for the data (also an unprivileged TCP port), and then the client connects to the specified port.

The server responds to commands from clients with three-digit status codes accompanied by a text message (all in ASCII format) that is readable by humans (e.g., for administration purposes). For example, status code "200" means "OK," while status code "500" means "Syntax error, command unrecognized." The same approach for responses and status codes (although extended with new codes) is later used for other types of Internet communications that appeared after FTP, such as email protocols and HTTP.

Most FTP implementations use the so-called stream mode for data connections (i.e., data is sent as a continuous TCP stream, without any processing by the FTP), and establish a new TCP connection when the server needs to send new data to the client. However, data transfer can also be done in two other modes: block mode (FTP breaks data into blocks and then passes them to the TCP) and compressed mode (data are compressed using a given algorithm). These two data transfer modes do not close the TCP connection to indicate the end of a transferred file, which is required in the stream mode. In all cases, if the control connection is terminated, then the FTP session is terminated, resulting in termination of all data transfer processes.

4.2.1 Trivial FTP (TFTP)

FTP is a demanding application that includes two processes running in parallel (control process and data process). When it is not necessary to exploit the full potential of FTP, but only do a simple transfer of a file (e.g., for a diskless workstation or when booting a router) then Trivial FTP (TFTP) can be used [7]. TFTP uses the UDP/IP protocol stack and single process on the host. TFTP servers listen on UDP port 69. TFTP can be used in conjunction with DHCP to initialize given devices by providing easy download of configuration files from a TFTP server.

The sending host typically sends the data in blocks of 512 bytes (as a default value; it is later increased to 1,468 bytes [8]), and the last block must be less than 512 bytes to indicate the end of a file. If the file size is a multiple of the block size, then the source sends the last data packet containing zero bytes of data. However, because TFTP uses UDP on the transport layer, the reliability is accomplished with acknowledgments (ACKs) on the application layer (i.e., by TFTP application).

4.2.2 FTP Discussion

FTP was one of the first important applications used over the Internet. It provides and efficient means for the transfer of files between heterogeneous types of hosts. The approaches that were standardized for FTP influenced most of the Internet services that appeared after FTP, including HTTP, considered the most important successor of FTP. In fact, most Internet applications include transfer of files between two hosts in a manner similar to that used by FTP. However, FTP's successors (e.g., HTTP) are simpler in terms of the number of processes that need to be run on a given host.

In general, FTP was a crucial protocol for file transfer in the 1980s and to some extent in the 1990s. However, after the invention of the WWW, FTP usage over the Internet has significantly decreased because client–server file transfer is performed mainly by HTTP.

4.3 Electronic Mail

Electronic mail (email) is one of the "flagship" Internet services. It provides for the exchange of digital messages between the sender and one or more recipients. Although email is one of the oldest Internet services, it is very important from different aspects (from business and personal communication aspects, to authentication of users to other Internet services), as well as one of the most frequently used in the past several decades (e.g., it has replaced the traditional fax communication in enterprise environments). Early systems (predecessors of email) appeared in the 1960s and 1970s. However, in 1982 ARPANET published proposals for electronic mail, RFC 821 and RFC 822, that defined the Simple Mail Transfer Protocol (SMTP) [9] and the Internet message format [10] as foundations of the email service that have remained up to the present day.

The email architecture, shown in Figure 4.5, consists of the following main standardized components called agents:

- Message user agent (MUA): This is a process that is used to create an email message (hence, it has a user interface), and to process the delivered email messages (at the receiving side, also through the given user interface). Examples of MUAs are Thunderbird and Outlook.
- Message transfer agent (MTA): The MTA acts as an SMTP server, and accepts messages from the MUA, message submission agent (MSA), or another SMTP server (i.e., another MTA). In cases where the email recipient is not hosted locally, the message is forwarded to another MTA (i.e., another SMTP server). It uses as a default TCP port 25, while secure SMTP uses TCP port 465.
- Message submission agent (MSA): The MSA acts as a submission server to accept messages from user agents (i.e., MUAs), and it either delivers them or acts as an SMTP client to relay them to an MTA. It is typically integrated with MTA (i.e., MSA/MTA), but also can be separated (i.e., used for message

4.3 Electronic Mail

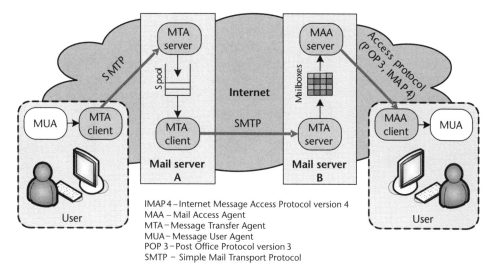

Figure 4.5 Email communication.

relaying to MTAs). When an MUA sends email to an MSA, it must use TCP port 587 [11]. In such a case the MSA acts as an SMTP client to the MTA, which is the SMTP server. Separation of the MSA and MTU is beneficial when there is a need to apply certain security and policy requirements (e.g., when ISP limits the ability to connect to remote MTAs on port 25, to limit automatically generated spam email traffic). MSA requires authentication by default (unlike MTAs).

Further, in email communication there are two main scenarios:

- Sender and receiver are on the same mail server: In this case there are two MUAs (a sender and a receiver) and a mail server.
- Sender and receiver are on two different mail servers: In this case there is a pair of MTAs and a pair of MUAs. It is also possible to have an MSA between the MUA (on sender's side) and the MTA.

SMTP is used for sending email messages from an MUA (on the user side) to a mail server, and sending email messages from one mail server (i.e., MTA) to another mail server.

For access to email messages, two defined mail access protocols (also sometimes referred to as mail access agents) are used:

- Post Office Protocol, version 3 (POP3) [12], and
- Internet Message Access Protocol version 4 (IMAP4) [13].

Additionally, popularization of free email accounts for Internet users (Figure 4.6) increases the portion of web-based access to email (when compared to standalone applications on a user's host, either using POP3 or IMAP4 as mail access protocol). In such cases the protocol that is used for access to email is HTTP.

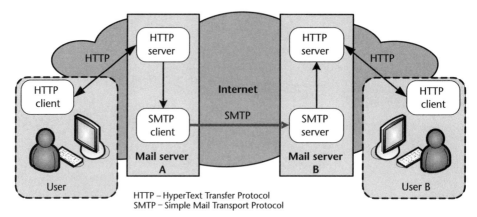

Figure 4.6 Web-based email.

4.3.1 Simple Mail Transfer Protocol (SMTP)

Historically, SMTP has its roots in the first mail sending programs created for ARPANET at the beginning of the 1970s, when it relied on FTP. The standardization of SMTP at the beginning of the 1980s (RFC 821) removed its reliance on FTP, but FTP influenced SMTP design regarding the command and response mechanism. The SMTP and Internet message format were last updated in 2008, with RFC 5321 [14] and RFC 5322 [15], respectively.

SMTP is the protocol that implements the MTA. It uses commands and responses to accomplish transfer of messages between an SMTP client and an SMTP server. In most cases (when the sender and receiver are on different mail servers), SMTP is used two times: (1) for sending the email from the sender's user agent to the mail server and (2) during message transfer between two mail servers.

How is an email message addressed so that it will reach its recipient? It is accomplished by an email address, which is a character string that identifies the user to whom the message should be sent or a location into which the message should be deposited. The depository is referred to as mailbox, and the email address is in fact a reference to a given destination mailbox. Typically, an email address is in the form "username@domain" where "username" uniquely identifies the mailbox of the email user within a given "domain" mail server ("domain" refers to domain name of the mail server, and "@" is a delimiter between the "username" and the "domain"). The MTA (SMTP client) that has to transfer the message to another MTA (SMTP server) uses DNS to look up the mail-exchanger record (i.e., MX record) for the recipient's domain. DNS replies to the requesting mail server with the IP address of the target mail server.

Each SMTP communication has three main phases (as shown in Figure 4.7):

- Connection establishment (i.e., greeting): This is the so-called "handshake" between the SMTP client (i.e., the sender) and SMTP server (i.e., the recipient). The server sends code 220 (if it is ready) or code 421 (if service is not available), the client sends "EHLO," and the server responds with code 250 (request command completed).
- Mail transfer: This phase begins when the client sends a "MAIL FROM" command to introduce itself as the sender of the message. Data transfer is

```
S: 220 example.com Simple Mail Transfer Service Ready
C: EHLO feit.ukim.edu.mk
S: 250-example.com greets feit.ukim.edu.mk
S: 250-8BITMIME
S: 250-SIZE
S: 250-DSN
S: 250 HELP
C: MAIL FROM :<tonij@ feit.ukim.edu.mk>
S: 250 OK
C: RCPT TO :<dario@ example.com>
S: 250 OK
C: RCPT TO :<abc@ example.com>
S: 550 No such user here
C: RCPT TO :<antonio@ example.com>
S: 250 OK
C: RCPT TO :<jasmina@ example.com>
S: 250 OK
C: DATA
S: 354 Start mail input; end with<CRLF >.<CRLF >
C: Hello...
C: ...etc.etc.etc.
C: .
S: 250 OK
C: QUIT
S: 221 example.com Service closing transmission channel
```

Figure 4.7 SMTP command and response example.

initialized by a client with the "DATA" message, and the client sends the email message in consecutive lines until the message is terminated by a line containing a single full stop (for successful message delivery the server returns code 250).

- Connection termination: For termination of the SMTP connections, the client sends a QUIT command and the server responds with code 221 for successful termination (or other appropriate codes in other cases).

The default email format defined in SMTP is 7-bit ASCII text [10]. Each email message consists of two main parts: a header and a body. The header is separated from the message body by a null line called carriage-return/line-feed (CRLF). Defined multipurpose Internet mail extensions (MIMEs) allow attachments to be added to an email message [16]. The MIME does not replace SMTP, but it is a supplementary protocol that allows arbitrary binary data to be encoded in ASC form and then transmitted as a standard email message. For that purpose MIME defines five additional headers, which include MIME-Version, Content-Type, Content-Transfer-Encoding, Content-Id, and Content-Description.

Content types include text (plain or HTML, no transformation by MIME is needed), image (JPEG, GIF), video (MPEG), audio (single channel encoding of voice at 8 kHz), application files (postscript, or general binary data), message (i.e., message body), or multipart combination (e.g., message body contains files of different data types). Content transfer encoding further defines how to encode the message into binary format (i.e., into ones and zeros). There are five encoding types:

- Seven-bit: This is 7-bit ASCII encoding, with less than 1,000 characters per line.
- Eight-bit: This is 8-bit encoding in which ASCII and non-ASCII characters can be sent, but the line length is limited to 1,000 characters. It relies on underlying SMTP for transfer of 8-bit non-ASCII characters, and does not use MIME. Therefore, it is not used in practice.
- Binary: This is very similar to 8-bit encoding, and relies on SMTP for transfer of binary data. Again, it does not use MIME and hence it is not used in practice.
- Quoted-Printable: In this case if the character is in ASCII format it is sent as is. Otherwise, non-ASCII characters are sent as three characters where the first character is an equals sign (=) and the remaining two characters are a hexadecimal representation of the byte.
- Base64: This type encodes 6-bit blocks of data into 8-bit ASCII characters.

Overall, the most used MIME encoding types in practice are Quoted-Printable and Base64. The table for binary-to-text encoding with Base64 is shown as Table 4.1. It is based on a simple principle that replaces each 6 bits (6 bits have maximum $2^6 = 64$ values) of a continuous bit stream of message data into the ASCII string format.

So, MIME provides the possibility for the transfer of multimedia email messages, as an extension of the SMTP and Internet message format, which originally supported text only. In general, MIME changes multimedia data into ASCII characters that are transferable through email systems.

4.3.2 Post Office Protocol Version 3 (POP3)

The most popular protocol for transfer of email messages from user's mailbox to the user's host device (computer, laptop, smartphone, and so on) is POP3 [17]. When using POP3, the user has a POP3 client (in the mail application, such as Outlook or Thunderbird) that creates a TCP connection to the POP3 server (it is typically colocated with the SMTP server on the same machine).

Table 4.1 Base64 Encoding Table

Base64 Value	ASCII Character	Base64 Value	ASCII Character	Base64 Value	ASCII Character
0	A	26	a	52	0
1	B	27	b	53	1
2	C	28	c	54	2
3	D	29	d	55	3
4	E	30	e	56	4
5	F	31	f	57	5
6	G	32	g	...	...
7	H	33	h	61	9
...	...	...	...	62	+
25	Z	51	z	63	/

4.3 Electronic Mail

When POP3 is used as mail access protocol, the mail server needs to have two active servers: (1) an SMTP server for incoming messages from other SMTP servers or messages sent by users through their email client applications and (2) a POP3 server for communication with the POP3 clients. The SMTP server accepts the message delivered to the user's permanent mailbox. The POP3 server allows the user to list and retrieve messages from the mailbox, and optionally to delete them after retrieval (this is typically set up via the user's mail application). A POP3 client creates a TCP connection to the server on TCP port 110. On the other side, POP3 secure uses TCP port 995.

POP3 uses commands (with arguments) and responses, as shown in Figure 4.8. It has the following states:

- Authorization state: This is the first state and it starts with a greeting from the server and follows with authentication of the client by means of a username and password.
- Transaction state: This state is reached after successful authentication of the client, and the client has access to the mailbox to list, retrieve, or delete mes-

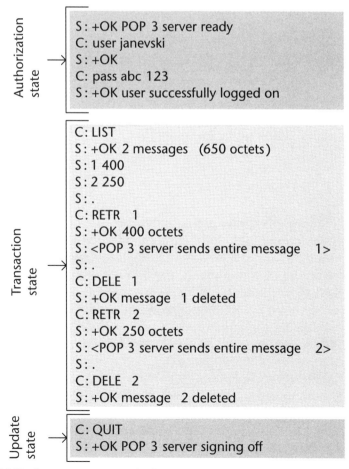

Figure 4.8 POP3 client–server communication.

sages. Messages are stored and transmitted as text files in RFC 822 standard format.
- Update state: When the client sends a QUIT command to the server, the POP3 session enters this state and the connection is terminated.

Overall, POP3 is one of the most popular mail access protocols, used in almost all email applications in fixed or mobile Internet hosts.

4.3.3 Internet Message Access Protocol Version 4 (IMAP4)

IMAP4 is another email access protocol [18]. It permits manipulation of remote message folders (in mailboxes) in a way that is functionally equivalent to local folders, including creating, renaming, and deleting mailboxes, then checking for new messages and deleting them, setting and clearing different flags, searching messages, and so forth. IMAP4 also provides the capability for an offline client to resynchronize with a server when the client goes online. An IMAP server listens on TCP port 143. Additionally, IMAP over the Secure Socket Layer (SSL), referred to as IMAPS, uses TCP port 993.

Overall, IMAP4 is a more complex and more powerful protocol than POP3. For example, with IMAP4 a user can have different message folders on the mail server, something that is not provided with POP3. Also, IMAP4 provides a means to check email prior to downloading (which is not possible with POP3), to search messages for a particular string (also prior to downloading), and to partially download messages. Also note that IMAP4 manages messages on the server side (local copies are possible, but they are considered as cached ones), whereas POP3 provides local management of the email messages (after retrieval). Hence, most of the email client applications have both email access protocols (POP3 and IMAP4), but fewer mail servers have implemented IMAP4.

4.4 World Wide Web (WWW)

WWW development was started in 1989 by Tim Berners Lee at CERN in Switzerland. He created a protocol for the distribution of documents in which a text document could be linked to other documents or objects (e.g., images, videos, audio files) located on the same or other server machines connected to the Internet. The WWW is made up of a large number of documents called web pages. Each web page is a hypermedia document ("hyper" because it contains links to other websites, and "media" because it can contain other media objects besides text).

4.4.1 Hypertext Transfer Protocol (HTTP)

The main protocol for access to web pages is HTTP, which is based on the client–server TCP/IP model. HTTP servers listen on TCP port 80. The web application on the user side is called a browser (e.g., Internet Explorer, Chrome, Firefox, Opera), which has an HTTP client. HTTP servers offer web pages to web clients. The main markup language for creating web pages is Hypertext Markup Language (HTML).

HTTP is used in two different modes:

- Nonpersistent mode: In this mode, a new HTTP connection is created for each object transfer (HTTP 1.0 [19]).
- Persistent mode: In this mode, one HTTP connection is used for transfer of multiple objects between a client and a server (HTTP 1.1 [20]). It was standardized with RFC 2616 in 1999, and updated in 2014 with RFCs 7230 through 7235.

HTTP works on request/response principle. A client sends a request, and obtains a response from a server. There are several defined HTTP requests, as listed in Table 4.2.

Each response consists of a three-digit response code followed by a blank space and readable description of the response code. HTTP response codes (i.e., status codes) are divided primarily into five groups: Informational (1xx), Successful (2xx), Redirection (3xx), Client error (4xx), and Server errors (5xx). For example, status code "100 Continue" means that the initial part of the request was received and the process can continue; status code "200 OK" means success; status code "302 Moved Temporarily" means that the requested URL has moved temporarily; "404 Not Found" denotes that the requested document was not found; "500 Internal Server Error" informs the HTTP client about an error at the server site; and so forth. Figure 4.9 shows an example of HTTP communication.

Each website has a unique identifier, the uniform resource identifier (URI). A URI generally refers to a Web address and it is a combination of a uniform resource locator (URL) and a uniform resource name (URN). The URL identifies a document, but it depends on the name of the server and other details of where the document is stored. URNs identify a document regardless of its storage location (i.e., host server). Note that there are lots of URLs, but few URNs. In that respect, URI is an "umbrella" category of URLs and URNs.

Table 4.2 HTTP Request Types

Request	Action
GET	Requests a document from an HTTP server.
HEAD	Identical to a GET request, but without the body of the response message.
POST	Used to send certain information from the client to the server (annotation of an existing resource, submitting a web form, and so on).
PUT	Requests a certain file to be stored under the supplied Request-URI. If the URI refers to an existing resource, it is modified.
TRACE	Echoes the received request to determine if changes/additions have been made by intermediate servers.
CONNECT	Reserved for use with HTTP proxy that dynamically switches to an encrypted tunnel (e.g., SSL tunneling).
DELETE	Requests the origin server to delete the resource identified by the Request-URI.
OPTION	Requests information about the communication options available on the request/response chain identified by the Request-URI.
PATCH	Request to modify an existing HTTP resource (this method was added later, with RFC 5789 [21]).

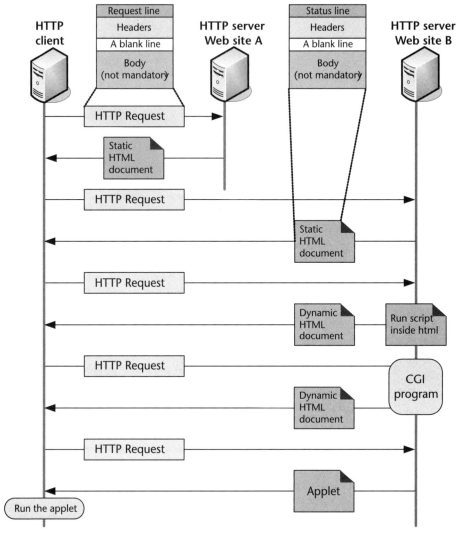

Figure 4.9 HTTP communication.

URL Example

http://www.example.com/itech/book.html
The first part specifies the protocol: http
The host server is: www.example.com
The local path to the web page (i.e., file) on that server is: /itech/book.html
The general form for an HTTP URL is as follows:
http_URL = http://host[:port][abs_path[?query]]
For example: http://www.example.com:8080/login.html?username=abc&password=123

In the general form of an HTTP URL [20], if the port is empty then port 80 is assumed for HTTP; otherwise, the specified port is used). The Request-URI for the resource is abs_path, which must start with "/" (an empty abs_path is equivalent to an abs_path of "/").

HTTP is a stateless protocol, which means that it does not remember the state of a connection. To be able to track transactions, HTTP uses cookies. A cookie is a piece of data sent from a web server to a web client and stored locally by the user's client (i.e., browser). Later the server can retrieve the cookie to track the user's previous activity.

HTTP also uses caching for higher efficiency (e.g., for reducing web traffic intensity on certain links or networks). Caching is implemented either locally in the user's machine or on the network side via proxy servers.

4.4.2 Web Documents

The various web multimedia documents can be classified into three main groups [22]:

- Static documents: In these types of documents, the contents are fixed and stored in a server. Typical examples are Hypertext Markup Language (HTML), Extensible Markup Language (XML), Extended Hypertext Markup Language (XHTML), and Extensible Style Language (XSL).
- Dynamic documents: This type of document is created by a server only at a given browser request. These web documents are handled by a set of standards called the Common Gateway Interface (CGI). The main benefit of CGI is that CGI programs are executed at the server site, and results are sent to the browser on the client's side. Several technologies are involved in the creation of dynamic documents such as a hypertext preprocessor (PHP) based on the Perl language, Java server pages (JSP) based on the Java language, active server pages (ASP) from Microsoft, and so forth.
- Active documents: This type of document is a copy of a program retrieved by the client and run at the client site. Typical examples include:
  - *Java applet:* A Java applet is a program written in Java and put on the server (in binary form, ready to run) that can be retrieved and run by the client's browser.
  - *JavaScript:* This is a small active part of a document, written in the JavaScript scripting language and given in a source code (i.e., in text form) that is interpreted and run by the client at the same time.

The WWW was initially associated with HTML, which is a simple semantic markup language that is intended for creating platform-independent hypertext documents [23]. HTML can represent various types of hypertext mail, news, documents, hypermedia, menus of options, database query results, structured documents with in-line graphics, and hypertext views of existing bodies of information. It has been in use by the WWW since the 1990s. HTML is an application of the ISO standard for Standard Generalized Markup Language (SGML) [23].

The markup language is written in the form of HTML elements, which consist of tags enclosed in angle brackets. The tags are used in pairs (start tag, end tag); for example, <html> is the start tag and </html> is the end tag. The documents involve a structure of nested HTML elements. The general form of an HTML element is:

<tag_name attribute1="value1" attribute2="value2">content</tag_name>

The structure of an HTML document is generally browser independent (i.e., all existing browsers have to support HTML), although it depends on the implementation. Also, an HTML page typically includes links to other objects and other HTML pages, including local content (on the same web server) and contents located on other web servers.

Simple HTML Example

```
<!DOCTYPE HTML>
<html>
  <head>
    <title>HTML example</title>
  </head>
  <body>
    <p>This is simple HTML text example.</p> <!-- This is a comment -->
    <p>This is <a href = "http://www.example.com"> HTML link</a>.</p>
  </body>
</html>
```

HTML content can be further styled by means of Cascading Style Sheets (CSS), which is a style sheet language for description of the look and provision of formatting of documents that are written in markup languages. Currently, W3C maintains the HTML and CSS standards, which are used by most websites on a global scale.

Overall, HTML, as a simple programming language for presentation of hypertext documents, together with HTTP, as a web communication protocol on the application layer, are the fundamental WWW technologies.

4.5 Multimedia Streaming

Multimedia streaming refers to the delivery of synchronized audio and video media in real time (i.e., with limited delay) to the end user by a service provider (e.g., an ISP or third-party service provider).

Several different application layer protocols and codecs have been standardized for multimedia streaming over packet-switched networks such as the Internet. One of the most well-known from the 1990s is the Real-Time Streaming Protocol (RTSP) [24]. Encoding of audio/video content and creation of streams is maintained by the Moving Pictures Experts Group (MPEG), which has created two main standards, MPEG-2 and MPEG-4, for multimedia streaming over packet-switched networks.

To keep delays as minimal as possible, the UDP is typically used as a transport layer protocol for real-time services, including multimedia streaming and VoIP, because TCP introduces additional delays due to retransmissions of lost or damaged segments. UDP, however, does not provide any traffic control mechanisms (end to end), so RTP is used over UDP within the transport protocol layer (i.e., RTP/UDP) to provide standardized mechanisms for synchronization of streams and feedback control communication between the receiver and the sender.

Broadband access to the Internet as well as high data rates in core and transport networks in the Internet reduce the delay, so multimedia streaming is also being provided over TCP/IP, especially for cases of video-on-demand (VoD) over best-effort Internet (e.g., YouTube). In the same manner, HTTP live streaming is also possible with a certain playout delay at the receiving side (to compensate for the packet delay variations, i.e., jitter, due to different network conditions at different times, such as level of congestion, packet error ratio, available bit rates for the given stream, and so on).

4.5.1 Real-Time Transport Protocol (RTP)

RTP is designed to provide end-to-end transport functions for provision of real-time services over the Internet, such as audio (e.g., VoIP), video (e.g., IPTV), and multimedia data, over unicast (e.g., for VoIP) or multicast networks (e.g., for IPTV or multimedia streaming services). It is used together with the RTP Control Protocol (RTCP), where RTP is used to carry media streams, and RTCP is used to provide control feedback between two end points of a given RTP session regarding the transmission statistics, QoS, and synchronization of multiple streams. Hence, the RTP standard [25] defines a pair of protocols, RTP and RTCP. Although RTP can run over TCP and UDP, it is typically used over UDP because it is targeted to be used by real-time applications that are sensitive to delay. RTP also provides support for translators (between different types of media streams) and mixers (for creation content from several media streams coming from different sources, such as video conferences).

RTP includes several important functionalities, such as timestamping, sequencing, and mixing of data from different sources, which are implemented via relevant fields in RTP headers as shown in Figure 4.10. The following fixed fields are defined in the RTP header:

- Version (the V bit): This identifies the current version of RTP (which is currently 2).
- Padding (the P bit): This bit indicates whether there are padding bits at the end of the packet that are not part of the payload. The number of padding octets (i.e., bytes) is written in the last padding octet of the packet.
- Extension (the X bit): When this bit is set, it indicates the presence of exactly one extension header between the RTP fixed header and the payload.
- Contributing sources count (4-bit field): This field denotes the number of contributors, where each contributor is identified by a contributor identifier. The maximum number of contributors is limited to 15 because the field has 4 bits.
- Marker (the M bit): This may be used for marking specific events such as a packet that contains a frame boundary. However, more than one or none

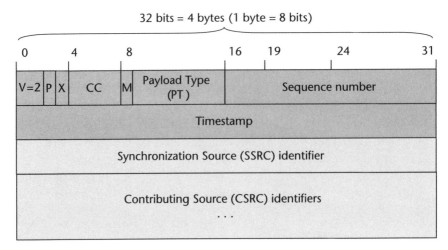

CC – CSRC Count; IC – Item Count; M – Marker; P – Padding; V – Version.

Figure 4.10 RTP packet format.

marker bits can be specified by changing the number of bits in the payload type field.

- Payload type (7 bits): This field indicates the format of the RTP payload and it is intended for use by the application that generates or processes the payload.
- Sequence Number (16 bits): This is used to identify a sequence of RTP packets between the sender and receiver. A sequence number starts with a random initial value and increments by one for each RTP packet; hence, it can be used to detect packet loss.
- Timestamp (32 bits): The timestamp field is used to provide information from the sender to the receiver to play back the received media samples at appropriate time intervals. The stamping rate must be derived from a clock (which is dependent on the application that uses RTP), which increments linearly and monotonically in time to allow synchronization as well as jitter calculations.

RTCP is used in parallel with RTP, and it is used for periodic transmission of control packets to all participants in a given session that uses RTP. RTCP uses the same mechanism as RTP, but on different ports. According to the standard [25], RTP ports should be even numbered, whereas RTCP ports should be odd numbered (in both cases they are unprivileged ports, between 1,024 and 65,535).

RTCP performs four functions: (1) provides feedback on the quality of data transfer (from the receiver to the sender); (2) carries a persistent transport-level identifier for a given RTP source called the canonical name (i.e., CNAME); (3) controls the rate at which packets are sent by the senders (which is individually calculated based on received RTCP packets); and (4) optionally carries a participant's information (e.g., email address).

To perform the given control functions, RTCP uses five packet types, which are defined by the packet type (PT) field in the RTCP header, given as follows:

- Sender report (SR): This packet type is sent periodically by all active senders to report transmission and reception statistics (to the recipients) from active participants in an RTP session, such as an absolute timestamp, sender's packet count, cumulative number of packets lost, highest sequence number received, interarrival jitter, delay since last sender report, and profile-specific extensions.

- Receiver report (RR): This packet type is used to carry reception statistics from receivers to senders.

- Source description items (SDES): This packet type provides a description of the sender, including mandatory CNAME item, and optional items such as name (i.e., personal name) and email address.

- End of participation (BYE): This packet type is used by a source to inform other participants that it is leaving the RTP session.

- Application-specific functions (APP): This packet type is used to design application-specific extensions (e.g., sending a document).

Generally, RTP can carry a wide range of multimedia formats such as MPEG audio and video (e.g., MPEG-2, MPEG-4), ITU-T H.261 encoded video, different voice codecs (e.g., ITU-T G.711, ITU-T G.723, GSM codec), and so forth. However, RTP and RTCP can provide adaptation of the stream (e.g., sending rate) to the available bandwidth end to end by using sender and receiver reports and by using mixing and synchronization of streams. However, they cannot provide QoS guarantees since the full RTP functionalities are based on the end hosts that are participants in the RTP session. Therefore, RTP is used for transport of real-time data over IP-based networks (e.g., VoIP, IPTV), while QoS guarantees (e.g., guaranteed bit rate) are provided by using control and signaling protocols (e.g., SIP, Diameter) prior to the data transfer.

4.5.2 Real Time Streaming Protocol (RTSP)

RTSP [24] is an application layer protocol for control of streaming media delivery from servers to clients via the Internet. One of the main features of RTSP is its videocassette recorder (VCR)-type control commands, such as "play," "stop," "pause," "fast forward," and "fast reverse," which provide real-time control of the playback of media files (Figure 4.11). So, RTSP provides "remote control" to end-user applications for multimedia servers on the Internet. However, it does not provide transfer of media data.

Although the streams controlled by RTSP typically use RTP/UDP/IP, the operation of RTSP does not depend on the transport layer protocol. Also, there is no notion of an RTSP connection; that is, an RTSP session is not tied to the transport layer protocol (e.g., TCP), and RTSP maintains a session by using a label called an identifier. So, an RTSP client may open and close many reliable transport connections (i.e., TCP) to the RTSP server with the goal of issuing requests for the media streams. RTSP can operate in multicast and unicast modes.

An RTSP server listens on port 554. The identified resource can be controlled by RTSP at the server listening on the dedicated port for TCP connections (in "rtsp" scheme) or UDP connections (in so-called "urtsp" scheme).

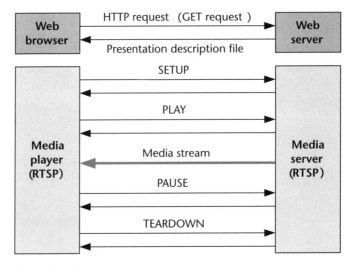

HTTP – HyperText Transfer Protocol
RTSP – Real Time Streaming Protocol

Figure 4.11 RTSP client–server communication.

The operations of RTSP can be grouped as follows:

- Media retrieval: Typically the client requests a presentation description via HTTP or some other method. In unicast communication, the client provides its IP address to the server; for a multicast presentation, the presentation description contains multicast addresses and ports for media delivery.
- Inviting of a media server to a conference: A media server can be invited to join a multimedia conference (e.g., for distributed distance learning applications).
- Adding media to an existing presentation: The server can inform the client about additional media becoming available (e.g., for live presentations).

In other words, RTSP provides the same services for streaming audio and video as HTTP does for text and graphics. Also, this protocol is designed to have a similar syntax and operation with HTTP, so all HTTP extensive mechanisms can be implemented to RTSP.

In RTSP each presentation and media stream is identified through RTSP URL. The full presentation and media properties are defined in the presentation description, which may include encoding type, URL destination, port number, and so forth. The presentation description file can be obtained from the client using HTTP, email, or some other means (typically HTTP is used).

Overall, RTSP is well designed to initiate and directly deliver streaming media content from media servers to clients.

4.5.3 MPEG-2

The first MPEG standard from ISO was MPEG-1, which was completed in 1993 [26]. However, it was designed for coding of moving pictures and associated audio for digital storage media (hard disk, CD, and so on). MPEG-1 was designed to fit into CD capacity, so it has a limited bit rate of up to about 1.5 Mbps (including video and audio streams) and was not intended for use over packet-switched networks such as ATM or the Internet.

The MPEG-1 standard supports the Standard Image Format (SIF) resolution of 352 × 240 pixels and 30 frame/sec (NTSC) or 352 × 288 and 25 frame/sec (PAL/SECAM). Also, the standard provided standardized mechanisms for synchronization of video and associated audio streams by using System Clock Reference (SCR) and Presentation Time Stamps (PTS).

Starting with MPEG-1, all MPEG standards use three types of video frames, named I, P, and B. The I frames are referent frames that are compressed independently of other frames; hence, they are the biggest (in bytes/frame). The P frames use data from previous I frames to achieve higher compression; hence, they are statistically smaller than the I frames. The B frames can use data from previous and forward frames (including I and P frames) for higher compression; hence, they are the smallest in size. All pictures of a given video stream are grouped into a so-called Group of Pictures (GoP), which is defined by two parameters: the number of P frames between two consecutive I frames and the number of B frames between two successive I or P frames.

MPEG-2 introduced the possibility of transmitting video and associated audio through packet-switched networks (standardized by ISO [27] and ITU-T [28]). It enables transmission in environments with a certain probability of packet losses (not possible with MPEG-1). Further, MPEG-2 includes more video resolutions, new image formats, and interactive multimedia services such as interactive television, remote control functions (i.e., VCR functions), and parallel transmission of multiple video and audio streams. Later MPEG-2 was revised to include high-definition television (HDTV) standardization, which was initially planned by ISO to be MPEG-3 (hence, there is no MPEG-3 standard).

MPEG-2 introduces and defines the transport stream. Two container formats are used for multiplexing in MPEG-2:

- Program stream: This format allows synchronization and multiplexing of multiple video and audio streams. The program stream format is compatible with the MPEG-1 standard and is intended for error-free transmission (e.g., for storage media such as hard disk drives, flash memory, and optical disks).
- Transport stream: This format combines one or more streams and can be used in real network scenarios (i.e., with a certain bit error probability). However, it does not provide error-control techniques; it leaves such tasks to the transport and network layers below it. The transport stream is positioned over transport layer protocols (e.g., over TCP and UDP in Internet).

The size of the MPEG-2 transport stream is 188 bytes, which equals 4 ATM cells with 47 bytes used for payload per cell (out of a total 48 bytes of payload size in 53-byte-long ATM packets), and it is designed to be transferred over the so-called ATM adaptation layer 1 (AAL1) based on a constant bit rate (CBR) transmission. ATM was an emerging technology in the first half of the 1990s (before Internet growth, after the invention and spread of the WWW) and hence MPEG-2 was initially adapted for transmission over ATM networks. However, in the second half of the 1990s, the Internet clearly won the packet-switching "battle" with the ATM, and multimedia streaming was further accommodated for transmission over IP networks.

4.5.4 MPEG-4

The most important standard to come out of the MPEG group of standards is MPEG-4. It was developed on the basis of three successful areas in the world of communications and computer systems: digital television, interactive graphics, and WWW. The main objective of MPEG-4 is to enable the integration of these three fields in terms of production, distribution, and content access.

One of the main features of the MPEG-4 standard is its interaction with the content of a particular scene in terms of the manipulation of objects therein. For that goal, MPEG-4 introduce audiovisual object (AVO), which can have video and/or audio components. Composition of different objects is defined by a so-called "scene description," which provides for the possibility of scene interaction and modification at the receiving side. The scene descriptions are coded independently of the streams (for different AVOs) in order to provide easier authorization in terms of the right to access, interact with, or manipulate the objects on the scene.

Coded AVOs are placed in one or more elementary streams. Each of these elementary streams is defined by a set of QoS parameters regarding the transmission (e.g., maximum bit rate, transmission error probability). The elementary streams are synchronized by adding timestamps in the synchronization layer (SL). The SL defines the packaging of elementary streams into access units, which are basic units for synchronization. One SL packet consists of a header and a payload. The header of the SL packet enables detection of packet losses and contains timestamps and associated information. Further, SL packets are multiplexed by using a so-called FlexMux layer (provides flexible multiplexing of SL streams, including interleaving) into FlexMux streams. The FlexMux layer, however, does not provide random access to the transport channel or error recovery. It relies on the underlying transport and network layer protocol stacks (e.g., RTP/UDP/IP, MPEG-2 transport stream, ATM adaptation layer).

Note, however, that MPEG-4 is an evolving standard, and it is divided into a number of parts. It was first standardized in 1998 [29], but in the past 15 years, parts of it have been standardized (currently there are more than 30 parts). Regarding the visual presentations, the most important MPEG-4 parts are Part 2, which includes widely used codecs such as DivX and Xvid, and Part 10, which defines advanced video coding (AVC), which is technically identical to the ITU-T H.264 standard. However, it is left to individual developers to decide which MPEG-4 features will be included in a given implementation.

In general, MPEG-4 is backwards compatible with MPEG-2 and MPEG-1, but includes much higher flexibility, higher resolutions, flexible bit rates, transmission in error-prone environments, and interaction with objects (which is dependent on the implementation). Nowadays, MPEG-4 is widely used in digital TV and IPTV implementations around the world.

Note that there are two other MPEG standards: MPEG-7, which is a multimedia content description interface, and MPEG-21, which is a multimedia framework for intellectual property management and protection. However, they are less important than MPEG-4, at least in terms of multimedia streaming over the Internet.

4.6 Session Initiation Protocol (SIP) for Internet Signaling

SIP is a widely accepted signaling protocol in IP networks. It was standardized by the IETF [30]. SIP is considered to be a replacement for SS7 in PSTN and PLMN, which was used mainly for signaling related to real-time services (e.g., voice) in fixed and mobile environments. It appeared in the second half of the 1990s (when the Internet started to grow globally), and it was finalized at the beginning of the 21st century.

SIP works on the application layer. Its main role is to establish, maintain, and terminate multimedia sessions. Hence, it is standardized as a signaling protocol in IMS, which is an essential part of the NGN for real-time multimedia services. In general, SIP can be used for creation of two-party sessions (e.g., VoIP), multiparty sessions (e.g., conference), or multicast sessions (e.g., IPTV). SIP can be implemented over TCP, SCTP, or UDP (as transport layer protocols). Additionally, SIP can be used as a component with other IETF standardized protocols for multimedia, such as HTTP, RTSP, and the Session Description Protocol (SDP) [31].

SIP is a client–server protocol. When it is used for VoIP, the caller acts as a client and the called party acts as a server. A call may involve several clients and servers, and a single host can be addressed as a client and as a server for a given call. In general, a caller may directly contact the called party (e.g., by using the IP address), which is a peer-to-peer communication approach. So, SIP communication can be used in both architectures, client–server and peer-to-peer.

4.6.1 SIP Messages

SIP is a text-based protocol similar to HTTP and SMTP. It uses messages in a request/response manner. Each SIP transaction consists of a client request and a response from a SIP server. SIP messages reuse most of the header fields, encoding

Table 4.3 SIP Messages

SIP Message	Description
INVITE	This message is used by the caller to initialize a SIP session.
ACK	This is a message with which the caller acknowledges the answer of the call (initiated with the INVITE message) by the called party.
BYE	This message is used to terminate an established session.
CANCEL	This message is used to cancel an already-started initialization process (e.g., client sends an INVITE and then changes its decision to call).
REGISTER	This message is used by a SIP user agent (located in the user's equipment) to register its current IP address and the SIP URIs for which the user would like to receive calls.
OPTIONS	This message is used to request information about the capabilities of a caller, without setup of a session. (A SIP session can be set up only with the INVITE message.)

rules, and status codes (reported in response messages) of the HTTP. This allows for easy integration of SIP with web servers and email servers.

SIP performs its operations with six messages used for session setup, management and termination, as listed in Table 4.3.

The given messages are used for performing SIP functions for different services, such as voice, video, messaging, and gaming. Each request SIP message is followed by a response SIP message from the other party in the given session. The response codes for SIP are the same as those defined for HTTP (i.e., three-digit response code followed by description in text format). All response codes are classified into six groups, with each group of codes being identified by the first digit, as given in Table 4.4.

The request and response SIP messages are exchanged between SIP components, which include network elements and user equipment that supports SIP.

4.6.2 SIP Addressing

A SIP entity must be named to be able to invite a given party to a session. Since the most used addressing format in the Internet at the time of SIP standardization was the email address format (e.g., username@FullyQualifiedDomainName) and URI schemes (e.g., http://www.example.com), the SIP also adopted the URI scheme based on general standard syntax (the same syntax used for WWW and email), as well as user addressing in a format similar to email addressing schemes.

So, each element in a SIP logical network (it is referred to as "logical," because a SIP network is usually an overlay network over an existing IP infrastructure) is identified by a SIP URI, which has the following general form:

```
sip:{user[:password]@}host[:port][;uri-parameters][?headers]
sips:{user[:password]@}host[:port][;uri-parameters][?headers]
```

The parts of the SIP URI in the brackets are not mandatory for the SIP URI, and their presence in the URI is dependent on the type of SIP entity. The format for

Table 4.4 SIP response codes

SIP Response Codes	Description
1xx	Provisional response, used by SIP servers to indicate progress, but they do not terminate a SIP transaction. *Example:* 100 Trying.
2xx	Success response, which means that the action was successfully received, understood, and accepted. *Example:* 200 OK.
3xx	Redirection response, which indicates that further action is needed for completion of a given request. *Example:*302 Moved Temporarily.
4xx	Client error, which usually means that the request contains bad syntax or cannot be completed by the server. *Example:* 404 Not Found.
5xx	Server error, which is a server-side error on a given valid request from the client side. *Example:* 504 Server Time-out.
6xx	Global failure, which means that the request cannot be fulfilled by any server. *Example:* 603 Decline.

4.6 Session Initiation Protocol (SIP) for Internet Signaling

SIPS (**SIP secure**) URI is the same as the used for a SIP URI, with the difference being the URI scheme; that is, "sip" is the URI scheme for SIP, and "sips" is the URI scheme for SIPS. The term "user" in the URI refers to a user in the given host (e.g., example.com). The password is associated with the given "user," but its presence in SIP URIs is not recommended due to possible security risks (e.g., exposing the password to others). The user and password are parts of the so-called user-info in the SIP URI. The host name in the URI contains either a fully qualified domain name (FQDN) or an IP address (including IPv4 and IPv6 addresses). The port number is used where necessary. If it is omitted, then the default SIP port numbers are used, which are 5060 and/or 5061 (for both, TCP and UDP) to connect to SIP servers and other SIP end-peers.

SIP URI Examples

```
sip:userABC;password123@example.com:3300
sip:userABC@example.com
sip:userABC@10.0.0.1
sip:server1.example.com
sip:1234567@example.com
```

The SIP user addresses of type "phone-number@gateway" are used to name PSTN telephone numbers to make them available through the "gateway" node. Also, some users may be able to use email addresses as their SIP addresses; this depends on the SIP implementation in a given domain.

4.6.3 SIP Network Elements

There are two general types of network elements for SIP called user agents (UAs):

- User agent client (UAC): A UAC creates SIP requests and sends them to servers.
- User agent server (UAS): A UAS receives requests from clients (i.e., UACs), processes them, and returns SIP responses.

A single UA can operate as both a UAC and a UAS. The roles of either UAC or UAS last only for the duration of a given SIP transaction. SIP UAs are logical end points in the SIP network architecture, which consists of the following types of SIP servers:

- Redirect server: This is a user agent that generates 3xx responses to redirect the request back to the client, indicating that the client needs to try a different route to reach the SIP recipient (e.g., when a recipient has moved to another location). With redirection, the servers push the routing information for a given request from the UAC in a response back to the client, thereby propagating URIs from the core network to its edges, which increases network scalability.

- Proxy server: This is the most common server type in SIP-based signaling networks. It usually works as both a UAC and a UAS, with the goal of providing a recipient's IP address to SIP clients. When a request is generated by a UAC, it typically does not know the recipient's IP address, so the client sends the request to its assigned SIP proxy server, which forwards the request to another proxy server or to the recipient. Proxy servers are also used to enforce network policy (e.g., checking whether a user is allowed to use given service), and also may rewrite certain parts of the SIP request message (according to the network policies).
- Registrar server: This is a logical end-point SIP entity that is used for registering the current location of the SIP clients (e.g., binding IP address of the client with one or more SIP URIs). A user registers its location by sending a REGISTER message to the registrar server (typically colocated with the SIP proxy server).
- Location service: This is a database that stores information about a user's location; that is, it stores bindings between a user network's location information (e.g., IP addresses) and SIP URIs. Such information is stored by registrar servers upon receiving REGISTER messages from SIP UAs.

Figure 4.12 shows three scenarios for using SIP signaling. Figure 4.12a shows a general view of SIP signaling between two SIP UAs when they are located in two different domains and hence are served by two different proxy servers. Figure 4.12b shows the so-called SIP triangle, which is the scenario in which both VoIP users (i.e., SIP UAs) are located in the same domain, and hence served by the same proxy server. The third case, shown in Figure 4.12c, presents a SIP scenario for peer-to-peer telephony, where the SIP network elements are excluded (i.e., peer-to-peer SIP scenario). Usually proxy servers are used to connect different network domains with each other, or to connect users (i.e., SIP UAs) with SIP network nodes (i.e., SIP servers deployed in the network).

SIP signaling for originating VoIP calls typically starts with a request message from a SIP UA that is located in a given host (e.g., personal computer, laptop, IP phone, smartphone). The exchange of SIP messages is very similar to those for HTTP, as shown in Figure 4.13.

Unlike other applications, the invitation to a call cannot immediately result in a response because locating the called party and waiting to answer the call takes several seconds. The call can be placed in the waiting line if the called party is busy. Responses from class 1xx are used to inform the calling party about the progress of the call, but they are always accompanied by other types of responses (e.g., "200 OK") for a given SIP request. While the responses with 1xx codes are only temporary, the answer messages with other classes of status codes determine the final status of the request, such as 2xx for success, 3xx for redirection or forwarding, as well as 4xx, 5xx, and 6xx in the cases of client, server, or global failures, respectively. For higher reliability of the SIP signaling, the server performs forwarding of the final response until the client or server confirms its receipt by sending an ACK message to the sender.

Finally, SIP is a well-standardized protocol (by the IETF). However, to replace SS7 in the telecommunication world, SIP needs to be implemented on a global scale

4.6 Session Initiation Protocol (SIP) for Internet Signaling

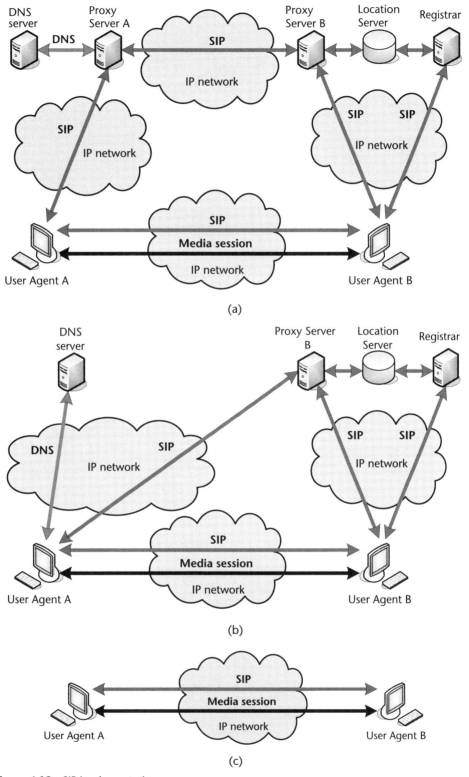

Figure 4.12 SIP implementations.

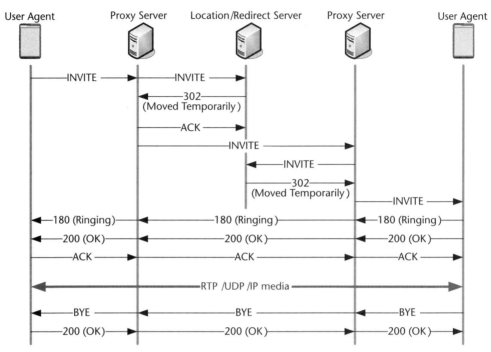

Figure 4.13 SIP signaling.

(e.g., for VoIP provided by telecom operators, including fixed and mobile ones). Such "umbrella" standardization is done by ITU-T (as was done for SS7 several decades before SIP) by accepting SIP globally as the main signaling protocol in NGN, together with the standardized common IMS. Overall, SIP is one of the most important signaling protocols in the converged Internet and telecommunication worlds, and is expected to remain so in the future.

4.7 Internet Security

Security in telecommunication networks is changing over time. Unlike the PSTN and PLMN, which are based on simple end-user devices and complex networks nodes, the Internet is based on smart and more powerful user devices (e.g., computers, smartphones) that pave the way for innovation of new applications and services without an explicit need for standardization and approval. On one side this drives the Internet in general (as a single networking platform for different types of services and different access networks), while on the other side it raises more security threats. The traffic can be intercepted or changed, users can be "tricked" into providing their personal data and information, servers can be under attack from malicious Internet hosts, and so forth. However, the threats have initiated development of appropriate Internet security solutions, which may be implemented on different protocol layers, as given in Table 4.5.

Several goals define Internet security, including the following:

Table 4.5 Security Solutions on Different Protocol Layers

OSI Protocol Layer	Security Solutions
Application layer	Firewalls, antivirus software
Presentation layer	
Session layer	
Transport layer	SSL/TLS
Network layer	IPSec, VPN
Data-link layer	Examples: 802.1X (for IEEE access networks), WPA (for Wi-Fi), LTE authentication, NAS (nonaccess stratum), and AS (access stratum) security
Physical layer	Different policies for user access to a given location with network elements (e.g., data centers, core routers, base stations)

- Confidentiality: Only the sender and targeted receiver should "understand" a message's contents. Hence, the sender encrypts the message, and the receiver should be able to decrypt it.
- Authentication: This operation is designed to allow the sender and receiver to confirm each other's identity.
- Message Integrity: This means that the message should not be altered in transit (or afterwards) on the way from the sender to the receiver and that changes should be made only by authorized entities. However, certain system failures (e.g., power failure, hardware failure) may violate the integrity of a message, which is not treated as a result of some security threats.
- Access and availability: The provided services must be accessible and available to users authorized to access the information.

Corporate networks emphasize data security more than ease of use, whereas in public IP networks simplicity of usage plays a more important role than the level of data security implemented in them. However, there should be always a balance between the security solutions and ease of use, especially in the public Internet (which consists of all hosts that are accessible by everyone who is connected to the Internet).

On one hand, the Internet is generally independent from underlying access networks (i.e., OSI layers 1 and 2); on the other hand, it provides the possibility (via the socket interface) for different applications (on the top protocol layers, i.e., OSI layers 5 through 7) to be installed and run. Hence, the main standardized Internet security solutions are given on the network layer (e.g., IPsec) and transport layer (e.g., SSL/TLS), which are therefore covered in the following sections. However, the application layer also has various security solutions, such as Pretty Good Privacy (PGP) for email encryption.

4.7.1 Internet Security on Network Layer (IPsec)

The lowest protocol layer that can provide end-to-end security in the Internet (across multiple routers on the way) is the network layer, which is IP (IPv4 and IPv6) in all Internet hosts and network nodes. The security architecture for the IP is

IPsec [32], which is a collection of protocols standardized by the IETF that provide security at the network layer. So, IPsec is an open standard.

In practice, the IPsec implementation operates in a given host or router as a security gateway (an intermediate system, such as firewall or router that is IPsec enabled) or as an independent device (with the goal of protecting IP traffic). In general, IPsec provides three processing actions for IP packets: (1) Protect the packet with IPsec, (2) discard the packet, or (3) allow IPsec protection to be bypassed.

What does IPsec do? It creates a boundary between unprotected and protected interfaces for a given host or network. In fact, IPsec provides encryption of the payload of each IP packet through selection of the proper security protocols, cryptographic algorithms, and keys. (An IP header cannot be encrypted across the Internet, because the routing is based on destination and source IP addresses located in the header.) In such a manner, IPsec can protect one or more paths between three different pairs of nodes: (1) between a pair of hosts, (2) between a pair of security gateways, or (3) between a security gateway and a host.

How does IPsec work? To provide security services IPsec uses two protocols, as shown in Figure 4.14:

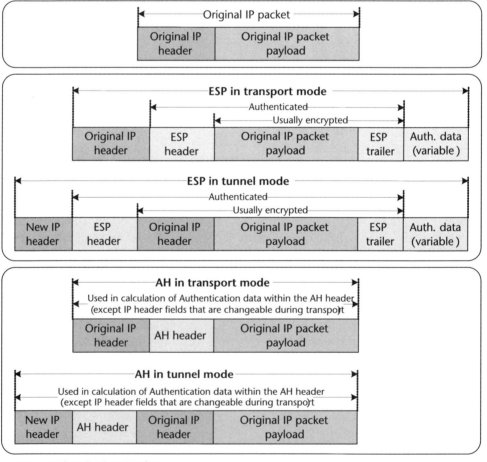

AH – Authentication Header
ESP – Encapsulating Security Payload

Figure 4.14 IPsec protocols AH and ESP.

- Authentication Header (AH) Protocol, [33]: AH provides source authentication, data integrity, but no encryption of IP packets. The AH header is inserted between the IP header and the IP payload field (the "next header" value for this protocol is 51). When AH is used, the intermediate routers process packets as usual. The AH header includes a connection identifier, authentication data (source-signed message digest calculated over original IP datagram), and a next header field (i.e., specifies next protocol header in the IP payload such as TCP, UDP, or ICMP).
- Encapsulating Security Payload (ESP) Protocol [34]: ESP provides secrecy (i.e., confidentiality), which is not provided with AH, as well as host authentication and data integrity. In this case data and so-called ESP trailer are encrypted. The next header field is placed in the ESP trailer (the next header value for this protocol is 51). On the other side, the ESP authentication field is similar to the AH authentication field.

Both AH and ESP offer access control that is enforced through the distribution of cryptographic keys and the management of traffic flows as dictated by a so-called security policy database (SPD). Each of them may be used individually or in combination; however, most security requirements in Internet networks can be met through the use of ESP only. Also, each of the two IPsec protocols (AS and ESP) can be provided in two modes:

- Transport mode: In this case AH and ESP provide protection for next layer protocols (i.e., transport and application layers, above the IP as the network layer).
- Tunnel mode: In this case AH and ESP are applied to tunneled packets; that is, IPsec encrypts the entire IP packet and then adds a new IP header to it that is used for tunneling. The added header has different information than the original IP header; the source and destination IP addresses in the added IP header correspond to the addresses of the routers (e.g., security gateways) that are at the start and end points of the tunnel, respectively.

The SPD management paradigm determines which security protocol (AH or ESP) should be employed, the type of the mode (transport or tunnel), various security service options, what cryptographic algorithms to use, in what combinations to use the specified protocols and services, and the granularity at which protection should be applied. Most of the security services require cryptographic keys, and IPsec can rely on manual and automatic distribution of keys such as the Internet Key Exchange (IKE) protocol.

Where is IPsec primarily used? One of the main applications of IPsec is in VPNs, which are used in core and transport networks to carry aggregated Internet traffic of a given type (e.g., VoIP, IPTV, best-effort, signaling and control traffic). A VPN is a network that is only logically private, not physically (i.e., it does not use real private transport networks, such as WANs). A VPN is typically used to connect a pair of routers to guarantee privacy of the transferred traffic between them (since traffic may be transferred across several network domains that belong to different organizations/enterprises). Typically, VPN technology uses the ESP protocol in the tunnel mode. In such a scenario each private datagram is encapsulated in an ESP packet and carried through the public Internet between the two end routers (for the given tunnel). The private datagrams are encrypted when they are transferred through VPN tunnels, so they cannot be decrypted during transit (e.g., by other parties that have access to the transport networks used for the given traffic). The router, which acts as a tunnel end point for the VPN, deciphers the packet and then routes it toward its destination. However, the deciphered IP packet can be put again into another VPN tunnel (e.g., different so-called bearer services in different networks), so a single packet may travel through several different VPNs until it reaches its destination.

Overall, VPNs are the most used technology for design and traffic management in core and transport networks, and they are typically used with certain QoS solutions in the Internet, such as MPLS.

4.7.2 Internet Security on the Transport Layer (SSL/TLS)

Internet security on the transport protocol layer is provided by two main protocols: SSL and Transport Layer Security (TLS). The older of the two protocols is SSL, which was initially developed by Netscape in the first half of the 1990s. TLS is IETF's standardized version of transport layer security. It is based in large part on SSL as its predecessor. In 2011 SSL version 3 was published by the IETF as a historical document (RFC 6101 [35]), because SSL never became an official Internet standard.

Both protocols, SSL and TLS, are placed between the transport and application layers in the protocol stack. Their purpose is to provide authentication between the client and server and ensure the integrity of transmitted data [36]. For example, when SSL is used, application programs such as HTTP actually store data in SSL (or TLS) packets. However, when talking about SSL, one has to keep in mind that TLS is a continuation of SSL, defined with RFC 2246 as TLS version 1.0, which is actually SSL version 3.1. The TLS standard was expanded with TLS version 1.1 (RFC 4346) [37], and later with TLS version 1.2 (RFC 5246) [38]. Therefore, SSL terminology is often used with regard to the IP transport security level that is defined by the IETF (which includes TLS), and it is used as such in the following sections of this chapter.

4.7.2.1 SSL Architecture

The protocol architecture of SSL is shown in Figure 4.15. SSL uses TCP as its transport protocol (i.e., TCP/IP), and provides end-to-end security service. On the other side, the most commonly used applications over SSL (e.g., HTTPS) are HTTP, SMTP, POP3, and FTP.

The SSL architecture has several protocols, including the following:

- SSL Record Protocol: This is the carrier between different entities in the SSL architecture; that is, this protocol carries messages from three other SSL protocols and receives data from the application layer. Finally, it creates the payloads to the underlying TCP. The protocol fragments the data it receives into units with a maximum size of $2^{14} = 16,384$ bytes, and optionally can

4.7 Internet Security

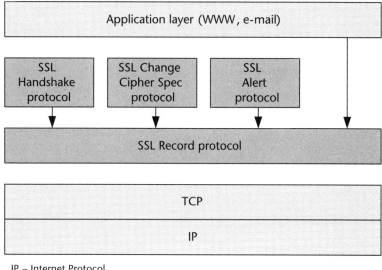

IP – Internet Protocol
SSL – Secure Sockets Layer
TCP – Transmission Control Protocol
WWW – World Wide Web

Figure 4.15 SSL protocol architecture.

compress the data. The MAC is added to the compressed message by using a negotiated hash algorithm. The compressed fragment and the MAC section are encrypted, and finally the SSL header is added to it. The SSL header consists of 5 bytes, where the first byte indicates a higher level protocol for that fragment (e.g., HTTP), the second byte contains the major version of SSL (e.g., in SSLv3.1 this byte contains a value of 3), the third byte contains the SSL subversion (e.g., in SSLv3.1 this byte contains a value of 1), and the last 2 bytes of the header contain the data size of the fragment (14 bits are used for that purpose).

- SSL Handshake Protocol: This protocol starts the SSL communication between the client and the server, and provides negotiation of security algorithms and parameters for the SSL Record Protocol, key exchange, server authentication, and optionally client authentication.

- SSL Change Cipher Spec Protocol: This protocol is based on a single message that indicates the readiness of cryptographic secrets, that is, the end of the SSL handshake. After that message, agreed parameters and keys are kept unchanged (as established during the handshake phase) and they are used by the Record protocol to sign/verify and encrypt/decrypt the messages.

- Alert protocol: This protocol is used for reporting fatal alerts and warnings via error messages.

SSL communication is based on client–server principles, which include the following two main approaches:

- SSL session: A session is an established association between a client and server, which means that both sides have exchanged general information such as the identifier of the session, certificates for authentication of the parties (e.g., X.509 certificates), security parameters, method of compression (if used), and the master key (it is a 48-byte secret key that is known to the client and server). A client–server pair can establish multiple sessions at the same time.

- SSL connection: This connection is created after an SSL session has been established. An SSL connection may belong to only one established SSL session. It provides the given type of service in a peer-to-peer manner. Hence, there can be multiple established SSL connections within a single session.

4.7.2.2 Use of SSL

In terms of use, SSL/TLS protocols can be used by any application layer protocol that uses TCP. In practice, SSL/TLS is used to protect data transmitted over the HTTP protocol (i.e., the Web) and for the email protocols SMTP and POP3. The implementation of HTTP over the SSL/TLS protocol is called HTTP Secure (HTTPS). Accordingly, a URL entry in the DNS system for web servers that use HTTPS begins with "https." However, there are no changes required in the HTTP protocol; the SSL/TLS protocol architecture is merely added between HTTP and TCP. HTTPS servers listen on port 443, while HTTP (without SSL/TLS) uses port 80.

Similar to HTTP, SSL/TLS can be used for email protocols. When using SMTP over SSL/TLS it is referred to as SMTP Secure and uses port 465 (SMTP uses port 25). If POP3 is used over SSL, then we have POP3 Secure, which uses port 995 (the default port for the POP3 protocol is 110).

Overall, using the SSL/TLS architecture for the two most important Internet native services, the WWW and email (represented via SMTP and POP3 protocols), provides a means for transparent use of such services with increased security and confidentiality when needed. It is widely used nowadays as a secure method of sending and receiving email messages (with STMP and POP3 over SSL/TLS) and for the use of web technology (with HTTPS) to access sensitive data, such as web-based email access, login data protection for various websites that offer certain types of services (e.g., social networks, portals), checking bank details, online payments (e.g., by credit cards), online transactions, and so forth.

4.8 AAA in Internet

AAA stands for authentication, authorization, and accounting. Briefly, authentication is used to identify a user, authorization manages what a user is allowed to do, and accounting measures the resources used by the user so that the user can be billed for services provided. Several systems are capable of performing AAA functions, such as Terminal Access Controller Access Control System (TACACS) [39], Remote Dial-In Remote Access Service (RADIUS) [40], and Diameter [41]. TACACS is not used much at all nowadays. RADIUS is the most widely used, but it is being replaced by Diameter.

4.8.1 RADIUS

RADIUS is an AAA application that is defined through RFC 2865 [40] and RFC 2866 [42]. It has specific application extensions that are defined in RFC 2869 [43]. Key features of RADIUS include the following:

- Client–server model: A network access server (NAS) operates as a client of RADIUS. The client is responsible for passing user information to designated RADIUS servers, and then acting on the response that is returned. RADIUS servers are responsible for receiving user connection requests, authenticating the user, and then returning all configuration information necessary for the client to deliver the service to the user. A RADIUS server can act as a proxy client to other RADIUS servers or other kinds of authentication servers.
- Network security: Transactions between the client and the RADIUS server are authenticated through the use of a shared secret, which is never sent over the network. Also, a user's password is encrypted to prevent snooping of the password.
- Variety of authentication mechanisms: RADIUS is able to support different authentication mechanisms, such as the Cleartext Password Authentication Protocol (PAP), the Challenge Handshake Authentication Protocol (CHAP), challenge/response procedures, and the Extensible Authentication Protocol (EAP) [44] and its extensions.
- Extensible protocol: RADIUS offers the ability to add new attribute values without disturbing existing implementations of the protocol.

RADIUS is UDP-based protocol. However, use of UDP instead of TCP is justified by higher reliability (with a secondary RADIUS server), simplification of the server (with UDP there is no need to maintain the connection between the client and the server), and simplified server implementations for multithreading (e.g., parallel authentication of users). Exactly one RADIUS packet is encapsulated in one UDP packet. The RADIUS server listens on port 1812 for authentication, and on port 1813 for accounting. Early versions of RADIUS used ports 1645 and 1646, respectively.

A basic RADIUS communication between a client and a server is shown in Figure 4.16. For example, if a RADIUS client wants to authenticate a user, it sends an Access Request packet to the server, which replies with an Access-Accept or Access-Reject packet.

A summary of the RADIUS packet format is shown in Figure 4.17. The Code field is 1 byte long, and it is used to identify the type of RADIUS packet. Available code types are shown in Table 4.6.

The Identifier field is 1 byte long and it is used in matching requests and replies. The RADIUS server can detect a duplicate request if it receives the same client source IP address, source UDP port, and an identifier within a short span of time.

The Length field defines the number of bytes in the packet including all fields. The minimum length is 20 bytes, and maximum length is 4,096 bytes.

The Authenticator is 16 bytes long and it is used in a password-hiding algorithm. The value is used to authenticate the reply from the RADIUS server. The value of the Authenticator field is called a Request Authenticator in Access-Request

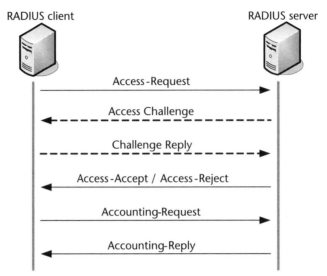

Figure 4.16 RADIUS communication.

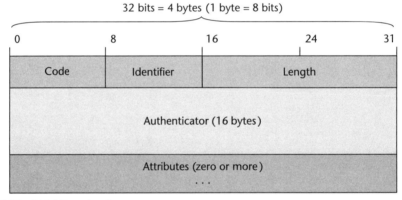

Figure 4.17 RADIUS packet format.

Table 4.6 Code Types for RADIUS

Code	Type
1	Access-Request
2	Access-Accept
3	Access-Reject
4	Accounting-Request
5	Accounting-Response
11	Access-Challenge
12	Status-Server (experimental)
13	Status-Client (experimental)
255	Reserved

packets, but it is called a Response Authenticator in Access-Accept, Access-Reject, and Access-Challenge packets. The Authenticator is based on a Message Digest algorithm MD5 value of a random number and the secret shared between the server and the client.

The last field, the Attribute field, is variable in length and contains the list of attributes that are required for the type of service, as well as any desired optional attributes. More than 60 attributes have been defined in RFC 2865, and additional attributes can be defined by the vendor. However, in the latter case there is no guarantee that RADIUS communication between different vendors will work.

When a client is configured to use RADIUS accounting at the start of the service, it will generate an Accounting Start packet, with information about the service and the user. The server acknowledges the packet upon the receipt. The RADIUS client can send update messages at configurable intervals (e.g., in range from 1 to 4,096 min). At the end of the service delivery, the client will generate an Accounting Stop packet describing the type of service being delivered and optional information such as elapsed time, input and output octets (i.e., bytes), or input and output packets. The RADIUS server sends an Accounting-Response to the client only when it has successfully recorded the accounting packet.

Note that some vendors may provide only RADIUS authentication and authorization clients, without the RADIUS accounting option.

4.8.2 Diameter

Diameter is a protocol that has been standardized by the IETF [41]. Its goal is to replace its predecessor, the RADIUS protocol. Diameter is an AAA framework protocol that works on the application layer. It is targeted for use by applications for network access as well as in IP mobility (including local and roaming scenarios).

Diameter uses TCP or SCTP on the transport layer, which provides reliable data transmission between end hosts that use the Diameter protocol. This is different than the RADIUS protocol, which uses UDP on the transport layer and provides reliability by retransmissions on the application layer. In general, Diameter introduces several important improvements over RADIUS, to suit different requirements from heterogeneous fixed and mobile networks. These improvement include the following:

- Failover: RADIUS does not have a failover mechanism since it uses UDP as the transport protocol, whereas Diameter defines application layer acknowledgments and failover methods.
- Transmission-level security: RADIUS may use the EAP framework [44] and may optionally use IPsec [45]. Diameter applies per-packet confidentiality by using IPsec and TLS [38], either as TLS/TCP or Datagram TLS for SCTP (i.e., DTLS/SCTP) [46].
- Reliable transport: On the transport layer RADIUS uses unreliable UDP, whereas Diameter uses reliable protocols (TCP and SCTP).
- Agent support: RADIUS does not have explicit support for agent nodes, such as relays, proxies, and redirect nodes. In contrast, Diameter defines explicitly the behavior of each agent that processes a Diameter message.

- Server-initiated messages: Support for this type of message in Diameter is mandatory, whereas in RADIUS such messages are defined, but support is optional, and hence it is difficult to function in heterogeneous network environments (e.g., server-to-client initiated unsolicited disconnect or reauthentication/reauthorization on demand).
- Transition support: Since Diameter's PDU is different from that of RADIUS, backward compatibility is needed with the goal of providing real-world deployments of Diameter, including networks that have deployed RADIUS.
- Capability negotiation: RADIUS does not support capability negotiations, so RADIUS clients and servers are not aware of each other's capabilities. However, Diameter includes error handling support and capability negotiations.
- Peer discovery and configuration: Whereas RADIUS requires names or addresses of servers and clients to be manually configured, by using shared secrets, the Diameter is based on peer discovery. There are two main cases: (1) when a Diameter client needs to discover a first-hop Diameter agent, and (2) when a Diameter agent needs to discover another Diameter agent for completion of a Diameter operation.

4.8.2.1 Diameter Message Structure

Diameter is a message-based protocol (one message per packet), with a message structure like that shown in Figure 4.18. There are two types of Diameter messages: request messages and answer messages. Each Diameter message contains a packet

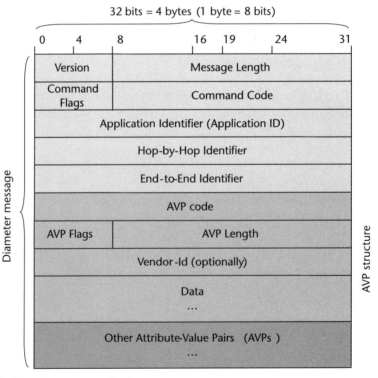

Figure 4.18 Diameter message structure.

header followed by one or more attribute-value pairs (AVPs). An AVP is used to encapsulate protocol-specific data, such as routing information, as well as AAA information.

The Version field in the Diameter header denotes the version of the Diameter protocol. In the current version of Diameter, which is version 1, it is set to 1. Message length contains the total length of the packet including the header and the AVPs.

The command flags are 8 bits, from which the first 4 bits are assigned and the last 4 bits are reserved. The four command flag bits are as follows:

- Request (R bit): If set, the message is a request; otherwise, it is an answer.
- Proxiable (P bit): If set the message may be proxied, relayed, or redirected; otherwise, it must be locally processed.
- Error (E bit): If set, the Diameter request caused a protocol-related error; otherwise, it is cleared. A message with a set E bit must not be sent as an answer.
- Potentially retransmitted message (T bit): This bit is set when resending requests are not yet acknowledged (e.g., in failover procedures). However, this bit must be cleared when a request is sent for the first time.

In Diameter each command has an assigned command code, which is stored in the given 24-bit field in the Diameter header. For backward compatibility with RADIUS, the values 0 through 255 of the command codes are reserved for RADIUS. The command code is used in both types of Diameter messages, requests and answers.

Diameter is an extensible protocol, a framework for different applications to use it with different approaches. Therefore, the Application ID is a 32-bit field in the Diameter packet header that carries the ID assigned to a given application by the IANA. However, the base Diameter protocol does not require an Application ID because its support is mandatory in all entities that use Diameter. Currently, the following application ID values are defined for Diameter: "0" for a Diameter common message; "3" for Diameter base accounting; and "0xffffffff" (i.e., all bits set to 1) for relay.

The Hop-by-Hop Identifier is a 32-bit unsigned integer field that is used for matching requests and answers (i.e., replies). The answer message for a given request must contain the same Hop-by-Hop Identifier. Its start value is randomly generated and it is a monotonically increasing number. Diameter answer messages without the specified Hop-by-Hop Identifier must be discarded.

Another identifier in the Diameter packet header is the End-to-End Identifier. This is a 32-bit integer field that is used to target duplicate messages. For that purpose this identifier must stay locally unique for a period of at least 4 min, even in the case of rebooting.

The AVP codes field is followed by an AVP flag, which is an 8-bit field. Note, however, that only 3 bits are defined as flags (the other bits are reserved): If the V bit is set, then the Vendor-Id field is present in the AVP; otherwise, it is not. The P bit is set when an end-to-end security mechanism is used (its default value is 0). The M (mandatory) bit is set to indicate that Diameter entities must handle the given

AVP, and each Diameter entity that does not understand the AVP (with the M bit set) must return an error message.

4.8.2.2 Diameter Communication

Diameter protocol communication is based on connections and sessions. A connection refers to a transport-level connection between two peers that is used to send and receive Diameter messages. A session refers to a logical connection at the application layer that is established between the Diameter client and server. Each Diameter Session-Id begins with DiameterIdentity of the sender, followed by 64-bit value consisting of a high 32 bits and low 32 bits and an optional value (which is implementation specific, i.e., it may be MAC address, timestamp, and so on), in the following format:

```
<DiameterIdentity>;<high 32 bits>;<low 32 bits>[;<optional value>]
```

DiameterIdentity is used to uniquely identify a Diameter node, and it must be the FQDN of the Diameter node (e.g., FQDN = diameter1.example.com). The redirect-host AVP is of type DiameterURI, which follows the URI syntax rules [47]. In DiameterURI possible values for transport protocol are "tcp," "sctp," and "udp"; possible values for the protocol are "diameter," "radius," and "tacacs+." Examples for different Diameter IDs include the following:

DiameterURI Examples Without Transport Layer Security

```
aaa://diameter1.example.com:3868;transport=tcp;protocol=diameter
aaa://diameter1.example.com:3868;transport=sctp;protocol=diameter
```

DiameterURI Example with Transport Layer Security (i.e., "aaas")

```
aaas://diameter1.example.com:5658;transport=tcp;protocol=diameter
aaas://diameter1.example.com:5658;transport=sctp;protocol=diameter
```

In general, there are two types of Diameter sessions: (1) authorization sessions that are used for authentication and/or authorization, and (2) accounting sessions that are used for accounting.

A given session can be either stateful or stateless, depending on the application. In a stateful Diameter session, multiple messages are exchanged. However, Diameter is an application layer protocol that uses reliable transport protocols; therefore, we can virtually distinguish between two connection establishments, the transport layer connection and the Diameter connection.

The communication between Diameter peer nodes starts with establishment of a transport connection by using either TCP or SCTP as the transport protocol on port 3868. When DTLS is used, the Diameter peer that initiates a connection must establish that connection on port 5658. Typically, TLS runs on the top of TCP, while DTLS runs on top of SCTP.

After proper establishment of the transport connection, the application sender initiates a Diameter connection with a Capabilities-Exchange-Request (CER) message as a first message to the other peer. The other peer sends a Capabilities-Exchange-Answer (CEA) message as a response. If the CEA result code is set to "success," then the Diameter connection is established and ready for exchange of application messages. When a secure transport path is established, then all messages (including CER and CEA) are exchanged on that secured transport path. However, if no messages are exchanged over an established Diameter connection for a certain time, then either side may send a Device-Watchdog-Request (DWR). In such a case, the other peer of the Diameter connection must respond with a Device-Watchdog-Answer (DWA) message. So-called watchdog messages are used to probe a given Diameter connection. For termination of a Diameter connection, either side may send a Disconnect-Peer-Request (DPR) message, which is followed by a Disconnect-Peer-Answer (DPA) message from the other peer. After the DPR/DPA exchange, the transport connection can be closed.

4.8.2.3 Diameter Architecture

The Diameter network architecture consists of the end peers and Diameter agents. The four types of Diameter agents are as follows:

- Relay agent: This agent is used to route a message to other Diameter nodes based on routing information in the received message (e.g., Destination-Realm AVP), hence it may be connected with multiple IP networks. A Diameter relay does not change the message, but it must advertise its Application ID (i.e., 0xffffffff).
- Proxy agent: This agent is similar to a relay agent, and routes Diameter messages by using the Diameter routing table. The proxy agent does the same function as the relay agent, but it can also change Diameter messages. Proxies must maintain the states of their downstream peers (i.e., devices), and must advertise Diameter applications they support.
- Redirect agent: This agent is used in centralized Diameter architectures, so every Diameter node that needs Diameter routing information gets it from the redirect agent. The redirect agent does not route or forward messages to any other nodes, it just replies to requests by giving the requested routing information in the response. Redirect agents must advertise their Application ID, which is the same as the one for relay agents (i.e., 0xffffffff).
- Translation agent: This agent provides translation of RADIUS messages to Diameter messages, and vice versa.

The routing of Diameter messages is also a part of the Diameter base standard, and it is based on the Application ID and destination-host or destination-realm AVP. The routing rule is based on the following Diameter routing algorithm:

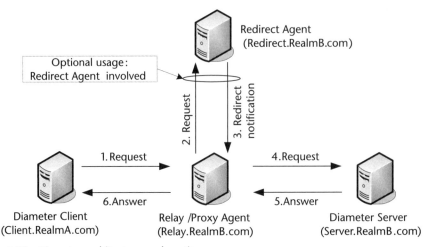

Figure 4.19 Diameter architecture and routing.

```
If local identity == destination-host AVP then process locally
else if peer identity == destination-host AVP then send message
to that peer node
else send unable to deliver answer message back to the sender.
```

Diameter routing is illustrated in Figure 4.19. Each Diameter request message is followed by an appropriate answer message in the reverse direction. In the case of Diameter routing management, the message follows the same path as the request message.

Each node that supports the Diameter protocol maintains two tables: (1) a peer table and (2) a realm-based routing table. The Diameter peer table is used for maintaining the identity of the next peer node (e.g., relay, proxy) and its state (e.g., idle, open, closed), how the peer was discovered (statically or dynamically), expiration time of the peer entries, and type of transport protocol and associated security with the peer (e.g., whether TLS/TCP or DTLS/SCTP is used). The peer table is also referenced by the Diameter routing table, which contains a list of realm routing entries that are used for realm-based routing lookups.

Overall, Diameter has been in development for more than a decade, and it is finally being deployed in NGN environments and future fixed and mobile networks, such as 4G mobile networks.

References

[1] P. Mockapetris, "Domain Names—Concepts and Facilities," RFC 1034, November 1987.
[2] P. Mockapetris, "Domain Names—Implementation and Specification," RFC 1035, November 1987.
[3] Root Servers, https://www.iana.org/domains/root/servers, accessed July 2015.
[4] D. Barr, "Common DNS Operational and Configuration Errors," RFC 1912, February 1996.
[5] R. Arends et al., "DNS Security Introduction and Requirements," RFC 4033, March 2005.
[6] J. Postel, J. Reynolds, "File Transfer Protocol (FTP)," RFC 959, October 1985.

[7]   R. Finlayson, "Bootstrap Loading Using TFTP," RFC 906, June 1984.
[8]   G. Malkin, A. Harkin, "TFTP Blocksize Option," RFC 2348, May 1998.
[9]   J. Postel, "Simple Mail Transfer Protocol," RFC 821, August 1982.
[10] D. Crocker, "Internet Message Format," RFC 822, August 1982.
[11] R. Gellens, J. Klensin, "Message Submission for Mail," RFC 6409, November 2011.
[12] J. Myers, M. Rose, "Post Office Protocol—Version 3," RFC 1939, May 1996.
[13] M. Crispin, "Internet Message Access Protocol—Version 4," RFC 1730, December 1994.
[14] J. Klensin, "Simple Mail Transfer Protocol," RFC 5321, October 2008.
[15] P. Resnick, "Internet Message Format," RFC 5322, October 2008.
[16] D. N. Freed, N. Borenstein, "Multipurpose Internet Mail Extensions (MIME) Part One: Format of Internet Message Bodies," RFC 2045, November 1996.
[17] J. Myers, M. Rose, "Post Office Protocol—Version 3," RFC 1939, May 1996.
[18] M. Crispin, " Internet Message Access Protocol—Version 4rev1," RFC 3501, March 2003.
[19] T. Berners-Lee, R. Fielding, H. Frystyk, "Hypertext Transfer Protocol—HTTP/1.0," RFC 1945, May 1996.
[20] R. Fielding et al., "Hypertext Transfer Protocol—HTTP/1.1," RFC 2616, June 1999.
[21] L. Dusseault, J. Snell, "PATCH Method for HTTP," RFC 5789, March 2010.
[22] B. A. Forouzan, "TCP/IP Protocol Suite," McGraw-Hill, 2010.
[23] T. Berners-Lee, D. Connolly, "Hypertext Markup Language—2.0," RFC 1866, November 1995.
[24] H. Schulzrinne, A. Rao, R. Lanphier, "Real Time Streaming Protocol (RTSP)," RFC 2326, April 1998.
[25] H. Schulzrinne et al., "RTP: A Transport Protocol for Real-Time Applications," RFC 3550, July 2003.
[26] ISO/IEC 11172, "Coding of Moving Pictures and Associated Audio for Digital Storage Media at Up to About 1.5 Mbit/s" (MPEG-1), 1993.
[27] ISO/IEC 13818, "Generic Coding of Moving Pictures and Associated Audio Information" (MPEG-2), 1995.
[28] ITU-T Recommendation H.222, " Information Technology—Generic Coding of Moving Pictures and Associated Audio Information: Systems," October 2014.
[29] ISO/IEC 14496, "Coding of Audio-Visual Objects" (MPEG-4), 1998.
[30] J. Rosenberg et al., "SIP: Session Initiation Protocol," RFC 3261, June 2002.
[31] M. Handley, V. Jacobson, "SDP: Session Description Protocol," April 1998.
[32] S. Kent, K. Seo, "Security Architecture for the Internet Protocol," RFC 4301, December 2005.
[33] S. Kent, "IP Authentication Header," RFC 4302, December 2005.
[34] S. Kent, "IP Encapsulating Security Payload (ESP)," RFC 4303, December 2005.
[35] A. Freier, P. Karlton, P. Kocher, "The Secure Sockets Layer (SSL) Protocol Version 3.0," RFC 6101, August 2011.
[36] William Stallings, *Data and Computer Communications*, 8th ed., Upper Saddle River, NJ: Pearson Education, 2007.
[37] T. Dierks, E. Rescorla, "The Transport Layer Security (TLS) Protocol Version 1.1," RFC 4346, April 2006.
[38] T. Dierks, E. Rescorla, "The Transport Layer Security (TLS) Protocol Version 1.2," RFC 5246, August 2008.
[39] C. Finseth, "An Access Control Protocol, Sometimes Called TACACS," RFC 1492, July 1993.
[40] C. Rigney et al., "Remote Dial-In User Authentication Service (RADIUS)," RFC 2865, June 2000.
[41] V. Fajardo et al. , " Diameter Base Protocol," RFC 7075, October 2012.

[42]   C. Rigney, "RADIUS Accounting," RFC 2866, June 2000.
[43]   C. Rigney, W. Willats, P. Calhoun, "RADIUS Extensions," RFC 2869, June 2000.
[44]   B. Aboba et al., "Extensible Authentication Protocol (EAP)," RFC 3748, June 2004.
[45]   B. Aboba, G. Zorn, D. Mitton, "RADIUS and IPv6," RFC 3162, August 2001.
[46]   M. Tuexen, R. Seggelmann, E. Rescorla, "Datagram Transport Layer Security (DTLS) for Stream Control Transmission Protocol (SCTP)," RFC 6083, January 2011.
[47]   T. Berners-Lee, R. Fielding L. Masinter, "Uniform Resource Identifier (URI): Generic Syntax," RFC 3986, January 2005.

CHAPTER 5

Internet Standardization for Telecom Sector

Telecommunication networks appeared for the first time in the form of telegraph networks in the first half of the 19th century, and then in the form of telephony networks in the second half (i.e., after invention of the telephone in 1876 by Bell, and then automation of telephone networks via the introduction of automatic exchanges). In 1865 the ITU was formed. It was initially targeted to telegraphy since the telephone had not been invented at that time) and the acronym ITU stood for International Telegraph Union. Nowadays, the same organization has the same acronym, ITU, but it stands for International Telecommunication Union, thus representing the whole ICT world. In fact, ITU is currently the United Nations' specialized agency for ICTs, and a type of umbrella group for all of the different standardization organizations (SDOs) such as the IETF, 3GPP, and IEEE.

The telecommunication world has changed during the past several decades given the success of the Internet as a global packet-switching network. Hence, the Internet technologies, as standardized by the IETF (the main standards from the network protocol layer up to the application layer), became the major technologies on the telecommunication scene for all types of networks—LANs, MANs, WANs, and so on—and all applications/services provided through them. As discussed in previous chapters of this book, the native Internet was best-effort based, providing network neutrality to services running over the top (i.e., independently provided by service providers, which generally are different than telecom operators). However, as all major players, including operators and equipment vendors, began to move toward an all-IP world, the native Internet architecture needed to change. Changes were driven by the introduction of the following:

- Standardized QoS over the Internet is required for the main legacy telecommunication services, telephony and television, as well as main business services (e.g., leased lines superseded by VPNs). However, best-effort OTT services continue to exist in parallel as the main driving force for innovative applications.
- A standardized signaling framework is also needed for certain types of services that require delivery of termination calls/connections across the Internet (e.g., VoIP, provided by telecom operators). For such purposes signaling overlay networks have been created in the existing Internet.

We can conclude that globally standardized QoS and signaling frameworks were required for transition of traditional digital telecommunication networks and services to all-IP networks and services. Although the standardization of the main protocol standards between the network and application layers is under the jurisdiction of the IETF, the oldest global SDO, the ITU, has the responsibility for harmonization of different Internet technologies (standardized by IETF) in a well-defined framework. The goal is to provide a smooth transition to the all-IP ICT world that began to happen at the beginning of the 21st century. Such framework standardization is titled next-generation networks (NGNs) and is standardized by ITU-T, the ITU Telecommunication standardization sector.

According to the ITU-T, an NGN is a packet-based network that is able to provide telecommunication services to users and for that purpose is able to make use of multiple broadband, QoS-enabled transport technologies. In addition, its service-related functions are independent of the underlying transport-related technologies. Also, an NGN supports generalized mobility, which allows for consistent and ubiquitous provision of services to users [1].

5.1 Next Generation Networks (NGNs) by ITU

The work on the NGN started in 2003 at several different SDOs. ITU-T in 2003 formed the Joint Rapporteur Group on NGN (JRG-NGN), which developed a general overview of the initial NGN recommendations [1] and NGN general principles and a reference model [2]. The initial JRG-NGN was superseded in 2004–2005 by the Focus Group on NGN, which was further superseded by the NGN Global Standardization Initiative (NGN-GSI) in 2006. The NGN-GSI carried out the main standardization of NGN through ITU-T recommendations. The NGN standardization timeline is illustrated in Figure 5.1.

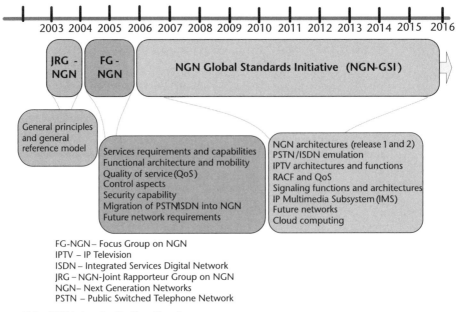

Figure 5.1 NGN standardization timeline.

5.1.1 Cooperation for NGN Development

Several other SDOs had important roles in the NGN's development, either directly or indirectly. Besides the ITU, other important global SDOs that have influenced the creation of the NGN are IETF, 3GPP, and IEEE. Other organizations, such as the Alliance for Telecommunications Industry Solutions (ATIS) and the GSM Association (GSMA), have also contributed. However, in terms of the development of the technical specifications for NGN, the most significant role was played by the European Telecommunication Standardization Institute (ETSI).

Since the NGN's main target was to provide a kind of "traditional telecommunication approach" for the initially best-effort and network-neutral Internet design approach, the IETF indirectly had the most significant role in NGN standardization. For example, NGN is based on IP protocols on the network layer (IPv4 and IPv6), Internet transport protocols on the transport layer (e.g., TCP, UDP, SCTP), and Internet applications (e.g., email, WWW, peer-to-peer). Also, NGN uses standardized routing protocols and QoS solutions for IP networks, then fundamental Internet technologies such as DHCP, DNS, and VPN. However, different Internet standards (e.g., protocols) are specified in NGN in a given context, with defined roles in network and service functional architectures. Note, however, that because NGN was targeted for the introduction of QoS and signaling into IP-based networks in a standardized manner, the protocols having the highest influence on the NGN have been SIP [3] and Diameter [4].

ETSI contributed to NGN standardization through its Telecommunications and Internet Converged Services and Protocols for Advanced Networking (TISPAN) technical committee, created in 2003. In fact, TISPAN developed the initial IMS specification, targeted to be used for implementation of fixed VoIP by telecom operators, as a replacement for PSTNs. However, TISPAN later transferred the IMS standardization process to 3GPP, with global acceptance by other SDOs (e.g., ITU) with the goal of developing a common IMS standard for all different networks (fixed and mobile ones) and services. Regarding the ETSI work, TISPAN's NGN release 1 was finalized in 2005, and it focused primarily on VoIP based on an IMS standard for SIP-based applications. NGN release 2 was targeted to IPTV including non-IMS and IMS-based IPTV, and it was finalized in 2008. So, the major work of NGN standardization was carried out by ETSI and adopted by the ITU-T. It was focused on the use of the main traditional telecommunication services (telephony and television) in all-IP environments. However, ETSI closely collaborated with 3GPP on IMS specifications. As a result, TISPAN NGN release 1 specifications correspond to 3GPP release 7 specifications, and TISPAN NGN release 2 specifications correspond to 3GPP release 8 specifications (regarding the IMS standardization by 3GPP).

Further, 3GPP has been playing an increasingly important role in NGN standardization. It standardized the common IMS in 3GPP release 8, which was finished at the beginning of 2009 and is currently used in all VoIP implementations by fixed and mobile telecom operators. Also, 3GPP has introduced Internet technologies that will be mandatory in the all-IP fourth-generation (4G) mobile systems, thus eliminating the circuit-switched approach in the PLMN that was the main design concept in 2G mobile systems (GSM) and also existed in the circuit-switched part of 3G mobile systems (used mainly for mobile telephony). In fact, 3GPP created

LTE (releases 8 and 9) and later LTE-Advanced (releases 10–13 and beyond) according to the NGN framework.

Finally, the IEEE has not been directly involved in NGN standardization, but it has a certain indirect influence on it, something similar to the IETF role. The importance of the IEEE to NGN emanates from their standards for access networks that are IP native, such as Ethernet (IEEE 802.3 standards group) and wireless networks including Wi-Fi (IEEE 802.11 standards group) as a WLAN and WiMAX (IEEE 802.16 standards group) as a WMAN. In that manner, Ethernet-based access is standardized in several NGN specifications by ITU-T, which is targeting QoS provisioning over such access networks.

5.1.2 NGN Network Concept

NGNs are all-IP networks that are being established as PSTN/PLMN converge with the Internet as a single networking platform for all types of services, including existing ones as well as future ones. Note, however, that traditional telecommunication networks have influenced the NGN network concept, particularly for QoS-enabled end-to-end VoIP and IPTV.

Traditional telecommunication networks (before the NGN appearance) were designed primarily to carry circuit-switched digital voice, by using unified 64-Kbps bit stream in each direction for a single voice connection (ITU-T G.711 voice codec standard). Such networks were based on simple telephone devices (without any computational capabilities), circuit-switched transport networks based on time division multiplexing (TDM) with 64-Kbps "capacity slots" (e.g., SDH/SONET transport networks), complex and expensive digital exchanges located in telecom operators' networks, and use of the globally unified SS7 (standardized by ITU-T). As shown in Figure 5.2, the evolution of PSTN and PLMN toward NGNs is performed by the gradual replacement of TDM-based transport systems and interfaces, circuit-switching network nodes (e.g., telephone exchanges), and SS7 signaling (which is packet based and separated from user traffic, but is not IP based) with IP/MPLS transport networks, media gateways, and IP-based signaling (e.g., SIP), respectively.

NGN specifies IP on the network layer end to end. Such a requirement defines the all-IP network concept. That means that all access, core (or backbone), and transit networks in NGN are IP based. On one side, IP hides the lower protocol layers (i.e., OSI-1 and OSI-2) from the upper layers, so an NGN can have a single core and transport network infrastructure for different fixed and wireless access networks. In the all-IP network principle, the interconnection between different networks is also done via IP links established between pairs of gateway routers or controllers. On the other side, IP hides the network interfaces from the upper protocol layers (e.g., the transport and application layers), thus providing single network abstraction (through the socket interface) to the application running on the top of the Internet protocol stack (even for cases in which several network interfaces reside on a given host). So, the Internet was initially designed with separated transport and services parts, with IP (including both IPv4 and IPv6) as the main protocol between them in each host and network node connected to the Internet. NGNs put that in a given context with the goal of specifying the required additional functionalities in given network nodes and end-user hosts.

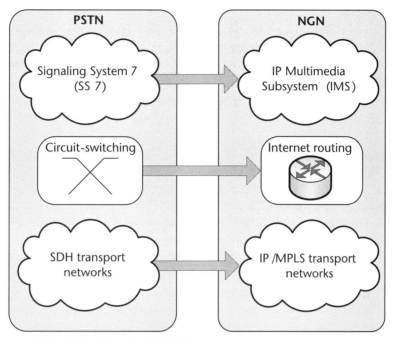

MPLS – Multi-Protocol Label Switching
NGN – Next Generation Networks
PSTN – Public Switched Telephone Network
SDH – Synchronous Digital Hierarchy
SS 7 – Signaling System No.7

Figure 5.2 Network concept evolution from traditional telecom networks to NGNs.

5.2 Transport and Service Strata of NGN

The separation of services from transport by allowing them to be offered separately and to evolve independently (e.g., invention of new services, changes to existing services) is the key characteristic of the Internet and that characteristic has been used in NGNs also [1]. The separation is allowed for by defining two distinct blocks (or strata) of functionalities, which define the basic reference model for NGN [5]:

- NGN service stratum: The service stratum provides user functions that transfer service-related data (related to voice, video, data and multimedia services) to network-based service functions, which manage service resource and network services with the goal of enabling user applications and services. So, the service stratum provides originating and terminating calls/sessions between end peers, which is different than the client–server model in the best-effort Internet, such as WWW, where clients always initiate connections while servers always receive connection requests.
- NGN transport stratum: The transport stratum provides user functions that transfer user data. On the network's side, the transport stratum provides functions that control and manage transport resources for carrying the data end to end. The related transferred data may carry user, management, and/or control information. To carry the data between the end-peer entities, the transport stratum may provide different static or dynamic associations.

Figure 5.3 shows the separation of NGN service and transport strata, as well as the architectural concept for the user plane (refers to data generated or received by the user, such as voice, video, web pages, and emails), control plane (refers to call/session control, such as signaling), and management plane (refers to communication among entities related to configuration, accounting, fault management, performance, and security management needs). Both strata consist of different types of functions.

5.2.1 Transport Stratum

Transport stratum functions are divided into two groups: transport functions and transport control functions.

Transport functions are related to physical network resources. The physical architecture of the NGN (e.g., interconnected routers) is divided into two main network parts: the access network and the core network. The routers on the edges between access networks and the core network are called edge routers. The routers on the edge of the core network in NGNs are referred to as gateways. Accordingly, the transport functions are grouped into functions for each network part and interconnection router as follows:

- Access network functions: These functions are used for user traffic aggregation and control from the access network toward the core (and vice versa). Further, these functions provide QoS support in the access networks (e.g., scheduling, buffer management, traffic classification, packet filtering) as well as mobility support in certain mobile networks. In general, the access network functions are related to four different types of access networks: xDSL (digital subscriber lines over twisted pairs), cable (over coaxial cables), wireless/mobile access (e.g., IEEE 802.11, IEEE 802.16, UMTS, LTE), and optical access (e.g., FTTH).

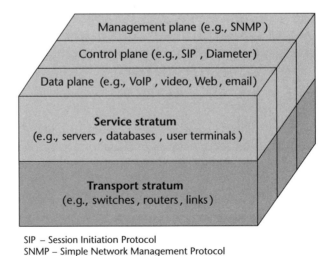

SIP – Session Initiation Protocol
SNMP – Simple Network Management Protocol
VoIP – Voice over IP

Figure 5.3 NGN reference model.

- Edge functions: These functions are used for traffic aggregation from several access networks to a core network, as well as for interconnection of different core networks.
- Core transport functions: These functions provide transport of the information in the core networks, including QoS provisioning for user traffic.
- Gateway functions: The NGN has two types of gateways. One is located at the user premises (for interworking with end-user functions), and the other at the edge of the core network (for interworking with other NGN and non-NGN networks such as PSTN and best-effort Internet). Generally, gateway functions may be controlled either by transport control functions in the transport stratum or service control functions in the service stratum.
- Media handling functions: These functions provide media-related functions, such as transcoding. They are located in the transport stratum only.

Transport control functions are covering functions related to network attachment and control, mobility management in mobile networks, and resource and admission control.

The main entity for control of physical resources in the transport stratum is the network attachment and control functions (NACF), which provide functions needed for user attachment to the NGN and use of its services. They include dynamic allocation of IP addresses, autodiscovery of user equipment capabilities (e.g., voice/video codecs supported by the user device), mutual authentication (between the user and the network) and then authorization of the user, access network configuration, and location management (at the IP layer).

Mobility management and control functions (MMCF) provide support for seamless IP mobility in the transport stratum, including horizontal handovers (between the cells that use the same radio access technology) and vertical handovers (between different types of radio access networks). However, it does not guarantee the persistence of QoS provisioning before the handover and after its execution. MMCF manages mobility as a service, independent from the type of access technology.

The main QoS entity in the NGN is the resource and admission control functions (RACF), which is a "link" between the service control functions and transport functions in the NGN. Their main goal is to provide admission control regarding given call/session to/from end users. So, a user connection can be admitted in the transport network or rejected (or modified) according to transport subscription information, service-level agreements (SLAs), different policies and service priorities, and so forth. The RACF hides the network topology from the service stratum, thus providing an abstract view of the transport architecture to service control functions. On the other side, RACF is not mandatory in the NGN, for example, when admission control is not needed (e.g., best-effort Internet traffic). On the other side, RACF interacts with service control functions (SCFs) for applications that require certain QoS support, which results in a need for admission control (e.g., QoS-enabled VoIP or IPTV). Besides that, RACF also provides network address and port translation (NAPT) traversal functions, which are necessary when an NGN operator uses private IP addresses for addressing end-user terminals. Finally, RACF interacts with NACF for exchange of user subscription information, as well as with RACFs in other NGNs for delivering services over multiple network or service providers.

5.2.2 Service Stratum

The service stratum has two main groups of functions. The first group consists of application and service support functions, which interface with applications. The second group of functions in the service stratum consists of service control and content delivery functions (including a service user profile database), which interact with the transport stratum and its functions via the control plane. Also, service stratum functions interact with management functions including identity management functions. So, the following groups of functions are defined for the NGN service stratum:

- Service control and delivery functions: These consist of the SCFs used for authentication and authorization as well as interaction with RACF in the transport stratum, and also content delivery functions (CDFs), which are used for delivering content (e.g., video, TV content) to users.
- Application and service support functions: These include functions for registration, authentication, and authorization at the application layer and provide services to the end users in cooperation with the service control functions.
- Management functions: These are allocated in each functional entity in both strata, with the goal of managing faults, configurations, accounting, performances, and security [6].
- Identity management functions: This is a set of functions and corresponding capabilities (e.g., administration of users, authentication, binding of different identifiers) that provide assurance of the identity of an entity, as well as its storage, usage, and distribution [7].
- End-user functions: These refer to functions located in end-user equipment (i.e., user terminals), which may be fixed (e.g., a desktop computer) or mobile (e.g., a mobile phone or device). These functions can interact with all functional groups in the service and transport strata.

Service stratum usage is not mandatory for network-neutral Internet services (e.g., WWW and email). In such a case, however, the service provided by the telecom operator is referred to as "Internet as a service," which includes the service and transport strata functions (with their described roles in NGN). For example, to be able to access OTT services, the end user first must be authenticated and authorized for use of the Internet as a service.

5.3 NGN Architectures

Overall, NGN has two main releases that influence the architecture and its capability sets, namely, NGN release 1 [8], and NGN release 2 [9]. The fundamentals of the NGN architecture 1 are set in release 1 (although its main applicability is used

5.3 NGN Architectures

for transition of digital telephony to VoIP with QoS support end to end), while release 2 has added architecture functionalities that are related to IPTV provisioning end to end.

The generic NGN architecture is shown in Figure 5.4. All functional entities in the NGN framework are built by using Internet technologies. However, different NGN networks may have different physical network topologies and different access networks, including non-NGN architectures.

Specific architectures related to VoIP and IPTV as services provided by telecom operators are covered in Chapter 7 of this book. One of the main advantages of NGN (besides transition of telephony and TV to QoS-enabled VoIP and IPTV, respectively) is its ability to provide ubiquitous networking. We can define it as networking that provides the following features [10]:

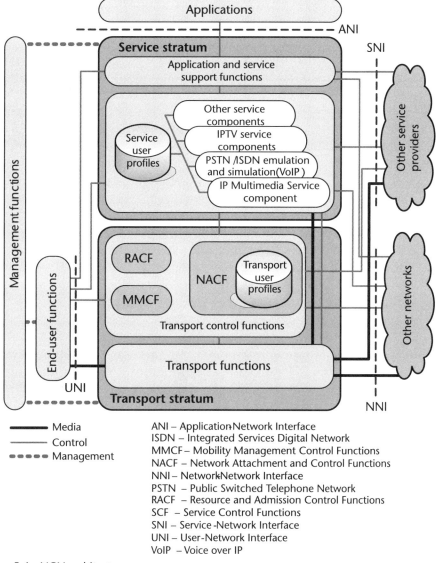

Figure 5.4 NGN architecture.

- Connectivity of whenever, whoever, wherever, whatever types of communications,
- Pervasive reality for effective interface to provide connectable real-world environments, and
- Ambient intelligence allowing for innovative communications and providing increased value creation.

Generally speaking, high-level capabilities for ubiquitous networking include IP connectivity of different objects to communicate with each other within a network (i.e., NGN) or when objects have to be reachable outside the network. Due to many possible objects that have to be accessible, IPv6 deployment is a prerequisite for this.

Another characteristic is personalization of services by delivering appropriate contents and services to users (what the user needs, either explicitly chosen, or indirectly by using past user experience). To provide user-centric and context-aware personalized services, the need for intelligence in network entities becomes essential (e.g., introduction of artificial intelligence in the network).

Ubiquitous networking in NGNs has the ability to support tagging objects and smart devices. One of the tag-based solutions for real-time identification and tracking of objects are radio-frequency identifiers (RFIDs). Tag-based solutions in an ubiquitous environment should allow users to get or retrieve information from anywhere in the network. Smart devices have the ability to support multiple functions (e.g., phones, cameras, TVs, music players). In that manner, specific environments (e.g., homes, offices, cars, buildings) will also require adaptive smart devices, including many very small devices (sometimes referred to as "smart dusts") that provide status and/or sensory information and interconnect with other devices connected to the network [10].

However, one of the main drivers for convergence to Internet technologies and NGN deployments by telecom operators is fixed and mobile broadband access, which provides the bit rates needed to support a wide range of services.

5.4 Next-Generation Broadband Access

In the telecom sector broadband has been discussed since the 1980s and it was mainly targeted to digital video transmission over telecommunication networks. However, the growth of the Internet since the second half of the 1990s redefined the term broadband to refer to broadband access to the Internet. So, the diffusion of broadband, defined as a technology that enables high-speed transfer of data, is inseparably linked to the appearance of the public Internet.

What does broadband refer to? There is no single exact definition (although certain regulatory bodies have tried to define it at certain times), but generally we can note that broadband refers to all access technologies, whether fixed or mobile, that provide enough bandwidth (e.g., in megabits per second) end to end for all existing applications to run smoothly. So, broadband refers to different bit rates (or data rates, in other words) at different times? Yes, exactly that. Individual broadband access in the 1990s referred to access bit rates of hundreds of kilobits per second; in the 2000s broadband referred to several megabits per second; in the 2010s it refers to several tens of megabits per second; and in the 2020s it is likely to refer to several hundreds of megabits per second.

Regarding the technologies for broadband access to the Internet, they can be grouped into the following two main types:

- Fixed broadband Internet access: This includes all broadband networks that use either copper such as twisted pair (first used for telephony, while another type is used for copper-based Ethernet) or cable networks (initially used for analog and digital TV broadcast), and fiber including passive and active optical networks.
- Wireless and mobile broadband Internet access: This includes broadband access via broadband wireless and mobile networks, which include 3G and 4G mobile networks, and Wi-Fi as WLANs.

However, broadband access to services is not related only to technology, but also to society in general. It has important business and regulation impacts. Deployment of broadband networks influences the development of the economy of a given country, and regulation is required in many cases when resources are limited (e.g., spectrum for mobile broadband access is a limited resource).

5.4.1 ITU's Role in Development of Broadband Internet Access

Broadband access is globally promoted and supported by ITU. The main prerequisite for broadband development is building a telecommunication infrastructure for transport of huge amounts of data generated to/from end users (including residential and business users) and network entities, and providing fixed and mobile broadband access to the Internet.

The main fixed broadband technologies, such as ADSL, which has sped up tremendously the introduction of global broadband Internet access, have been standardized by the ITU-T. The ITU has also standardized the main passive and active optical access networks and their interfaces, sometimes referred to as next-generation access networks (NGANs).

Another sector of the ITU, ITU-R, has an important role related to global synchronization for the allocation and use of frequency bands for mobile broadband. ITU-R has developed the International Mobile Telecommunications (IMT-2000) umbrella for 3G mobile standards. Similarly, ITU-R set the requirements for 4G mobile systems within the umbrella called IMT-Advanced. In that manner all mobile standards that comply with IMT-2000 requirements are referred to as 3G mobile networks (e.g., UMTS/HSPA, Mobile WiMAX release 1), whereas those mobile technologies that satisfy IMT-Advanced requirements are referred to as 4G (e.g., LTE-Advanced from 3GPP, and IEEE 802.16m, also known as Mobile WiMAX 2.0). In a similar manner, the requirements for the fifth-generation (5G) mobile networks will be defined by ITU-R in an umbrella called IMT-2020, which is expected to be finished a couple of years before the year 2020.

The ITU sector for development, ITU-D, works with different organizations all over the world, including individual users and enterprises, to examine the factors and key aspects for success of broadband deployments. The role of ITU-D also includes assisting countries in the creation of broadband strategies and policies, which are essential for the deployment of broadband Internet access on a global scale.

5.4.2 Fixed Broadband Internet Access Technologies

Fixed broadband access is provided via copper lines (e.g., twisted pairs and coaxial cables) and fiber-optic cables. The main fixed broadband access technologies over copper lines are DSL technologies and cable networks.

5.4.2.1 Digital Subscriber Line (DSL)

For broadband access for residential users, the most expensive part is building the access network. To provide gradual and less expensive development of broadband, the initial approach was (and still is) based on use of the existing access networks of telecom operators that were deployed during the 20th century for telephony. Such technologies were called DSL. Generally, ADSL was the first global technology that provided broadband access to the Internet to residential users, with bit rates many times higher than the maximum 56 Kbps provided via modem dial-up over a telephone line [11] or the maximum 144 Kbps provided via ISDN. The initial standard for ADSL was created by ANSI [12], but the most commonly used standard for ADSL on a global scale was developed by the ITU-T [13]. It is known as G.DMT because it uses DMT modulation. For longer lengths for a subscriber's local loop, the ITU-T standard known as G.Lite [14] is used. (Lower data rates result from the trade-off for longer lengths of the subscriber's line.)

The OSI-1 physical layer of ADSL is designed to be able to coexist with standard POTS (plain old telephone service) spectrum. The two services can coexist because ADSL is using a higher-frequency spectrum than the baseband spectrum used for telephony (the frequency band dedicated to analog voice is 0.3 to 3.4 kHz).

The successor of ADSL is ADSL2 [15] and its splitterless version [16]. Further, ADSL2+ provides two times higher bit rates than ADSL2. The main ADSL standards are shown in Table 5.1, with their specified maximum bit rates in the downstream and upstream directions. However, for longer distances (i.e., longer subscriber lines in the local loop), such as 3 to 5 km, there is no significant difference between ADSL and ADSL2/ADSL2+ bit rates due to higher attenuation of the

Table 5.1 Main ADSL Standards and Bit Rates

ADSL Name	Maximum Downstream Bit Rate (Mbps)	Maximum Upstream Bit Rate (Mbps)	Standard Name
ADSL	8	1	ANSI T1.413-1998
ADSL (G.DMT)	12	1.3	ITU G.992.1
Splitterless ADSL (G.Lite)	1.5	0.5	ITU G.992.2
ADSL2	12	1.3	ITU G.992.3
Splitterless ADSL2	1.5	0.5	ITU G.992.4
ADSL2+	24	3.3	ITU G.992.5

signals, which influences the type of modulation and coding scheme used for the line.

Besides ADSL, several other DSL technologies exist. However, the next DSL technology that may have an impact on the reuse of installed copper wires in the last kilometer is very-high-bit-rate DSL (VDSL), which is intended for use on shorter distances in the local loop, that is, several hundreds of meters. VDSL is also asymmetric (like ADSL) and it provides the highest bit rates for shorter distances of the local loop, as found in so-called fiber-to-the-building (FTTB) environments in which fiber deployment reaches the building and the existing twisted-pair lines are used inside the building. The next-generation broadband access over old telephony copper lines (based on VDSL) is the G.fast standard by ITU-T [17]. In this standard, for copper line lengths of less than 100m bit rates in the range of 500 to 1,000 Mbps are possible (i.e., up to a 1-Gbps downstream bit rate per line). Further, G.fast provides downstream rates of 500 Mbps at 100m, 200 Mbps at 200m, and 150 Mbps at 250m lengths of subscriber line. One may wonder why G.fast was developed in the time when fiber technologies are considered more powerful regarding the bit rates they provide. However, G.fast was developed to provide hybrid-FTTB solutions (besides pure FTTB) with the goal of it being used where it is attractive.

Various DSL technologies have been defined, as listed in Table 5.2. The technology that will be applied depends on the service to be provided. For example, ADSL is suitable for asymmetric services such as VoD and WWW. For symmetrical services (e.g., video telephony) more convenient DSL technologies that have symmetric flows in both directions, upstream and downstream, are used. However, an important factor for the choice of DSL technology is the supported length of the telephone line (i.e., the local loop length). Nowadays, of all of the standardized DSL technologies, ADSL has the incomparably greatest success on a global scale, with the possibility of VDSL and G.fast following next.

Figure 5.5 shows a typical deployment architecture for an ADSL network. In the access network, the digital subscriber line access multiplexer (DSLAM) is located

Table 5.2 Comparison of Different DSL Technologies

DSL Type	Maximum Downstream Bit Rate (Mbps)	Maximum Upstream Bit Rate (Mbps)	Approximate Maximum Local Loop Length* (m)
ADSL2+	8	3.3	5500
HDSL	1.54	1.54	3650
MSDSL	2	2	8800
RADSL	7	1	5500
SDSL	2.3	2.3	6700
VDSL	52	16	1200
VDSL2	200	200	500
G.fast	Maximum aggregated downstream and upstream = 1 Gbps		250

ADSL, asymmetric DSL; VDSL, very-high-bit-rate DSL; RADSL, rate adaptive DSL; HDSL, high-bit-rate DSL; SDSL, symmetric DSL; MSDSL, multirate symmetric DSL.

*Bit rates are several times lower at the approx. max. local loop length (for every DSL type), i.e., they decrease by increasing the length of the subscriber line.

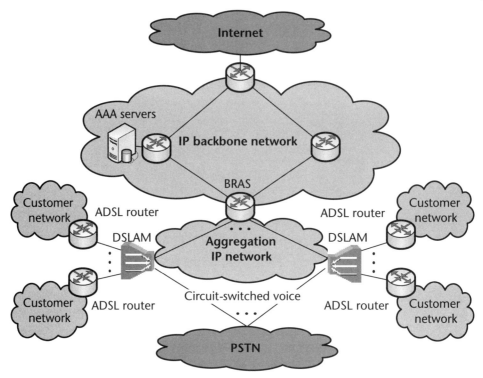

AAA – Authentication Authorization and Accounting
ADSL – Asymmetric Digital Subscriber Line
BRAS – Broadband Remote Access Server
DSLAM – Digital Subscriber Line Access Multiplexer
PSTN – Public Switched Telephone Network

Figure 5.5 ADSL network architecture.

on the network side. Several local loops end in a single DSLAM, which provides aggregation of the traffic from users (and vice versa). The network node between the access part and the core network is the broadband remote access server (BRAS), which is a router responsible for routing the traffic between the DSLAM-enabled access network and the core IP network of the telecom operator. It performs aggregation of user sessions from the access network, and vice versa.

5.4.2.2 Cable Broadband Networks

Internet access technologies for data transmission over cable networks have been developed by CableLabs. In fact, CableLabs developed architectures for integrated cable network for different services, such as analog TV and radio, digital TV, HDTV, VoD, Internet access, and telephone services by using VoIP. The architecture of the integrated cable network is shown in Figure 5.6.

The integrated network with cable access can be divided into several segments, including the following: home IP network, DOCSIS, PacketCable, and core IP network.

DOCSIS exists in three versions. DOCSIS 1.0 was the first standard for Internet broadband access via a cable access network, and DOCSIS 2.0 enhanced the upstream direction by providing higher bit rates, which became necessary due to

5.4 Next-Generation Broadband Access

Table 5.3 DOCSIS Bit Rates Including Overhead

Version	Downstream DOCSIS (Mbps)	Downstream Euro DOCSIS (Mbps)	Upstream Both (Mbps)
DOCSIS 1.0 and 1.1	42.8	55.62	10.24
DOCSIS 2.0	42.8	55.62	30.72
DOCSIS 3.0 (4 channels)	171.52	222.48	122.88
DOCSIS 3.0 (8 channels)	343.04	444.96	122.88

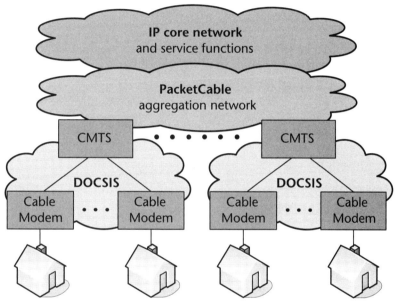

CMTS – Cable Modem Termination System
DOCSIS – Data-Over-Cable Service Interface Specifications

Figure 5.6 Cable network architecture.

competition from the DSL and FTTB technologies. The next-generation Internet access over cable networks is based on DOCSIS 3.0, which includes support for IPv6 as well as higher bit rates than previous versions. The bit rates for the different versions of the DOCSIS standards are summarized in Table 5.3. In the downstream direction, we can distinguish between DOCSIS (6-MHz bandwidth per TV channel) and Euro DOCSIS (8-MHz bandwidth per TV channel). The given bit rates are aggregated ones on a given coaxial cable, so they are shared between users with a single coaxial cable in the DOCSIS access network (e.g., in a residential building).

The PacketCable specification defines the interface used to enable interoperability of equipment for the transmission of packet-based voice, video, and other broadband multimedia services over a hybrid optical-coaxial network using the DOCSIS. The main reason for the development of PacketCable was provisioning of packet-based voice communication for users connected to cable networks. Its architecture includes call signaling, accounting, configuration management, security, and a PSTN interconnection. To guarantee QoS over the DOCSIS access network (which is also used for access to best-effort Internet services, such as WWW and email), PacketCable services are delivered with guaranteed priority in the DOCSIS access part, which ensures guaranteed bit rates and controlled latency (i.e., packet delay).

5.4.2.3 Optical Broadband Access Networks

Next-generation broadband access is developing toward all-optical access due to the much higher throughputs that can be achieved over fiber rather than over copper media (such as twisted pairs and coaxial cables).

Various architectures are used for fiber implementation in the last mile. They are generally denoted as FTTx (fiber-to-the x), where x stands for cabinet or curb (FTTC), building (FTTB), home (FTTH), premises (FTTP), and desk (FTTD). With FTTH, the optical connection reaches the home of the end user, whereas FTTP and FTTD are targeted for use by small enterprises.

The speed of deployment of FTTH is driven by the economics, especially by comparison of its cost to competitive technologies such as cable access networks and xDSL technologies. The cost of an FTTH network drives the chosen architecture. Different FTTH architectures can be characterized by location of electronics (on the user's side or the operator's side), location of bandwidth aggregation on the operator's side of the network, bandwidth allocation to the end user, and applied protocols. Regarding the topology, the fiber access networks can be point-to-point or point-to-multipoint. Based on these characteristics, there are four basic architectures for the FTTx access network, also called an optical distribution network (ODN):

- Point-to-point (P2P) architecture: P2P uses a separate direct fiber link between the central office of the operator and the home of the user. In this architecture the number of fiber links is equal to the number of homes.
- Active optical network (AON) architecture (i.e., active star): AON uses a shared point-to-point fiber link (also called a feeder fiber) between the optical line terminal (OLT) and an active remote switch (i.e., curb switch), and P2P links between the remote terminal and the optical network terminals (ONTs) at users' homes.
- Passive optical network (PON) architecture (i.e., passive star): PON uses a shared point-to-point fiber link between the OLT and passive splitters (splitting is performed only by using optics, without any power supply to the unit) and between the shared fiber and ONTs. This is a point-to-multipoint (P2MP) architecture regarding the optical path between the central office and end users.
- Wavelength division multiplexing (WDM) PON architecture: This scheme uses the PON architecture (listed above) with WDM (uses several wavelengths per fiber) with the goal of further increasing the bandwidth to the end user. This is also a P2MP architecture.

The main optical network architectures for broadband access to the Internet are illustrated in Figure 5.7. Nowadays the most important P2MP configuration optical access network is the PON, which is based on TDM. Legacy TDM-PON

5.4 Next-Generation Broadband Access

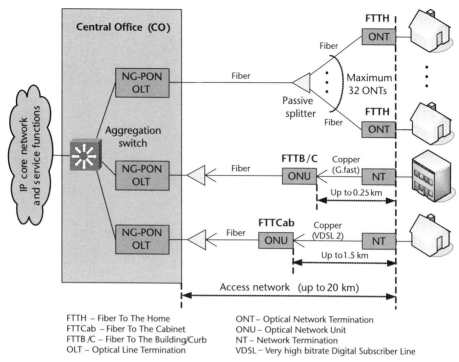

Figure 5.7 Optical network architectures.

standards define line bit rates of up to 2.5 Gbps and a maximum length of links of 20 km. The splitter used for PON is typically up to 1/32 (one feeder fiber is split into 32 connections to end users), which is limited by the optical power available at ONT. The PON equipment does not require an electric power supply in the optical part, which makes it easier to deploy and power supply independent. Hence, the first deployments of FTTH are based on PON networks. The main PON standards were created by ITU-T and IEEE and are listed in Table 5.4.

The standardization of next-generation PON is also carried by ITU and IEEE. The ITU continues with further standardization toward 10-Gbps-capable PON called XG-PON (also referred to as next-generation PON 1 or NG-PON1), the G.987 standard [18], and the next-generation 40-Gbps-capable PON (XLG-PON; also referred to as NG-PON2), the G.989 standard [19].

Table 5.4 Main PON Standards

	Standard	Maximum Downstream Bit Rate	Maximum Upstream Bit Rate	Transmission
BPON	ITU-T G.983	1.244 Gbps	622 Mbps	ATM
EPON (IEEE 802.3ah)	IEEE 802.3ah	1.25 Gbps	1.25 Gbps	Ethernet
GPON	ITU-T G.984	2.4 Gbps	2.4 Gbps	Ethernet, ATM, TDM
XG-PON	ITU-T G.987	10 Gbps	2.5 Gbps	TDM, TWDM
XLG-PON	ITU-T G.989	40 Gbps	10 Gbps	TDM, TWDM
10G-EPON	IEEE 802.3av	10 Gbps	1 or 10 Gbps	Ethernet

ATM, asynchronous transfer mode; TWDM, time and wavelength division multiplexing.

The IEEE contribution to next-generation optical broadband networks is a 10-Gbps Ethernet PON (10G-EPON). It provides asymmetric access with 10 Gbps downstream and 1 Gbps upstream, as well as symmetric access with 10 Gbps in both directions. The standard was approved by the IEEE in 2009 as IEEE 802.3av [20].

The topologies of AONs are very similar to those of PON. The main difference is the replacement of passive elements (i.e., splitters) in PON with active elements (i.e., switches) in AON. Today AONs are end-to-end Ethernet-based networks. AON switches located at the central office in the access network are almost the same as aggregation switches in the aggregation part of the network.

For the purposes of better traffic engineering (e.g., QoS provisioning) and a common control plane in access, distribution, and core optical networks, next-generation AON may incorporate the Generalized Multi-Protocol Label Switching (GMPLS) architecture [21]. GMPLS differs from the legacy MPLS by adding support for different types of switching, such as TDM switching, wavelength switching, and fiber link switching. GMPLS is used as a control plane in wavelength-switched optical networks that are initially used as transport networks. With next-generation AON the GMPLS may be also used in access networks. It can provide the control functions for multilayered Ethernet networks (e.g., Ethernet tunneling), such as Metro Ethernet.

5.4.2.4 Metro Ethernet

With the transition of digital circuit-switched networks to all-IP networks, the Ethernet has extended its target area from LANs to MANs with the Metro Ethernet standard, especially for the front-haul networks of mobile and fixed telecom operators, thus replacing TDM bit streams from the legacy telecom era [22].

The Metro Ethernet Forum (MEF) has been driving the development of Ethernet-based access in the metropolitan areas (i.e., up to several tens of kilometers). From the birth of the Internet in its present form (i.e., based on TCP/IP protocol stack), the Ethernet has become the main choice for local access technology including users' LANs (e.g., home, office access networks), as well as LANs for connecting servers and databases in data centers. It is also the technology used for interconnection of switches and routers in the local access part. End users who use Ethernet (in the last mile or miles) are demanding the same level of performance that they had in the past with TDM-based networks. This evolution of the Ethernet was defined as "Metro Ethernet" network or "Carrier Ethernet" (both names are interchangeable).

Because Carrier Ethernet is replacing traditional telecom access technologies, it must provide the same level of service [21], especially regarding scalability and QoS support. (It should provide the same performance guarantees regarding throughput, delay, and so on, as TDM-based access such as leased lines for business users.) Further, telecom operators have to guarantee the SLAs; therefore, it needs to have tools for performance monitoring of different parameters of the supported services toward end users via the Metro Ethernet. Finally, TDM emulation is required for Carrier Ethernet to seamlessly interwork with TDM-based leased line services (e.g., PDH or SDH/SONET bitstreams).

The subscriber network (e.g., computers, switches, base stations) attaches to the Metro Ethernet network (Figure 5.8) at a user/network interface (UNI), which can belong to one of the Ethernet standards such as 10 Mbps, 100 Mbps, 1 Gbps, or 10 Gbps.

Generally, MEF has defined Ethernet-based services in metropolitan areas by using so-called Ethernet virtual connections (EVCs). The three defined types of EVCs are shown in Table 5.5 [23].

The EVC-based services are provisioned by using several IEEE standards, the most important of which for Carrier Ethernet is IEEE 802.1Q for virtual LANs [24], IEEE 802.1ad for provider bridges [25], and IEEE 802.1ah for provider backbone bridges [26].

5.4.3 Wireless and Mobile Broadband Internet Access Technologies

The development of wireless and mobile networks has influenced the Internet, and vice versa. Their convergence started to happen in the second half of the 1990s in so-called 2.5G mobile systems, such as GPRS from 3GPP.

Overall, the mobile evolution has gone through these generations:

- 1G: The first generation included analog mobile systems, based on frequency division multiple Access (FDMA), without global roaming; used in the 1980s.

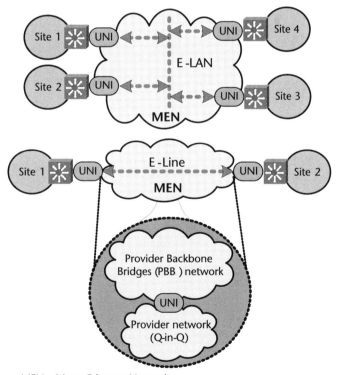

MEN – Metro Ethernet Network
UNI – User Network Interface

Figure 5.8 Metro Ethernet architecture.

Table 5.5 Metro Ethernet Services

Type of Ethernet Service	Port-Based Ethernet Service	VLAN-Based Ethernet Service
E-Line	Ethernet Private Line (EPL)	Ethernet Virtual Private Line (EVPL)
E-LAN	Ethernet Private LAN (EP-LAN)	Ethernet Virtual Private LAN (EVP-LAN)
E-Tree	Ethernet Private	Ethernet Virtual Private Tree (EVP-Tree)

- 2G: The second generation included digital mobile systems, based mainly on TDMA and FDMA (e.g., GSM), CS access and core networks, with global roaming, telephony, and short message service (SMS) as the main services; started at the beginning of the 1990s.
- 3G: The third generation was the first generation of mobile systems that included by default the PS domain [for Internet access and multimedia messaging service (MMS)] in parallel with CS (for voice and SMS). It is based on wideband code division multiple access (WCDMA) with TDMA/FDMA in the radio part. The 3G standardization and deployment processes started at the beginning of the 2000s.
- 4G: The fourth generation is the first generation of mobile systems that are all-IP by default including radio access and core networks. It is based on orthogonal frequency division multiple access (OFDMA) with TDMA/FDMA in the radio part. It started at the beginning of the 2010s.
- 5G: The fifth generation is the next generation of mobile systems that is expected to have its first standards around 2020 and to be deployed in the 2020s. 5G is expected to be all-IP, mainly based on IPv6, with higher bit rates than its predecessors, and targeted to exploit heterogeneous radio access networks.

So, although in the past couple of decades there were also wireless standards that had success on national or regional levels (e.g., WiBRO in Korea and iMode in Japan in the 2000s), the wireless and mobile broadband networks have converged mainly to standards from two SDOs: (1) 3GPP mobile broadband technologies and (2) IEEE wireless and mobile broadband technologies.

5.4.3.1 3G Mobile Broadband Standard by 3GPP: UMTS/HSPA

Since GSM has been the most successful 2G technology worldwide and the Internet has become winning packet-switching technology, their integration in the form of GPRS has made possible a breakthrough for mobile Internet access on a global scale (although GPRS bit rates at that time, in the late 1990s, were comparable with dial-up modem bit rates, i.e., several tens of kilobits per second). In fact, the packet switching in the core network was introduced in GPRS by adding two new network nodes in the GSM core architecture: SGSN and GGSN.

The 3G standardization started with 3GPP release 99, completed in 2000. The 3GPP standardized coexistence of two different domains in its 3G standard called UMTS:

- CS domain: This domain is used for GSM-alike services (mobile telephony and SMS).
- PS domain: This domain is used for accessing the Internet (i.e., for third-party Internet services, such as websites) and MMS.

The UMTS network architecture is shown in Figure 5.9. It has the same organization as the GPRS core network (i.e., with SGSN and GGSN as main network nodes), but with higher bit rates in the radio access part, due to different access technology (WCDMA) and the fact that more spectrum is dedicated to 3G. The typical frequency carrier width for the 3G radio interface is 5 MHz, whereas in 2G the spacing between frequency carriers was 200 kHz for GSM.

Note that in parallel with 3GPP mobile standards (led mainly by the European countries), there were also 3GPP2 standards (led mainly by American countries) such as CDMA2000 and accompanying wireless IP standards. However, further developments have shown global domination of the 3GPP mobile standards (starting with GSM as 2G, via UMTS as 3G, to LTE/LTE-Advanced as 4G), hence the main focus is given to 3G mobile technologies from the 3GPP.

In general, three segments have been standardized by the 3GPP:

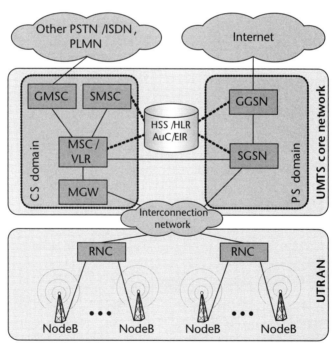

AuC – Authentication Center
EIR – Equipment Identity Register
ISDN – Integrated Services Digital Network
GGSN – Gateway GPRS Support Node
GMSC – Gateway MSC
HLR – Home Location Register
HSS – Home Subscriber Server
MGW – Media Gateway
MSC – Mobile Switching Center
PLMN – Public Land Mobile Network
PSTN – Public Switched Telephone Network
RNC – Radio Network Controller
SGSN – Serving GPRS Support Node
UMTS – Universal Mobile Telecommunications System
UTRAN – UMTS Terrestrial Access Network
VLR – Visitor Location Register

Figure 5.9 UMTS 3G network architecture.

- High-speed access: This includes radio access technologies.
- IP core network: This includes all controllers/gateways and databases in the networks as well as their interconnection.
- Mobile services: This part includes the service overlay networks, which are implemented in a given mobile network architecture (e.g., IMS).

The move toward higher data rates in the radio interface was made with 3GPP release 5, which defined high-speed downlink packet access (HSDPA). The enhanced bit rates in the uplink were introduced in 3GPP release 6 as high-speed uplink packet access (HSUPA). HSUPA merged with HSDPA into high-speed packet access (HSPA) in 3GPP release 7, thus making UMTS/HSPA truly mobile broadband technology.

The available bit rates for the different 3G mobile broadband standards by 3GPP (i.e., UMTS/HSPA) are given in Table 5.6. Note that these bit rates are the highest possible with a given mobile technology. In practice, the bit rates per individual mobile user are much lower, because the given bit rates are aggregate values, which are shared among all users who are using the mobile network in the same geographical area (e.g., in the same cell). Also, achievable bit rates are highly dependent on the mobility of the users (higher mobility means lower bit rates due to the less efficient MCS that has to be used) and on the distance between the mobile terminal and serving base station (longer distance results in worse signal-to-noise ratio, and hence a less efficient MCS with smaller bit rates, and vice versa). Also, different obstacles that influence radio propagation may further decrease the bit rate (e.g., signal attenuation due to urban buildings or walls).

Generally, 3G mobile networks were standardized and implemented worldwide in the first decade of the 21st century. The goal of all-IP mobile networks, higher bit rates, and NGN implementation in the mobile network is being accomplished by means of 4G mobile standards (from 3GPP) called LTE/LTE-Advanced.

Table 5.6 3G Mobile Broadband Technologies by 3GPP

3GPP Mobile Broadband Technology	Maximum Downlink Bit Rate (Mbps)	Maximum Uplink Bit Rate (Mbps)
UMTS/HSPA release 6	14.4	5.76
UMTS/HSPA+ release 7 (64 QAM downlink, 16 QAM uplink)	21.1	11.5
UMTS/HSPA+ release 8 (2 × 2 MIMO, 64 QAM downlink, 16 QAM uplink)	42.2	11.5
UMTS/HSPA+ release 9 (2 × 2 MIMO, 64 QAM downlink, 16 QAM uplink, dual carrier i.e., 2 × 5 MHz)	84.4	23
UMTS/HSPA+ release 10 (2 × 2 MIMO, 64 QAM downlink, 16 QAM uplink, four carriers i.e., 4 × 5 MHz)	168.8	46
UMTS/HSPA+ release 11 (4 × 4 MIMO, 64 QAM downlink, 16 QAM uplink, up to eight noncontiguous carriers i.e., 8 5 MHz)	675.2	184

5.4.3.2 4G Mobile Broadband by 3GPP: LTE/LTE-Advanced

The evolution of all three parts of a mobile network (radio access technology, core network, and services) resulted in standardization of the next step in the evolution of 3GPP in its release 8, which uses LTE in the radio part, system architecture evolution (SAE) for the core network, and IMS for the services. The next release (release 9) standardized the leftovers from release 8. However, LTE does not satisfy all requirements set by ITU-R for the IMT-Advanced systems, particularly requirements for bit rates above 1 Gbps in the downlink for nomadic mobile users. Therefore, LTE is usually referred to as 3.9G, but in reality it is noted as a 4G technology, due to its similarity with LTE-Advanced.

The road to LTE continued toward LTE-Advanced in 3GPP release 10, which was followed by releases 11, 12, and 13. Both LTE and LTE-Advanced use OFDMA in the radio interface in the downlink, and single-carrier FDMA (SC-FDMA) in the uplink. As its name denotes, LTE-Advanced is an advanced version of LTE that has more similarities than differences with it, such as the same core network, the same radio access technology (on the physical layer), and the same IMS for the services.

With the convergence of 3GPP networks toward all-IP, certain changes were needed in the mobile network architecture to provide less delay for delay-sensitive traffic such as voice when it is transferred over IP end to end. That led to a change in the hierarchical network architecture [with, e.g., the radio network controller (RNC) until release 7] to a flat architecture, as shown in Figure 5.10. The SAE defines the core network for LTE/LTE-Advanced mobile networks. (Additional details on SAE are given in Chapter 6.)

The radio access network in LTE and LTE-Advanced is called the Evolved UMTS Terrestrial Radio Access Network (E-UTRAN) and it consists of eNodeBs (i.e., base stations) and interfaces on the links to a centralized gateway (MME and S-GW) on one side and to mobile terminals (i.e., user equipment) on the other side. The radio interface between the mobile terminals and the base stations is referred to as Evolved UMTS Terrestrial Radio Access (E-UTRA). Evolved packet core (EPC), E-UTRAN, and E-UTRA form the IP connectivity layer in LTE/LTE-Advanced mobile networks, called overall the Evolved Packet System (EPS).

The main difference between LTE and LTE-Advanced is the bit rate. LTE-Advanced has higher bit rates due to the possibility for frequency carrier aggregation on the OSI-2 MAC layer of up to five noncontinuous frequency carriers (each with up to 20 MHz). The maximum bit rates for LTE and LTE-Advanced are given in Table 5.7.

So, the SAE/EPC approach provides an all-IP core architecture for 3GPP mobile broadband networks, including the evolved 3G network (UMTS/HSPA+) as well as 4G mobile broadband technologies like LTE and LTE-Advanced. Each new 3GPP release enhances the bit rates of the radio access part and decreases the total delay budget. All 3GPP networks in certain periods considered use of Wi-Fi for Internet traffic offload in areas with heavy traffic (e.g., in urban areas).

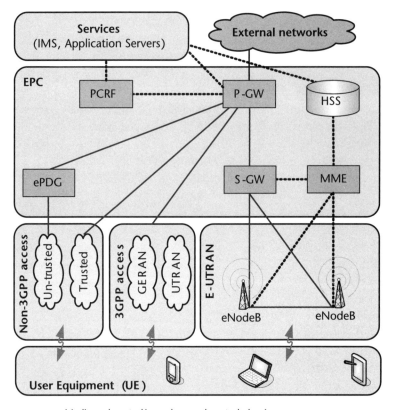

Figure 5.10 Evolution of 3GPP 4G mobile broadband network architecture.

GERAN – GSM EDGE Radio Access Network
HSS – Home Subscriber Server
E-UTRAN – Evolved UTRAN
ePDG – Evolved Packet Data Gateway
MME – Mobility Management Entity
P-GW – Packet data network Gateway
PCRF – Policy and Charging Rules Function
S-GW – Serving Gateway
UTRAN – UMTS Terrestrial Radio Access Network

Table 5.7 Bit rates of 4G Mobile Broadband Technologies by 3GPP

3GPP Mobile Broadband Technology	Maximum Downlink Bit Rate (Mbps)	Maximum Uplink Bit Rate (Mbps)
LTE, 2×2 MIMO, 64 QAM in downlink, 16 QAM in uplink, 20-MHz carrier width	172.8	57.6
LTE, 4×4 MIMO, 64 QAM in downlink, 16 QAM in uplink, 20-MHz carrier width	326.4	86.4
LTE-Advanced (five aggregated LTE carriers with total 100-MHz spectrum)	3000	1500

5.4.3.3 Wi-Fi and Fixed WiMAX for Wireless Local Broadband Access by IEEE

The main wireless and broadband technologies from the IEEE standards track are Wi-Fi (IEEE 802.11 standards group) as a WLAN, and WiMAX (IEEE 802.16 standards group) as a WMAN.

Currently Wi-Fi is the most widely spread wireless LAN for business, home, and public environments. Both Ethernet and Wi-Fi today create unified access to the Internet in any environment that is not a mobile network.

Wi-Fi has neither "explicit" QoS nor mobility support because it has no TDMA, which is needed for managed radio resource allocation. Also, Wi-Fi operates in unlicensed frequency bands (2.4 and 5 GHz), which may be congested by the large number of users using their Wi-Fi networks (e.g., in residential or enterprise buildings).

The IEEE 802.11 standards for the physical layer define the bit rates by specifying the spectrum bands, modulation schemes, coding rate, number of antennas, and so forth. The first IEEE 802.11 standard was published in 1997. It provided bit rates up to 2 Mbps. It was followed by IEEE 802.11b with up to 11 Mbps (in 2.4 GHz), then IEEE 802.11g appeared (2.4 GHz) and the IEEE 802.11a (5 GHz) with up to 54 Mbps. The continuous development of the popular IEEE 802.11 technology continued with IEEE 802.11n, which provides up to 600 Mbps and can be used for both 2.4 and 5 GHz, and IEEE 802.11ac with up to 6.9 Gbps (in the 5-GHz band). Wi-Fi development goes even further with the next amendment (IEEE 802.11ad), and we may expect it to continue to advance in such a manner in the future. However, the IEEE 802.11ad operates in the 60-GHz band, which is completely different than previous standards in the 2.4- and 5-GHz bands. An overview of Wi-Fi maximum bit rates is given in Table 5.8.

With regard to the OSI-2 layer, there are Wi-Fi standards for different purposes, such as application support (e.g., 802.11p/z/aa), network convergence (e.g., 802.11u), network management (e.g., 802.11 k/v), QoS (e.g., 802.11e), security (e.g., 802.11i), and so forth.

All Wi-Fi networks follow a simple architecture, based on access points (APs), which work on the OSI-2 layer, although they also can be implemented in routers (e.g., Wi-Fi routers). In general, Wi-Fi architectures can be systematized into two main modes:

- Infrastructure mode: In this case Wi-Fi APs are connected via switches to a router connected to the Internet.

Table 5.8 Wi-Fi Maximum Bit Rates

IEEE Standard	Maximum Bit Rate
IEEE 802.11	2 Mbps
IEEE 802.11b	11 Mbps
IEEE 802.11g/a	54 Mbps
IEEE 802.11n	600 Mbps
IEEE 802.11ac	6.9 Gbps
IEEE 802.11ad	6.9 Gbps

- Ad-hoc mode: In this case Wi-Fi is used for direct communication between two or more hosts (i.e., wireless terminals such as laptops).

In the infrastructure mode a single AP coverage area is referred to as the basic service set (BSS), while multiple APs in the same IP network (on the same side of the router through which they are connected to the Internet) define the extended service set (ESS), as shown in Figure 5.11.

Let's continue toward fixed WiMAX and its architecture as well as its position on the world ICT scene. Fixed WiMAX appeared in 2004 and was defined by the IEEE 802.16 group of standards for the OSI-1 and OSI-2 protocol layers, while IP is assumed on the network layer (similar to Wi-Fi and other IEEE standards for network access, either fixed or wireless).

The WiMAX Forum defines the WiMAX network architecture for fixed and mobile WiMAX. The WiMAX architecture consists of two main network parts: the connectivity service network (CSN) and access service network (ASN). The ASN is the access network consisting of base stations (BSs) and an ASN gateway that is used for connection with the CSN. In contrast, the CSN is the core network part, which includes nodes for AAA, application servers, and so forth.

Overall, fixed WiMAX is a good "piece" of technology that has been developed further as a mobile version. However, sometimes that is not enough. The penetration of broadband is happening at the highest speed in the developed countries that have a deployed network infrastructure (typically based on optical networks), whereas in the developing parts of the world (which often lack a fixed network infrastructure for various reasons), the most prominent way to provide broadband access to the population of end users is based on fixed wireless broadband access solutions (e.g., fixed WiMAX) or mobile broadband access.

Contrary to WiMAX, we might say that Wi-Fi is currently the most widespread wireless technology in the world, taking into account the number of deployed Wi-Fi APs in homes, offices, and public places via Wi-Fi hotspots (at hotels, airports, malls, restaurants, and so on). Also, Wi-Fi is used as a wireless extension of fixed broadband access and for traffic offloading in mobile broadband networks (due to

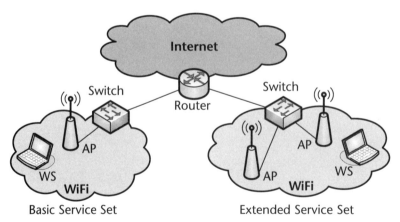

Figure 5.11 Wi-Fi network architectures.

the lower costs of Wi-Fi equipment and the free use of unlicensed frequency bands). Simply said, Wi-Fi is great for its purpose, whereas WiMAX is still lagging behind.

5.4.3.4 4G Mobile Broadband Standard by the IEEE: Mobile WiMAX 2

Mobile WiMAX is a "younger" technology compared with the 3GPP technologies. Its first release was standardized as IEEE 802.16e-2005 [27]. Its network infrastructure was standardized by the WiMAX Forum and it was called Mobile WiMAX release 1.0. The radio part of Mobile WiMAX 2.0 (which is based on IEEE 802.16m for the radio part [28, 29]) has many similarities with LTE-Advanced. Also, both technologies were approved as 4G by ITU at the same time period (around 2010). However, this is a different type of progress compared to the 3G development where Mobile WiMAX entered the 3G umbrella later (in October 2007), and had no significant impact on the 3G mobile networks on a global scale. At the same time, however, it was an indication of a "serious intention" on the part of another mobile technology to provide more competition and more possibilities in the mobile broadband world. The wireless interfaces of Mobile WiMAX 2.0 and LTE-Advanced as well as the all-IP core architecture in both standards will lead to a convergence of these technologies in some way. However, that is dependent on the regulation (e.g., spectrum management) and business strategies of vendors and mobile operators.

The Mobile WiMAX architecture is shown in Figure 5.12. It consists of three main parts (similar to fixed WiMAX): ASN, CSN, and mobile stations (MSs).

The ASN is the radio access network of Mobile WiMAX, consisting of BSs and an ASN gateway (ASN-GW). The CSN provides the means for IP connectivity between the ASN (and mobile stations connected to the ASN) and the Internet. It is an all-IP based core network, which includes different servers for support of the necessary functionalities to mobile stations in the ASN. The main parts of the CSN include AAA servers (e.g., RADIUS, Diameter), a DHCP server, DNS servers, a home agent (for Mobile IPv4 and Mobile IPv6), a foreign agent (for MIPv4), application servers, and so forth.

The main difference in network architectures of Mobile WiMAX release 1.x and 2.x is in the ASN part, where Mobile WiMAX release 2 introduces the advanced mobile stations (AMSs) and advanced base stations (ABSs), which implement the new functionalities of the IEEE 802.16m radio interface.

Overall, 4G Mobile WiMAX is very similar to LTE-Advanced, but it is lagging behind due to (1) better positioning of 3GPP mobile technologies on the side of mobile operators, users, and vendors; (2) 3GPP success with 2G (i.e., GSM) and 3G mobile technologies (i.e., UMTS/HSPA); and (3) being able to provide a 4G evolution (with LTE/LTE-Advanced).

Overall, all-IP networking and Internet technologies are now fundamental prerequisites for all existing and future mobile (and fixed) broadband standards.

5.5 Naming and Addressing

The Internet has exactly two naming and addressing spaces: (1) domain names and (2) IP addresses. The mapping between Internet addresses and domain names is done by DNS. In telephone networks the addressing is based on telephone numbers

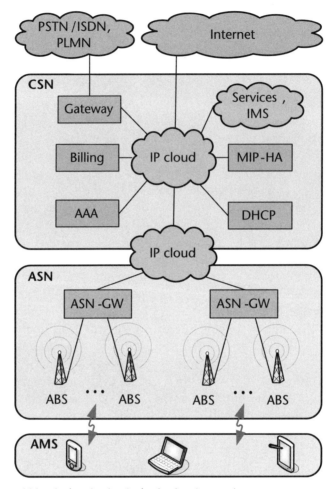

AAA – Authentication, Authorization, Accounting
ABS – Advanced Base Station
AMS – Advanced Mobile Station
ASN – Access Service Network
BS – Base Station
CSN – Connectivity Service Network
DHCP – Dynamic Host Configuration Protocol
IMS – IP Multimedia Subsystem
MIP-HA – Mobile IP Home Agent

Figure 5.12 Mobile WiMAX network architecture.

as defined by ITU-T E.164 standard, which is used globally. Because NGN is a convergence of the two worlds, it includes both approaches for numbering, naming, and addressing (for Internet services and telephony).

The E.164 standard is the dominant numbering scheme for voice users, so we might expect that it will continue to be used further, at least for some time (e.g., short or medium term). However, E.164 is expected to adapt to the IP environments with the newer numbering (E.164), naming (DNS), and addressing (IP addressing) scheme called ENUM, which is standardized by IETF [30] and ETSI (for the NGN usage) [31]. ENUM was created to translate digits from E.164 telephone number to labels (in terms of DNS), written in the reverse order, with the root "e164." For example, if an E.164 telephone number is "+389-70-1234567," then the full domain name for this telephone number will be "7.6.5.4.3.2.1.0.7.9.8.3.e164.arpa."

In fact, the main approach for naming and addressing in converged Internet and telecom worlds is defined by NGN. Three types of IDs specified in NGN [32]:

- Home domain name: This identifier is used to identify the operator for routing the SIP registration requests to the home operator's IMS network (IMS is in the service stratum of NGN). It is specified in a domain name format (e.g., "TelecomOperator123.com") and it is resolved by using DNS.
- Private IDs: These identifiers are used to identify a user's subscription. So, each user should have in the NGN at least one private ID. The syntax of a private ID is the form "username@realm" and it is a permanent ID (not related to a given session or a call). It is stored locally in the ISIM. The "realm" in the private ID is the home domain name. Private IDs are network (operator) aware.
- Public IDs: These identifiers are used for message routing (e.g., SIP message routing to the end user for initiation of a call/session). These IDs may take the form of tel URI (e.g., E.164 based) or SIP URI (e.g., "sip:user@domain"). Public IDs are user aware.

Numbering, naming, and addressing are related to public IDs (i.e., user-aware IDs). Regarding the NGN strata, all identifiers may be divided into two groups: IDs in the service stratum, and IDs in the transport stratum. An overview of some of the public identifiers in NGN is given in Table 5.9 [32].

Besides being used to identify users, network interfaces, and services/applications, the identifiers are inseparably related to QoS frameworks in all telecommunication/Internet environments.

5.6 Quality of Service (QoS) Framework

The network layer in the Internet is IP and it does not contain mandatory QoS mechanisms. However, from the beginning it has had the type-of-service (ToS) field defined to specify QoS requirements on precedence, delay, throughput, and reliability. In a similar manner, IPv6 has the DHCP field and can support QoS per flow on the network layer (e.g., by using the flow label and next header options). However, IPv6 does not guarantee the actual end-to-end QoS because there is no reservation of network resources (this should be provided by some other mechanism).

In contrast, the traditional telecommunication approach includes end-to-end QoS support by default in the network. For that purpose there is required signaling

Table 5.9 Public Identifiers in NGN

Type of Identifier	Public ID	Format of Identifier
User/service identifiers (service stratum)	Name	SIP URI
	Number	tel URI, SIP URI
Network identifiers (transport stratum)	Address	Routing number, IP address

end to end (e.g., SS7 signaling) and admission control (due to reservation of resources per call or session, only a limited number of calls/session can be admitted in the network with the goal of maintaining the desired QoS). The convergence of telecommunications toward all-IP networks and Internet technologies resulted in standardization of the NGN by ITU-T (as discussed at the beginning of this chapter).

The main purpose of the NGN framework was to standardize the end-to-end QoS support in all-IP networks (including all needed functions in transport and service strata), which is essential for real-time services, such as VoIP and IPTV. Such services have strict requirements regarding the QoS (e.g., guaranteed bit rates, losses, delay, and delay variation—jitter). So, NGN provides standardized implementation of QoS (instead of proprietary case-by-case implementations), which is mandatory for the transition from PSTN and PLMN to all-IP networks.

The following subsections define QoS and quality of experience (QoE), give an overview of the Internet QoS mechanism, and cover QoS parameters in IP networks and end-to-end QoS provisioning.

5.6.1 Introduction to QoS, QoE, and Network Performance

The QoS is extensively used today in the telecommunication/ICT world in which it has its roots, including the current developments in the field regarding broadband, wireless/mobile, and multimedia services. However, QoS is moving from its initial definitions targeted to traditional telecommunication networks (e.g., PSTN/ISDN, broadcast networks) to QoS in all-IP networks.

Depending on what aspects of quality are examined and what kind of services and technologies are involved, different definitions and concepts of quality are often used. However, the main reference to a QoS definition in the telecommunication world is ITU-T recommendation E.800 [33]. According to E.800, QoS is the totality of characteristics of a telecommunications service that bear on its ability to satisfy the stated and implied needs of the user of the service. From the network's point of view, QoS can be defined as the ability to segment traffic or differentiate between traffic types in order for the network to treat certain traffic flows differently from others. QoS also can be defined as a criterion of the degree of user satisfaction of the offered service, which is a more subjective definition of QoS that depends on users' perceptions. However, in traditional telecom traffic engineering, the QoS refers to measurable parameters and techniques to select, control (e.g., via admission control), measure, and guarantee the required quality for a given service.

To provide QoS support in all-IP–based networks, standardization of QoS provisioning end to end is certainly needed for those services and applications that have specific QoS requirements. Such standardization is carried out by the ITU-T on a global scale, while Internet technologies are continuing to be standardized by the IETF and then they are incorporated (where needed) into framework standards (e.g., ITU-T recommendations).

Besides the network performances that can be measured (e.g., losses, packet delay and jitter, throughput in bits per second) end-user perceptions of telecommunication services (which also include all Internet native services such as WWW, email, and so on) are also influenced by various trends (e.g., what is "cool"), advertising (online, TV, various media), tariffs, and costs, which are interrelated with customer expectations of QoS. So, user perception of the quality is not limited to the objective characteristics at the man/machine interface (e.g., on a PC or smartphone). Typically, for end users what is important is the quality they personally experience during use of a given service (e.g., users at different locations on a mobile network can have different personal experiences for the same services provided with the same QoS functions by the network). The relationship between technical QoS and nontechnical QoS influence on customer satisfaction is given in Figure 5.13.

Network performance (NP) differs from the QoS, because the QoS is the outcome of the user's experience or perception, whereas the NP is determined by the performances of different network elements, or by the performance of the network as a whole (i.e., the combination of the performance of all single elements in it) [34]. However, the NP has an influence on the QoS and it represents a part of it. So, simply said, QoS consists of network performances and nonnetwork performances, as illustrated in Figure 5.13.

5.6.2 ITU-T Standardization on QoS and QoE

There are four aspects of QoS that should be considered according to the ITU-T G.1000 [35]. They are placed into two groups of customer viewpoints and service provider viewpoints:

- Customer's QoS requirements: This is the QoS level required by the subscriber.
- QoS offered by the service provider: This includes QoS criteria or parameters offered by the service provider, which may be used for creation of an SLA as a bilateral agreement between the customer and the service provider, public offering (i.e., declaration) of the service level that can be expected by

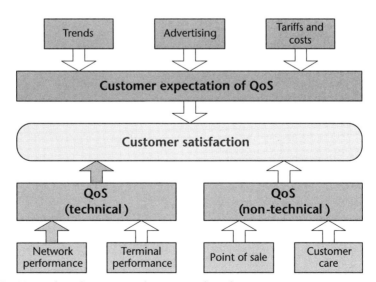

Figure 5.13 Network performance and nonnetwork performance.

the subscribers, as well as planning and maintaining the service at a given performance level.

- QoS achieved or delivered by the provider: This is the actual level of QoS achieved or delivered by the service provider, which can be used for checking of the delivered QoS (e.g., considering the SLA) or for any corrective action regarding the QoS.
- Customer perception of the QoS: This is the QoS level obtained by user ratings of the provided QoS by the service operator, which can be used for comparison purposes among QoS levels provided from different service providers as well as for corrective actions (e.g., when the perceived QoS level is below the QoS level offered by the provider).

The four viewpoints of QoS are interrelated. QoS offered by the service provider defines the limits for the QoS achieved or delivered (by the service provider), which influences the customer's perception of the QoS and customer's QoS requirements (e.g., demand for higher bit rates). Finally, the customer's perception has influence over the QoS offered by the service provider.

Besides the QoS as an objective measure (via performance parameters that can be measured at defined measurement points, such as network interfaces in hosts and network nodes), there is a need for subjective merit for the quality of the given service perceived by the end users. ITU-T defines so-called quality of experience (QoE) as the overall acceptability of an application or service, as perceived subjectively by the end user.

In general, QoE is influenced by all QoS criteria, which include speed, accuracy, service availability, reliability, security, simplicity of use, and flexibility. For example, speed (as a QoS criterion) influences the available throughput and latencies and it is of crucial importance for the QoE. That is why when moving toward broadband access and higher access bit rates (including fixed and mobile networks) the overall QoE improves. Availability and reliability are also very important and depend on the capability of the network to recover from a failure as well as appropriate planning and dimensioning of the network (to serve the expected number of users for given services). Also, security aspects, accuracy, simplicity of use, and flexibility regarding the services influence the QoE.

The most used measure for QoE is the mean opinion score (MOS). Initially, the MOS scale referred to voice service only (ITU-T P.800 [36]), but nowadays it is also used for other services such as video (e.g., IPTV). MOS is expressed as a single number in the range from 1 to 5, where a value of 1 corresponds to the lowest quality experienced by the end user and 5 is the highest quality experienced by the end user, as shown in Table 5.10.

QoE is quite different from QoS and NP, because it has a subjective feature in the definition, that is, a dependency on end-user perception of the services received. It is clear, however, that QoE is impacted by QoS and NP. Taking into account those differences given by the definitions of NP, QoS, and QoE [37], the key features of these three parameters can be summarized as given in Table 5.11.

So, the ITU-T QoS framework is based on three cornerstones: network performance, quality of service, and quality of experience. NP is network provider ori-

5.6 Quality of Service (QoS) Framework

Table 5.10 Mean Opinion Score Scale

MOS	Quality	Impairment
5	Excellent	Imperceptible
4	Good	Perceptible
3	Fair	Slightly annoying
2	Poor	Annoying
1	Bad	Very annoying

Table 5.11 Relationship Among Network Performance, QoS, and QoE

Quality of Experience (QoE)	Quality of Service (QoS)	Network Performance (NP)
User oriented		Provider oriented
User behavior attribute	Service attribute	Connection/flow element attribute
Focus on user-expected effects	Focus on user-observable effects	Focus on planning, development (design), operations, and maintenance
User subject	Between (at) service access points	End-to-end or network element capabilities

ented, whereas QoS and QoE are user oriented. However, today, all three elements are dependent on the Internet QoS mechanisms.

5.6.3 Overview of Internet QoS

The Internet was initially created for best-effort service, which means that every connection is admitted in the network and every packet is transferred from source to destination via the best effort by the network, without any QoS guarantees.

The whole telecommunication world is transiting to the Internet as a single networking platform for all telecommunication services, including native Internet services (e.g., WWW, email) and traditional telecommunication services (e.g., telephony, television). Therefore, understanding the Internet technologies and standardized mechanisms of the IETF has high importance for QoS issues and solutions.

Overall, Internet traffic characteristics influence the NP and QoS solutions. The Internet has the following typical types of traffic:

- Voice traffic: This type of traffic has a constant bit rate when sending, requires relatively small IP packets (e.g., 50 to 200 bytes), and is very sensitive to end-to-end delay (the recommended delay in PSTN is below 150 ms, while above 400 ms is not acceptable because people who are talking will start interrupting each other) and jitter. Such a delay budget is also valid in IP environments, although the delay in the Internet is always higher than in PSTN/ISDN, due to packetization, buffering in network nodes, and other differences. To compensate, the jitter voice must have a playback point. However, audio is more tolerant to errors due to the characteristics of the human ear.

- Video traffic: This type of traffic generally has high variable bit rates that are controlled by codec efficiency on picture. The message size is generally determined by the MTU and typically it uses the maximum possible MTU (e.g., 1,500 bytes). On the receiving side, the video player application buffers data to ensure consistency.
- Data traffic: This type of traffic is typically non–real-time traffic over the Internet (e.g., WWW, email) and is typically TCP based. Most Internet traffic is TCP while the rest is UDP. TCP slow start and fast retransmit mechanisms ensure maximum utilization of bottlenecks, but that results in long queues (at network nodes) without a control mechanism. On one side, Internet end-to-end flow control is based on TCP at the end users, while on the other side it is one of the main reasons for Internet congestion.

How can we provide QoS guarantees for the Internet? Well, we can add QoS mechanisms to the initially best-effort Internet. Overall, all such QoS mechanisms (or technologies) in the Internet can be grouped as follows:

- Best-effort services: This is the traditional Internet service model, without any QoS guarantees. The IP networks just route packets until they reach the destination.
- Integrated services (IntServ): This is the first mechanism to be standardized by IETF. It is based on resource reservation in routers on the path by using signaling (e.g., RSVP) [38]. Routers have to store traffic and QoS information per connection (i.e., flow).
- Differentiated services (DiffServ): This is the most commonly used method for traffic differentiation in the Internet. In this method all packets are classified into a limited number of classes, so routers have to store only information per class (not per connection/flow).
- Multiprotocol Label Switching (MPLS): This is the "default" approach for QoS provisioning in transport IP networks, which may be combined with DiffServ as well as other protocols such as BGP and VPNs.
- Policy-based QoS: This is typically used for QoS provisioning between network providers (e.g., BGP policies).

All protocols for implementation of the QoS framework in the Internet have been standardized by IETF, starting in the 1990s. However, regarding the transition of telecommunication networks and services to all-IP networks and services, other SDOs have created standardized frameworks for QoS implementations (e.g., ITU-T). As a result, the QoS architectural framework is organized into three planes (ITU-T Y.1291): a control plane (includes admission control, QoS routing, and resource reservation), data plane (includes buffer management, congestion avoidance, packet marking, queuing and scheduling, traffic classification, traffic policing, and traffic shaping), and management plane (includes SLA, traffic restoration, metering and recording, and policy).

5.6.3.1 Admission Control and QoS Mechanisms

Typical QoS mechanisms in network nodes (e.g., routers) are packet scheduling, metering, and classification. Regarding the scheduling, the default approach in all routers and switches is first in, first out (FIFO). For packet streams in a statistical multiplexing situation (i.e., there is queuing), FIFO means letting arrival characteristics determine the QoS. A single bursty flow (bursty refers to the high peak to mean bit rates of the flow) can cause high delays for many other flows. Typically scheduling is dependent on the traffic classification (if it is applied), which is used to sort packets into flows per class (e.g., separated classes for voice, video, and data traffic). After classification, scheduling is applied, such as priority scheduling or weighted fair queuing.

Usually the following fields are used for classification (5-tuple): source IP address, destination IP address, protocol number, source port, and destination port. So, IPv4 requires fields from the IP header (IP addresses and protocol) and transport headers (ports). In IPv6 networks classification can be performed with 3-tuple {Source address, Destination Address, Flow Label}, which is possible when the flow label in the IPv6 header is used (typically it is not used nowadays).

For QoS provision to telecommunication services, the key mechanism is the admission control. A general model for QoS implementation in routers is shown in Figure 5.14. No matter how good the scheduler works, one may still have no QoS in practice when traffic demand exceeds available resources. Admission control is used to reject a flow if the resources are insufficient. It is also different than policing, which is used to reject/drop a packet at a router, because it usually is applied to an established connection/session.

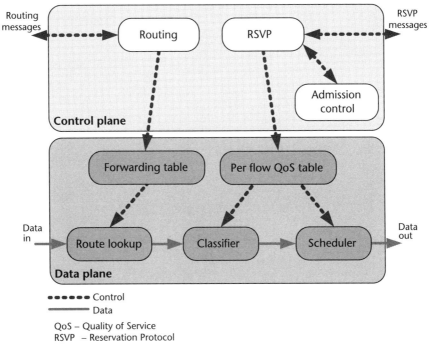

Figure 5.14 QoS implementation in routers.

In the traditional telephone networks (e.g., PSTN, PLMN), it is easy to do admission control due to circuit switching (allocation of a pair of channels per call, one per direction, during the call duration). In the Internet, flows have different QoS requirements and therefore admission control is much harder for implementation and requires standardization of applied network architectures and control protocols in different networks end to end.

5.6.3.2 QoS Technologies Standardized by IETF

IETF has standardized three main solutions for QoS support in the Internet: IntServ, DiffServ, and MPLS, which are the focus of this section.

IntServ with initial RSVP for QoS support per flow in the Internet was standardized in the 1990s [38]. It lacks scalability because each node (i.e., router) in the path of the flow should admit the flow with the given QoS requirements. With the standardization of aggregation of RSVP for IPv4- and IPv6-based flows [39], the scalability issue with RSVP has been solved. The version of this signaling protocol used in NGN core MPLS networks is RSVP-TE [40].

The other standardized solution for QoS support in IP networks, DiffServ, uses ToS and DSCP (i.e., DS) fields in the IPv4 and IPv6 headers, respectively [41]. DiffServ architectures are well suited to large-scale environments because they support classification of all packets into a limited number of classes based on the DS field in the IP packet header. Elimination of the per-flow state and per-flow processing in the network nodes has given an advantage to DiffServ over IntServ architectures in the past. However, DiffServ has no mechanisms for communication between an individual application and the network. Hence, in the cases where per-flow resource and admission control is required, the IntServ architectures have advantages over DiffServ. With the aggregation of reservations per individual flows at the edge nodes of the core network and their tunneling [42], there scalability of the RSVP-based network architectures for QoS support is much better.

Since the beginning of the 21st century, MPLS has been the main approach for implementation of QoS in the transport IP networks of carriers (e.g., telecom operators) because it provides a high degree of scalability due to aggregation of traffic [in label edge routers (LERs)] according to its type and its QoS and business requirements (e.g., type of subscribers). In the MPLS architecture short, fixed-length labels (inserted between the layer 2 header and IP packet header) are associated with streams of data in the ingress edge MPLS router (i.e., LER) of a given administrative domain (i.e., MPLS domain). Packets belonging to those streams are forwarded along the label switched paths (LSPs) based on their labels, without IP header examination in the label switching router (LSR) inside the MPLS domain. MPLS provides so-called traffic engineering (TE) that enables predetermined paths to specific destinations and uses connection-oriented switching (based on labels); therefore, it is referred to as MPLS-TE. Typically MPLS is combined with DiffServ for class-based QoS provision, as well as with RSVP (i.e., IntServ approach) for flow-aware QoS provision. DiffServ and MPLS together satisfy two necessary conditions for QoS: (1) DiffServ provides differentiated queue servicing treatment, and (2) MPLS forces flows into the paths with guaranteed bandwidth (i.e., bit rate).

The NGN puts the MPLS, IntServ with RSVP, and DiffServ in a defined QoS framework within the transport stratum and interconnects it with the service control functions (e.g., IMS) in the service stratum via the RACF entity, thus providing practical implementation of NGN core networks by using standardized and well-known Internet QoS solutions and technologies.

Generally, MPLS, DiffServ, and IntServ can be combined within a standardized QoS framework to provide end-to-end QoS support. For that purpose we also need a definition of certain QoS parameters.

5.6.4 QoS Parameters and Classes

QoS parameters characterize the quality level of a certain aspect of a service being offered and ultimately the customer satisfaction. They can be used by network and service providers to manage and improve how they offer their services, as well as by the customers (end users or other partner providers) to ensure that they are getting the level of quality for which they are paying. Hence, a subset of QoS parameters is being used to support commercial contracts such as SLA formulation and verification.

According to ITU-T I.350 [37], there are QoS parameters (which are user oriented, where service-related attributes are observed) and NP parameters (which are network provider oriented, and refer to network technology-related attributes).

Considering the Internet as a network, the commonly used parameters that may be used in specifying and assessing IP network performance (ITU-T Y.1540 [43]) are as follows:

- IP packet transfer delay (IPTD): This is the time difference between the occurrences of two corresponding IP packet reference events (i.e., packet transmission via a given measurement point in the network). There are several types of IPTD such as minimum IPTD (the smallest IP packet delay), median IPTD (the 50th percentile of the frequency distribution of IP packet transfer delays), and average IPTD.
- IP packet delay variation (IPDV): This is the difference between the one-way delay of an IP packet and the reference IP packet transfer delay.
- IP packet error ratio (IPER): This is the ratio of the total number of errored IP packets to the total number of transmitted IP packets in a given measurement.
- IP packet loss ratio (IPLR): This is the ratio of the total number of lost IP packets to the total number of transmitted IP packets in a given measurement.

Additionally, transfer capacity (in bits per second) is the QoS parameter that has the highest impact on the performance perceived by the end user, so higher bit rates (i.e., broadband access and transport) are normally better for all services.

The values of the defined IP network performance parameters vary, depending on different so-called network QoS classes. Based on the requirements of the key applications such as conversational telephony, reliable data applications based on TCP (e.g., WWW, email), and digital television, network QoS classes are specified by ITU-T Y.1541 [44], as given in Table 5.12. They define the upper bounds on the key performance parameters for end-to-end IP services. For example:

Table 5.12 IP Traffic Classes as Defined by ITU

QoS Class	Upper Bound on IPTD	Upper Bound on IPDV	Upper Bound on IPLR	Upper Bound on IPER
Class-0	100 ms	50 ms	10^{-3}	10^{-4}
Class-1	400 ms	50 ms	10^{-3}	10^{-4}
Class-2	100 ms	Unspecified	10^{-3}	10^{-4}
Class-3	400 ms	Unspecified	10^{-3}	10^{-4}
Class-4	1 sec	Unspecified	10^{-3}	10^{-4}
Class-5	Unspecified	Unspecified	Unspecified	Unspecified
Class-6	100 ms	50 ms	10^{-5}	10^{-6}
Class-7	400 ms	50 ms	10^{-5}	10^{-6}

IPDV, IP packet delay variation; IPER, IP packet error ratio; IPLR, IP packet loss ratio; IPTD, IP packet transfer delay.

- Class-0 and Class-1 are targeted to real-time jitter-sensitive applications (e.g., VoIP, video conferences), where class-0 provides higher interactions due to the lower bound on the IPTD parameter.
- Class-2 and Class-3 are targeted to transaction data, from which class-2 is intended for signaling traffic, while class-3 is for interactive applications.
- Class-4 is targeted for short transactions, video streaming, or bulk data.
- Class-5 is unspecified (regarding all performance parameters) and hence it is targeted to traditional best-effort Internet applications.
- Class-6 and Class-7 are provisional in the given table. However, such QoS classes are needed for new emerging applications with strict performance parameters.

The specified upper bounds on IP network performance parameters refer to end-to-end QoS provisioning, and that is not an easy task to implement in the existing Internet.

5.6.5 End-to-End QoS

The whole concept of QoS is attractive to the end user if the presentation at the user interfaces satisfies needs and expectations. This means that an overall approach to QoS provisioning is necessary (i.e., end-to-end QoS provisioning).

The end-to-end network model across the Internet is also referred to as UNI-to-UNI (UNI is a user/network interface) [44]. In that model, IP network sections may be represented as clouds with edge routers on their borders and a certain number of interior routers with various functionalities. Such IP clouds may support user-to-user connections, user-to-host connections, and other variations of the end points. In general, a path can have one or more network sections. Also, network sections that support the packets in a flow may change during their lifetime, and an IP connection may span national or regional boundaries. In all cases UNI–UNI performance of a given path can be estimated knowing the performance of its subsections across different networks.

5.6 Quality of Service (QoS) Framework

End-to-end performances of a path can be estimated if its subsections are known. In that manner, the following calculations apply for end-to-end estimation of the Internet QoS parameters:

- Mean IPTD: End-to-end performance is the sum of means contributed by network sections. For example, when a flow portion does not contain a satellite hop, its computed IPTD is:

$$\text{IPTD microseconds} \leq (R_{km} \times 5) + (N_A \times D_A) + (N_D \times D_D) + (N_C \times D_C) + (N_I \times D_I) \quad (5.1)$$

where R_{km} represents the route length assumption and $(R_{km} \times 5)$ is an allowance for "distance" within the portion; N_A, N_D, N_C, and N_I represent the number of IP access gateway, distribution, core, and internetwork gateway nodes, respectively; and D_A, D_D, D_C, and D_I represent the delay of IP access gateway, distribution, core, and internetwork gateway nodes, respectively.

- IPLR: End-to-end performance may be estimated by inverting the probability of successful packet transfer across n network sections:

$$IPLR_{UNI-UNI} = 1 - \{(1 - IPLR_{NS1}) \cdot (1 - IPLR_{NS2}) \cdot (1 - IPLR_{NS3})\ldots(1 - IPLR_{NSn})\} \quad (5.2)$$

- IPER: End-to-end performance can be calculated by inverting the probability of error-free packet transfers across n network sections:

$$IPLR_{UNI-UNI} = 1 - \{(1 - IPLR_{NS1}) \cdot (1 - IPLR_{NS2}) \cdot (1 - IPLR_{NS3})\ldots(1 - IPLR_{NSn})\} \quad (5.3)$$

- IPDV: The end-to-end estimation of this parameter is difficult because it may significantly vary in different network sections. In addition, detailed information about delay distributions that influence it may rarely be shared among operators, so there are limitations to the accurate estimation of end-to-end IPDV.

Generally, there is an existing difficulty for end-to-end QoS provision in the IP-based platforms and networks due to their heterogeneity. In other words, using IP as a transport technology does not mean networks and platforms are the same or compatible. An extreme case of configuration among different IP networks is a scenario where each network has a different mechanism and level of performance control and QoS provision (e.g., ISP-1 uses DiffServ, Backbone-1 uses IntServ, Backbone-2 uses MPLS-TE, ISP-2 uses over provisioning, as shown in Figure 5.15). In this case, it is hard to provide services with end-to-end QoS level guarantees, because each network has a different QoS mechanism.

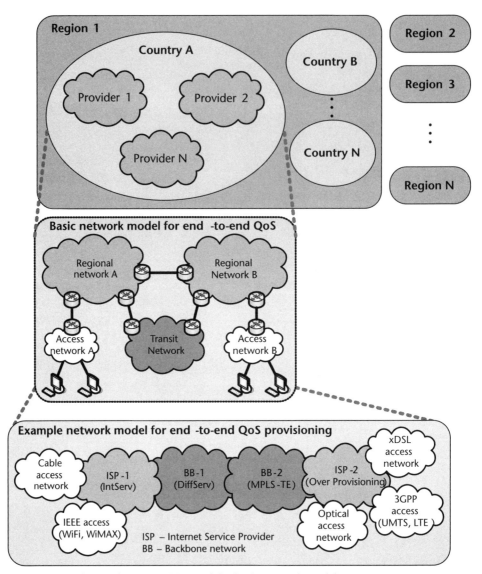

Figure 5.15 End-to-end QoS provision and the basic network model.

A basic network model (consisting of access networks, backbone networks, and transit networks) should be applied commonly to the providers, but the detailed technologies to compose such a network model would be different based on providers and countries. Also, various differences in the regulatory environments are fundamental to SLAs, QoS, and QoE (see Figure 5.15).

Because the most used LANs are Ethernet LANs, but the QoS in transport IP networks is implemented by using MPLS and DiffServ, the QoS mapping between the three technologies may be crucial. Such mappings are defined in ITU-T Y.1545 [45] and shown in Table 5.13.

Finally, end-to-end performance assessment and evaluation are essential for provision of end-to-end QoS provision, including for fixed and mobile networks.

Table 5.13 Mapping Between DiffServ, MPLS, and Ethernet

Packet Network QoS Class	Description	Layer 3 Packet Marking: DSCP (DiffServ Code Point)	Layer 2 Packet Marking		Applications
			MPLS (Class of Service)	Ethernet (Priority Code Point)	
Classes 0, 1	Jitter sensitive	EF (expedited forward)	5	5 (default) or 6	Telephony
Classes 2, 3, 4	Low latency	AF (assured forward)	4, 3, or 2	4, 3, or 2	Signaling, interactive data
Class 5	Best efforts	DF (default forward)	0	0	Web browsing, email

5.7 Fixed–Mobile Convergence (FMC)

FMC initially appeared in the form of unlicensed mobile access (UMA) and generic access network (GAN) concepts. However, in the first decade of the 21st century, all ideas about FMC have converged to an architecture with common core IP networks and common service functions, and specific access networks for fixed and mobile environments. In fact, FMC is based on convergence of different access networks to the same core and transport networks (regarding the transport stratum) and the same signaling protocols and overlay networks needed for service provisioning (regarding the services stratum), and finally to the same applications and content provided via different (i.e., heterogeneous) mobile and fixed networks. The FMC approach provides different services/applications to end users regardless of the access network.

Regarding standardization, the NGN provides generalized mobility and defined FMC requirements. Generalized mobility is based on separation of the transport stratum and service stratum and provision of the same services over different wireless and mobile access networks (including horizontal mobility within a given access network, and vertical mobility among different access technologies) and fixed networks.

Note that the evolution of core networks toward the generalized mobility and NGN principles (e.g., IMS in the service stratum, PCRF in the core network) has been realized in 4G mobile networks.

The first FMC objective is seamless service operation over heterogeneous fixed networks (e.g., PSTN/ISDN, Ethernet, cable networks) and mobile networks (e.g., GSM/GPRS, UMTS, Mobile WiMAX, LTE/LTE-Advanced). The second objective is seamless provisioning of services from the service operator's point of view. The third objective is support to different types of mobility, such as terminal mobility (i.e., one terminal with multiple IP addresses), user mobility (one person with multiple user terminals/devices), and session mobility (one user with multiple terminals in sequence or parallel). The fourth objective is ubiquity of service availability; that is, users should be able to enjoy virtually any application or service, anytime, on any end-user terminal, from any location. However, certain limitations are part of specific access networks, such as QoS, which is different in various fixed and mobile access networks (e.g., Ethernet, Wi-Fi, 3G, 4G). The last objective for FMC (according to the NGN framework) is support for multiple user identifiers and AAA mechanisms. Different identifiers are available for end users such as E.164 telephone numbers; international mobile subscriber identity (IMSI), which is directly related to mobile subscriber ISDN (MSISDN; i.e., a mobile E.164 number); URI for different types of protocols (e.g., HTTP and SIP); tel URI; SIP URI; and so forth. Finally, all these FMC objectives in NGNs are addressed via the specification, standardization, and implementation of IMS as the fundamental technology for the convergence of IP multimedia services to a single platform.

5.8 IP Multimedia Subsystem (IMS)

IMS is a crucial part of the service stratum of NGNs implemented by fixed and mobile telecom operators. Although different SDOs started the work on IMS at the beginning of the 2000s, the main work was carried out by ETSI and then transferred to 3GPP. So, IMS was first standardized within 3GPP release 5 as an application development environment, while 3GPP release 7 retargeted the IMS for telephony replacement in all IP-based networks. The common IMS was finally standardized in 3GPP release 8 as a unified standard that implemented all different requirements from other SDOs including ITU, 3GPP2, TISPAN, CableLabs, and so forth. Hence, the 3GPP standardization of IMS is included as an integral part of the NGN service stratum [46], and it also is considered to be a legacy approach for FMC deployments [47]. 3GPP continued with IMS development in the releases that followed release 8 [48].

IMS uses SIP [3] as the signaling protocol for different services, including multimedia session services and other services such as presence and message exchange services. IMS for NGN supports SIP-based services, but also supports PSTN/ISDN simulation and emulation services based on use of twisted-pair telephone access in the last mile or last meters [46].

The IMS architecture is shown in Figure 5.16. The main functional entities in the IMS architecture are called call session control functions (CSCFs): Proxy-CSCF (P-CSCF), Serving-CSCF (S-CSCF), and Interrogating-CSCF (I-CSCF).

Besides the CSCF entities, the "core" part of IMS [46] includes three additional functional entities. The first one is the breakout gateway control function (BGCF), which is used for processing of user requests for routing from an S-CSCF for the situations when it cannot use session routing based on DNS or ENUM/DNS (e.g., cases with PSTN termination of a given call from an IMS user). If a breakout occurs in the same IMS network, then BGCF selects a media gateway controller function (MGCF) for interfacing with the PSTN. If the routing in BGCF results in breakout into another network, then BGCF forwards the request to the I-CSCF node in the other IMS network.

The second functional entity is the MGCF, which performs signaling translation between SIP and ISUP (the ISDN user part, from SS7 signaling system in the PSTN). The third functional entity is the multimedia resource function controller (MRFC), which interprets the information coming from the application server and the S-CSCF and uses that information for control of the media stream. MRFP is a user plane node that provides mixing of media streams (e.g., for multiple receiving

5.8 IP Multimedia Subsystem (IMS)

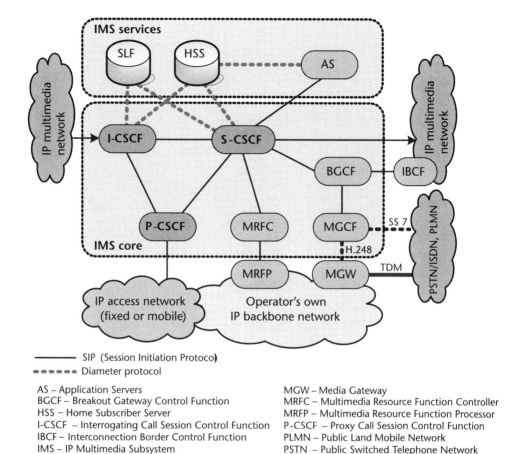

Figure 5.16 IMS architecture.

parties). It can also be a source for the media stream (e.g., for multimedia announcements) and provides media stream processing as well as floor control (i.e., management of access rights).

Several other entities are used for interconnection of the IMS to access networks and user information, including the home subscriber server (HSS), subscriber location function (SLF), application server (AS), and interconnection border control function (IBCF).

HSS is the main user database in the IMS architecture. It contains home subscriber profiles and subscription information, provides authentication and authorization of the users via CSCF entities, and contains information about the network location of the user (i.e., the IP address). SLF is a resolution function in IMS that provides information on HSS that has information for a given IMS user. It can be queried by I-CSCF, S-CSCF, or ASs. However, SLF is not required in networks that have a single HSS.

AS hosts and executes a given service by using SIP for communication with the S-CSCF. IBCF is used as a gateway to external networks. It is in fact the session border controller (SBC) that provides a firewall and network address translation (NAT) functionalities in the IMS architecture.

5.8.1 Proxy-CSCF (P-CSCF)

In terms of IMS-based signaling, the P-CSCF is the first contact point for the user equipment (UE; e.g., mobile terminal, fixed terminal). It acts as a SIP proxy and it interacts with the admission control subsystem to provide authorization only of media components that can be provided with an appropriate QoS level. The IP address of the P-CSCF is provided by the access network during the configuration phase of the user terminal or via DHCP. The P-CSCF may also behave as a user agent; that is, in certain conditions it may in fact terminate and independently generate SIP transactions (e.g., to handle the case of a mobile terminal that goes out of the mobile network coverage area). This entity is also an end point of security associations between the IMS and the UE (with IMS capabilities), with the goal of maintaining confidentiality of SIP sessions. The P-CSCF for a given terminal can be located in a home network or in a visited network (e.g., in the case of roaming).

5.8.2 Serving-CSCF (S-CSCF)

The central node in the IMS is the S-CSCF, which acts as a SIP node with several SIP functionalities. When acting as a SIP server, the S-CSCF performs session control services for the UE and maintains the overall sessions. Regarding the UE registrations, the S-CSCF behaves as a SIP registrar by accepting registration requests and making the registration information available via HSS (i.e., the location server). So, it binds the user location (determined with the IP address) with the SIP address. S-CSCF is always located in the home network of the subscriber and uses the Diameter protocol to access the HSS database to obtain user profiles (the user profile cannot be changed by S-CSCF). In the opposite direction, S-CSCF uploads user to S-CSCF associations to the HSS. Since it is a central node in the IMS architecture, the S-CSCF enforces the established policies of the telecom operator (e.g., provides forwarding of SIP messages between IMS nodes and performs routing services based on ENUM).

For the purpose of load balancing and higher availability of the IMS functions, the IMS architecture may include several S-CSCFs. In such a case, when the HSS is queried for a given user, it assigns one of the deployed S-CSCF nodes in the operator's network.

S-CSCF may also behave as a SIP proxy server or a SIP user agent (e.g., for independent initiation or termination of SIP sessions) and support interaction with service platforms via the SIP-based interfaces.

5.8.3 Interrogating-CSCF (I-CSCF)

I-CSCF is a SIP location function located at the edge of an IMS network. It is the first contact for all incoming messages from other IMS-based networks, and therefore its IP address is published in DNS. For each SIP request from another network, I-CSCF queries HSS to retrieve the address of the S-CSCF that is registered for that user. Further, I-CSCF performs forwarding of SIP messages to the S-CSCF address obtained from the HSS. It also generates charging records that are used in billing arrangements between IMS-based network operators.

5.8.4 User Identities in IMS

IMS is used in all-IP networks that are standardized according to the NGN framework. So, IMS as a main all-IP signaling platform for telecom operators must support different types of identities used in the PSTN/ISDN and the Internet. Generally, all user identities in IMS can be grouped into public and private user identities (i.e., IDs) [48].

Every IMS user must have an IP multimedia private user ID (IMPI), which is stored in the HSS and is globally unique. It is used to identify the user from the network perspective, and typically used for registration and AAA functions by the user's home operator. The IMPI must have the form of a network access identifier (NAI) [49]. For example, "user1@telecom1.example.com" is a valid IMPI. This ID is only authenticated by the network at network registration (or re-registration) of the user equipment. It is valid during the user's subscription (e.g., based on SLA) within its home network, and hence the user cannot modify it. Optionally, the IMPI may be present in charging records.

The IP multimedia public identity (IMPU) is the user ID that can be used by any user for requesting communications to other users (e.g., similar to E.164 telephone numbers in PSTN or email addresses in Internet). A single user may have multiple IMPUs per single IMPI. The IMPU can be shared with another phone, so two or more users (e.g., with a single subscription) can be reached with the same identity (Figure 5.17).

IMS users must be able to receive and initiate sessions from/to many different networks (e.g., GSM, PSTN/ISDN, UMTS, LTE, WiMAX, PON networks, cable networks), so their IDs they must be reachable from both CS and PS networks. Therefore, both user ID types in IMS (IMPI and IMPU) are URIs and typically have the form of a tel URI (used to express traditional E.164 telephone numbers in URL syntax) [50] or SIP URI (for communication with Internet clients) [3].

On the UE's side, an IMS subscriber identity module (ISIM) securely stores one IMPI. However, if neither an ISIM nor a USIM (UMTS SIM) is present in the UE, then IMPI is stored in the IMC (IMS Credentials) application. In practical IMS use, each public user ID may have one or more globally routable user agent URIs (GRUUs). The GRUU is an identity that identifies a unique combination of IMPU and established UE instance (to ensure that a SIP request is not forked to another UE registered with the same IMPU). One may distinguish two types of GRUUs:

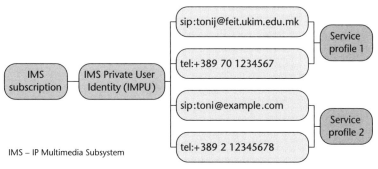

Figure 5.17 Relationship between different user IDs in IMS.

- Public-GRUU (P-GRUU): P-GRUU is a persistent ID that makes public the IMPU.
- Temporary-GRUU (T-GRUU): T-GRUU contains a URI that reveals the IMPU of the user, and it is valid until the contact is explicitly deregistered or the existing registration expires.

Each P-GRUU and T-GRUU pair is associated with one IMPU and single UE. During subsequent re-registrations of the UE, the same P-GRUU should be assigned to the UE, but each time a new and different T-GRUU should be generated and assigned. However, all previously generated T-GRUUs remain valid after a re-registration. For a given IMPU a current set of the P-GRUU and all T-GRUUs that are currently valid is referred to as the GRUU set. In the case when a UE registers with multiple IMPUs, then a different GRUU set is associated with each IMPU.

In general, IMS is used for signaling related to different services, such as presence, messaging, conferencing, and group service capabilities, for which it defines public service identities (PSIs). Such identities are different from the IMPUs in that they identify existing services (not users), which are hosted by ASs. For example, a messaging service may use a PSI (e.g., sip:messaging.list@example.com) to which the users have established a session to be able to send and receive messages from other participants in the given session of that service.

5.9 Discussion

The Internet developed separately from the traditional telecommunication technologies for voice and TV broadcast services. However, its design approach, which separates applications/services from underlying transport technologies, has led to success of the Internet paradigm on a global scale, consequently changing the telecommunication concept toward higher flexibility in applications innovation and convergence of all services to IP networks and Internet technologies. However, the native Internet was lacking end-to-end QoS support and established global standardization and harmonization. Therefore, ITU-T has taken the role of developing an NGN, which systematically introduced a globally standardized QoS framework and signaling systems (e.g., IMS) in the Internet. As a result, the Internet has become more complex than before, while telecommunications have become more flexible than before in network architectures as well as service creation and offerings, thus creating a win–win situation. Such development is possible due to the development and deployment of fixed and mobile broadband access technologies, and vice versa. The convergence of telecom and Internet sectors is a driving force for speedy development and deployment of broadband access, which positively influences all aspects of everyday life and society in general.

References

[1] ITU-T Recommendation Y.2001, "General Overview of NGN," December 2004.
[2] ITU-T, ITU-T NGN FG Proceedings, 2005.
[3] J. Rosenberg et al., "SIP: Session Initiation Protocol," RFC 3261, June 2002.

[4]   V. Fajardo et al., "Diameter Base Protocol," RFC 6733, October 2012.
[5]   ITU-T Recommendation Y.2011, "General Principles and General Reference Model for Next-Generation Networks," October 2004.
[6]   ITU-T Recommendation M.3060/Y.2401, "Principles for the Management of Next Generation Networks," March 2006.
[7]   ITU-T Recommendation Y.2720, "NGN Identity Management Framework," January 2009.
[8]   ITU-T Recommendation Y.2006, "Description of Capability Set 1 of NGN Release 1," February 2008.
[9]   ITU-T Recommendation Y.2007, "NGN Capability Set 2," January 2010.
[10] ITU-T Recommendation Y.2002, "Overview of Ubiquitous Networking and of Its Support in NGN," October 2009.
[11] ITU-T Recommendation V.90, "A Digital Modem and Analogue Modem Pair for Use on the Public Switched Telephone Network (PSTN) at Data Signaling Rates of Up to 56 000 bit/s Downstream and Up to33 600 bit/s Upstream," September 1998.
[12] ANSI T1.413 issue 2, "Network and Customer Installation Interfaces—Asymmetric Digital Subscriber Line (ADSL) Metallic Interface," 1998.
[13] ITU-T Recommendation G.992.1, "Asymmetric Digital Subscriber Line (ADSL) Transceivers," June 1999.
[14] ITU-T Recommendation G.992.2, "Splitterless Asymmetric Digital Subscriber Line (ADSL) Transceivers," June 1999.
[15] ITU-T Recommendation G.992.3, "Asymmetric Digital Subscriber Line Transceivers 2 (ADSL2)," April 2009.
[16] ITU-T Recommendation G.992.4, "Splitterless Asymmetric Digital Subscriber Line Transceivers 2 (Splitterless ADSL2)," July 2002.
[17] ITU-T Recommendation G.9701, "Fast Access to Subscriber Terminals (G.fast)—Physical Layer Specification," December 2014.
[18] ITU-T Recommendation G.987.1, "10-Gigabit-Capable Passive Optical Networks (XG-PON): General Requirements," January 2010.
[19] ITU-T Recommendation G.989.1, "40-Gigabit-Capable Passive Optical Networks (NG-PON2): General Requirements," March 2013.
[20] IEEE 802.3av-2009, "Local and Metropolitan Area Networks—Specific Requirements Part 3: Carrier Sense Multiple Access with Collision Detection (CSMA/CD) Access Method and Physical Layer Specifications Amendment 1: Physical Layer Specifications and Management Parameters for 10 Gbit/s Passive Optical Networks," October 2009.
[21] R. Sanchez, L. Raptis, K. Vaxevanakis, "Ethernet as a Carrier Grade Technology: Developments and Innovations," *IEEE Communications*, September 2008.
[22] T. Janevski, *NGN Architectures, Protocol and Services*, John Wiley & Sons, April 2014.
[23] MEF Technical Specification, MEF 6.1, "Ethernet Services Definitions—Phase 2," April 2008.
[24] IEEE 802.1Q-2005, "Virtual Bridged Local Area Networks," May 2006.
[25] IEEE 802.1ad-2005, "Virtual Bridged Local Area Networks—Amendment 4: Provider Bridges," May 2006.
[26] IEEE 802.1ah-2008, "Virtual Bridged Local 14 Area Networks—Amendment 6: Provider Backbone Bridges," August 2008.
[27] IEEE 802.16e-2005, "Part 16: Air Interface for Fixed and Mobile Broadband Wireless Access Systems Amendment 2: Physical and Medium Access Control Layers for Combined Fixed and Mobile Operation in Licensed Bands and Corrigendum 1," February 2006.
[28] IEEE 802.16m-2011, "Part 16: Air Interface for Broadband Wireless Access Systems, Amendment 3: Advanced Air Interface," May 2011.
[29] WiMAX Forum, "WiMAX Forum Network Architecture: Detailed Protocols and Procedures—Base Specification," 17 April 2012.

[30] P. Faltstrom, M. Meallingm, "The E.164 to Uniform Resource Identifiers (URI) Dynamic Delegation Discovery System (DDDS) Application (ENUM)," April 2004.

[31] ETSI Technical Specification 184 011, "Requirements and Usage of E.164 Numbers in NGN and NGCN," February 2011.

[32] ETSI Technical Specification 184 002, "Identifiers (IDs) for NGN," October 2006.

[33] ITU-T Recommendation E.800, "Definitions of Terms Related to Quality of Service," September 2008.

[34] ITU-T Recommendation E.803, "Quality of Service Parameters for Supporting Service Aspects," December 2011.

[35] ITU-T Recommendation G.1000, "Communications Quality of Service: A Framework and Definitions," November 2011.

[36] ITU-T Recommendation P.800, "Methods for Subjective Determination of Transmission Quality," August 1996.

[37] ITU-T Recommendation I.350, "General Aspects of Quality of Service and Network Performance in Digital Networks, Including ISDNs," March 1993.

[38] R. Braden et al., "Resource ReSerVation Protocol (RSVP)—Version 1 Functional Specification," RFC 2205, September 1997.

[39] F. Baker et al., "Aggregation of RSVP for IPv4 and IPv6 Reservations," September 2001.

[40] D. Awduche et al., "RSVP-TE: Extensions to RSVP for LSP Tunnels," RFC 3209, December 2001.

[41] K. Nichols et al., "Definition of the Differentiated Services Field (DS Field) in the IPv4 and IPv6 Headers," RFC 2474, December 1998.

[42] F. Le Faucheur, "Aggregation of Resource ReSerVation Protocol (RSVP) Reservations over MPLS TE/DS-TE Tunnels," RFC 4804, February 2007.

[43] ITU-T Recommendation Y.1540, "Internet Protocol Data Communication Service—IP Packet Transfer and Availability Performance Parameters," March 2011.

[44] ITU-T Recommendation Y.1541, "Network performance objectives for IP-based services," December 2011.

[45] ITU-T Recommendation Y.1545, "Roadmap for the quality of service of interconnected networks that use the Internet protocol," May 2013.

[46] ITU-T Recommendation Y.2021, "IMS for Next Generation Networks," September 2006.

[47] ITU-T Recommendation Y.2808, "Fixed mobile convergence with a common IMS session control domain," June 2009.

[48] 3GPP TS 23.228, "IP Multimedia Subsystem (IMS); Stage 2 (Release 13)," December 2014.

[49] B. Aboba, M. Beadles, "The Network Access Identifier," RFC 2486, January 1999.

[50] H. Schulzrinne, "The tel URI for Telephone Numbers," RFC 3966, December 2004.

CHAPTER 6
Internet Technologies for Mobile Broadband Networks

The Internet was initially created for fixed-access networks because mobile and wireless digital networks were not present until the 1990s. So, the main concept of Internet routing was based on IP addressing and routing schemes, in which each IP packet is routed to the destination IP network identified by the network prefix of the destination IP address in the packet header. In contrast, in mobile networks the hosts are moving within a given IP network (from one cell or base station to another one, performing handovers) or between different IP networks. When a host is moving within the same IP network, the process is referred to as *micro-mobility*. When a given mobile host changes the IP network to which it is attached, however, the process is referred to as *macro-mobility*. The main problem with macro-mobility management in all-IP networks comes from the roles of IP addresses. In particular, the current Internet architecture is based on using IP addresses in two distinctive roles:

- Locator role: From the network point of view, an IP address defines the current topological location of an interface by which a host is attached to the network. So, when a host or node moves and attaches its network interface to a different network location (IP network with different network prefix), the IP address associated with the interface changes.
- Identifier role: From an application point of view, an IP address identifies a host and its interface to the Internet. So, the IP address is used as an identifier for the peer host, and it is expected that this identifier does not change as long as the association is active. This is realized through the socket by using API.

The dual role of IP addresses causes problems with Internet mobility, because its locator role requires a change of the IP address as a mobile host moves to another IP network, whereas its identifier role requires the IP address to remain the same as long as the socket for a given connection (that is bound to the IP address) is open. But, when a mobile host moves from one wireless IP network to another wireless IP network (identified with another network prefix in IP addresses), the host must change its IP address to the new one that belongs to the new wireless IP network to which it has performed a handover. The goal is to receive incoming packets from the corresponding host (that is, the host on the other side of the connection). A changed IP address, however, implies that the open socket (bound to the IP address of the old wireless IP network) must be closed and another socket (bound to the new IP address that belongs to the new wireless IP network) must be created. However, closing a socket means termination of the connection, and hence there will be no seamless handover on the network layer (i.e., no mobility, because mobility means handovers without connection interruption or termination).

6.1 Mobile IP

With the development of digital mobile networks at the beginning of the 1990s, work began on standards for Internet mobility management, which resulted in standardization of Mobile IP (MIP), which was created for two different versions of the Internet protocol (IPv4 and IPv6): Mobile IPv4 (or simply Mobile IP) and Mobile IPv6.

6.1.1 Mobile IPv4

According to the introductory discussions in this chapter, Internet mobility faces the problem of the dual role of IP addresses. To solve this issue, Mobile IP (MIP) [1] uses two IP addresses for each mobile host:

- Home address: This is a permanent address assigned to the mobile node (MN), which belongs to its home IP network.
- Care-of-address (CoA): This is a temporary IP address that is "assigned" to the mobile node when it moves into another network (i.e., visited network). In particular, CoA is the termination point of a tunnel toward a mobile node, for packets forwarded to the mobile node while it is away from home.

With the goal of managing the home address and CoA, Mobile IP uses an approach similar to the one used in GSM mobile networks in the 1990s. It is based on the home location register (HLR), which stores information for all mobile subscribers of the home network, and a visitor location register (VLR), which contains information for all subscribers that are attached to the mobile network, including home subscribers and roamers. Similar to HLR and VLR in GSM, the Mobile IP protocol [1] introduces two agents:

- Home agent (HA): This is a router on a MN's home network that intercepts IP packets addressed to the mobile node's home address and then tunnels them to the CoA of the mobile node in the foreign, or visited, network (when mobile node is roaming into another IP network).
- Foreign agent (FA): This is a router on a MN's visited network that provides routing services to the MN while registered, by detunneling and delivery of datagrams to the MN that were tunneled by its HA. In the opposite direction, for datagrams sent by a MN, the FA may serve as a default router for

all registered mobile nodes in that IP network. Typically, the CoA is an IP address that is assigned to one of the interfaces of the FA node.

Generally, the protocol may use two different types of CoAs: (1) a CoA that is an address of a FA with which the mobile node is registered or (2) a colocated CoA that is an externally obtained local address that the MN has associated with one of its own network interfaces.

MIP has three main phases that define its operation:

- Agent discovery: This is the process of discovery of an agent by the MN (e.g., the FA in a foreign network), which determines whether it is connected to its home network or to a foreign network. With the goal of keeping the MIP on the network layer, the agent uses ICMP router advertisement messages and extends it with an agent advertisement message. An agent solicitation message is completely identical to an ICMP router solicitation message, but with the TTL field in the IP header set to 1 (only one hop between the agent and mobile nodes). However, agent solicitation and agent discovery are not needed when the link layer in the mobile access network provides such functionality.
- Registration: It refers to binding of CoA that is obtained in a visited network with the home address by using the registration message from the MNs to the HA in the home network. A registration request, as well as the reply to it, is sent by using UDP port 434. Each new CoA in a new foreign network (due to mobility of the host) requires a new registration request to bind the new CoA with the home address of the mobile host. So, registration is a method for registering/renewing CoA addresses and for requesting forwarding services by the HA when the mobile nodes are in foreign networks and deregistering them when they return home.
- Data transfer and routing: The HA and FA must support tunneling to provide data transfer to a MN located in a visited network. When the MN is in the home network, it operates without mobility support; that is, it operates in the same way as fixed hosts or routers. But, when the MN is registered on a foreign network, then there are two cases:
  - If the MN is using CoA, then it may use the FA node as a first-hop router, and the MAC address of the FA can be learned from the FA's advertisement messages. Then all messages from a correspondent node (CN) addressed to the MN reach the home network of the MN and are further tunneled from the HA to the FA, which finally forwards them to the MN. In the opposite direction, the MN sends packets directly addressed to the CN by using typical Internet routing. This case results in so-called triangle routing, as shown in Figure 6.1.
  - If the MN uses a colocated CoA (i.e., the MN is directly registered with its HA), then the MN should select its default router from among those advertised in any ICMP router advertisement messages that it receives for which its obtained CoA and the router address match the same network

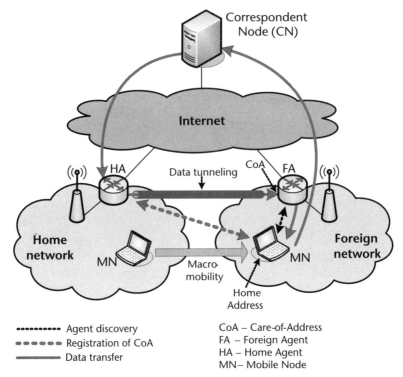

Figure 6.1 Mobile IPv4 operations.

prefix (obtained from router advertisement messages, if present, but also can be obtained through other mechanisms that are beyond the scope of MIP). In this case, the routing is also referred to as double-crossing.

Triangle routing and double-crossing are inefficient mechanisms. Several mechanisms have been created for solving such problems in MIPv4, such as micro-mobility solutions that hide the movement of the mobile node within a given foreign network by placing the FA at the gateway node of the mobile network (e.g., cellular IP, HAWAII) [2]. Also, MIPv4 may experience an ingress filtering problem (if it is applied at gateway nodes of the foreign network) because mobile nodes use as a source address the home address, which has a different network prefix than IP addresses in the foreign network (because the home address belongs to the home network). However, MIPv6 solves the inefficient triangle routing problem present in MIPv4, as well as the ingress filtering problem.

Note that the initial standard for Mobile IP was created in 1996 [1]; however, the last revision of it is done in 2010 [3], driven by the all-IP nature of 4G mobile networks. Note, however, that it was influenced by the fact that in 2010 around 0.5% of IP addresses were IPv6, whereas 99.5% were IPv4. So, mobility support in the second decade of the 21st century is mainly needed for IPv4 on the network layer (or it is provided on the OSI-2 data-link layer, as is typically done in 3G and 4G mobile networks for horizontal handovers within the same mobile network).

6.1.2 Mobile IPv6

The MIPv6 protocol was initially standardized in 2004 with RFC 3775 and updated in 2011 to its current standard, RFC 6275 [4]. It allows mobile hosts or nodes to remain reachable while moving within IPv6-based networks.

Similar to MIPv4, in MIPv6 a MN maintains two IP addresses: home address and CoA (when the MN is in a foreign network). While the MN is away from its home network, it sends information about its current location to the home agent. In the home network, the HA intercepts packets addressed to the mobile node and tunnels them to the MN's current location in a foreign network. However, only initial packets from a CN toward the MN are routed to the MN's home address and then tunneled to its CoA in the foreign network. MIPv6, however, provides the possibility for the IPv6-based CN to cache the binding of the MN's home address and CoA and then to send packets directly to the MN (i.e., to the CoA).

So the major differences between MIPv4 and MIPv6 are as follows:

- No FA: There is no FA in MIPv6, but to provide that feature MIPv6 requires every MN to support IPv6 decapsulation, address autoconfiguration, and neighbor discovery.
- Packet delivery: The MIPv6 MN uses CoA as its source address in foreign links (unlike MIPv4, which uses its home address), so there is no ingress filtering problem. The CN uses an IPv6 routing header rather than IP encapsulation; hence, "route optimization" is naturally supported.

So, MIPv6 uses several features present in IPv6 that help to optimize its operations. In that manner, the address autoconfiguration feature of IPv6 is used either in a stateless mode (i.e., based on the network prefix advertised by routers in the foreign network and interface ID, such as MAC address, of the mobile node) or stateful mode (i.e., with DHCPv6). For finding routers and discovery of each other's presence, the MN uses the neighbor discovery mechanism of IPv6. In that manner MN and routers in the mobile access network determine each other's link addresses and maintain reachability information. For route optimization MIPv6 uses IPv6 extension headers, such as the routing header and destination options header in IPv6. So, new messages in MIPv6 are defined with IPv6 destination options, which are used to carry additional information targeted to be used by the CN. Four new destination options are defined:

- Binding Update: This is used by the MN to inform the HA (or any CN) about its current CoA.
- Binding Acknowledgement: This is used to acknowledge the receipt of a binding update message from the other end (i.e., from CN).
- Binding Request: This is used by any node to request a MN to send a binding update with the current CoA.
- Home address: This is used in a packet sent by a MN to inform the other end (i.e., CN) about the home address of the MN.

To support its operations, the MIPv6 maintains several data structures, which include binding cache, binding update list, and home agent list.

MIPv6 operation starts with HA registration (similar to MIPv4). However, a MIPv6 MN may perform stateless autoconfiguration to obtain automatically its CoA in a foreign network. After that, the MN registers its CoA with the HA by using the Binding Update destination option. On the other side, the HA uses so-called "proxy" neighbor discovery and additionally replies to neighbor solicitations on the MN's behalf.

While a MN is attached to some foreign network (i.e., foreign link [4]), it may be addressable through one or several CoAs. In this case the CoA is an IP address that has the subnet prefix of a particular foreign network (the prefix is obtained through the periodically sent router advertisements). The association between the CoA and home address is referred to as "binding." So, the MN selects one of the addresses in the foreign network as a primary CoA and registers it with a router on its home network (i.e., home link) requesting that router to operate as the HA for the MN. With the goal of triggering the Binding Acknowledgment (ACK), the MN sets the ACK bit in the Binding Update message sent to the HA. The MN continues to periodically retransmit the Binding Update until it receives the ACK. Also, MN may send Binding Updates to the CN.

How is the HA discovered by the MN? The MN sends the Binding Update to the HA anycast address (which is an IPv6 address). Each of the HAs may accept or reject a registration request from the MN. However, Binding Updates may travel across different IP networks, so the MN must provide strong authentication (e.g., IPsec for Binding Update with HA, and cookie-based authentication with the CN).

Generally, MIPv6 provides two possible modes of communication between the MN and the CN:

- Bidirectional tunneling: In this case MIPv6 support is not required from the CN. Packets from the CN to the MN are routed through the home network and tunneled from the HA to the MN. The HA uses proxy neighbor discovery to intercept any IPv6 packets addressed to the MN's home address, and then each such packet is tunneled (with IPv6 encapsulation) to the MN's primary CoA. On the other side, packets to the CN are tunneled from the MN to the HA (this is noted as "reverse tunneling") and then routed normally (i.e., with well-known Internet routing principles) to the CN.

- Route optimization: This is the typical mode of operation in which the MN and CN are MIPv6 capable, so the CN node sends packets directly to the CoA (of the MN) using standard Internet routing (and vice versa, in the opposite direction from the MN to the CN). Even in the case of end-to-end QoS provisioning, this MIPv6 mode provides the same possibilities as for fixed nodes.

Bidirectional tunneling in MIPv6 is not efficient in terms of the routing of IP packets between the MN and CN (similar to MIPv4 in the CN-to-MN direction), but it is convenient for cases when CN does not support MIPv6 (or its IP network is based on IPv4). In all other cases, route optimization is the primary choice for

6.1 Mobile IP

MIPv6 operations due to its efficiency. In such cases, during the transfer of packets the MIPv6 provides route optimization, as shown in Figure 6.2.

The MIPv6 provides handover mechanisms on protocol layer 3, the IP layer. In the standard, the handovers are related to so-called movement detection. However, most of the handovers between cells (which cover a small geographic area by a single transceiver) are solved using link layer, or layer 2 (L2), techniques that are specific for a given radio access technology. In that manner, when the on-link subnet prefix changes (after the L2 handover), the L3 handover follows. The L3 handover could be, for example, a change to the access router subsequent to change in

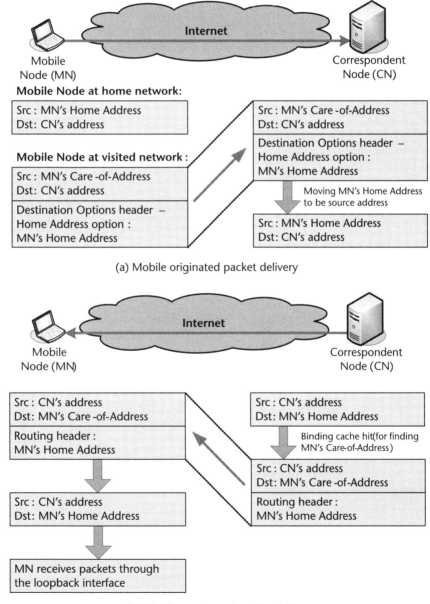

Figure 6.2 Mobile IPv6 operation: (a) MN terminated packet delivery; (b) MN originated packet delivery.

the wireless access point or base station. Generic movement detection uses neighbor unreachability detection to detect when the default access router is no longer reachable, so the MN must discover a new default router (that is typically over a new radio access link). But, such detection occurs only when the MN has packets to send; if there is no data to be sent and there is an absence of routing advertisements or certain indications from the link layer, the MN may become unaware that L3 handover has occurred. Therefore, the MN must use other information when it is available to it (such as information received on the link layer from the given radio access link).

Regarding the standardization aspects of MIPv6, several other standards from the IETF extend the mobility or implementation aspects of the protocol. While Internet mobility (on layer 3) in IPv6 networks is well standardized with the basic standard [4], several standardized enhancements are available for the following:

- Real-time mobility support (e.g., for VoIP): This includes Fast MIPv6 handovers (FMIPv6) [5] for fast detection of the change of the subnet link (i.e., new network prefix) and Hierarchical Mobile IPv6 (HMIPv6) [6], which uses mobility anchor points (MAPs) as domain-wise HA proxies (the MN uses bidirectional tunneling with MAP, so intradomain mobility, i.e., micromobility, is not visible to the basic MIPv6, whereas interdomain handovers use basic MIPv6 operations with Binding Updates).
- Carrier-grade mobility support for MIPv6 unaware nodes: The main standard is Proxy MIPv6 (PMIPv6) [7], which is standardized as in optional implementation in 3GPP core networks, but is targeted to be used in 4G mobile IPv6 environments. With PMIPv6 the network is responsible for managing mobility on behalf of the mobile host, because standard MIPv6 requires client functionalities in the protocol stack of the mobile nodes, which is hard to do in practice (e.g., there are many older mobile terminal types). So, a PMIPv6 node in the mobile core network (e.g., evolved packet core from 3GPP) operates signaling with the HA of the MN, thus hiding the micromobility handovers from the MIPv6.

MIPv6 is prepared for deployment and it is based fully on IPv6 features. It is an end-to-end paradigm, however, which directly creates a conflict with mobile operators based on 3GPP technologies (dominant globally) that use link layer handovers and technology-specific mobility management while transferring all data (voice, video, data) over end-to-end all-IP mobile networks, in which the main signaling protocol is SIP (referred to as SIP mobility). PMIPv6 may serve as a "mediation protocol" for deployment of MIPv6.

Note that there are also two other important issues. The other issue is the penetration of IPv6, which is not going to happen (at least to become dominant one) during the 4G era (i.e., 2010–2020). The full potential of IPv6 in the mobile world may be expected during the 5G era (i.e., after 2020). But, MIP was initially implemented in Mobile WiMAX from its first release and in some 3GPP2 standards from the 3G era such as the wireless IP architecture [2]. Finally, note that the main issue of MIPv6 implementation is to implement it in end devices, such as smartphones.

However, the 4G reality has shown that for 3G and 4G technologies from 3GPP (e.g., UMTS/HSPA, LTE/LTE-Advanced) MIP/MIPv6 is not considered to be first-choice technology (mobility is managed on the link layer, with specific signaling on defined interfaces in access and core networks), while for Mobile WiMAX the MIP is mandatory, but in reality the 3GPP technologies are dominant.

As a summary, unlike other Internet technologies that were developed separately from the legacy telecommunication networks, the main MIP philosophy (HA and FA) was influenced by GSM user databases (HLR and VLR), which were groundbreaking technology that provided global roaming and was deployed worldwide in the 1990s. So, whether MIP and especially MIPv6 will come into the focus for macro-mobility management worldwide (besides being extensively used in many research activities on all-IP mobility, and implemented in mobile WiMAX and 3GPP2 mobile networks) depends mainly on the further 3GPP evolution of the core network for 5G, due to the global dominance of 3GPP in the mobile world.

Several other mobility solutions for the Internet besides MIP have been researched and one of the most interesting (although it is experimental) is the Host Identity Protocol (HIP).

6.2 Host Identity Protocol (HIP)

The dual role of IP addresses limits the flexibility of IP environments in terms of mobility management. There are exactly two name spaces in the current Internet: IP addresses and domain names (there is currently no third one). The HIP [8] and its architecture [9] define a new, additional Internet name space. HIP proposes architectural changes to the IP protocol stack by placing a new protocol layer between the IP and TP, as shown in Figure 6.3.

The new name space introduced by the HIP architecture consists of host identities (HIs). Each HI is a public key (its ownership is proven by using a private key) that is presented as a hash value (obtained with hash functions that map data, such as a string, of an arbitrary size to data of a fixed size, and it is not reversible):

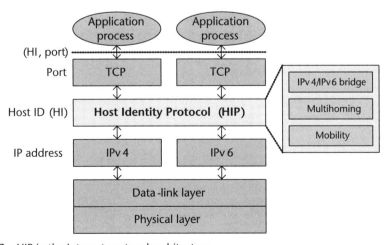

Figure 6.3 HIP in the Internet protocol architecture.

- IPv6: In this case the HI value is called the host identity tag (HIT) and, like IPv6 addresses, has a length of 128 bits.
- IPv4: In this case the HI value is referred to as the local scope identifier (LSI) and, like IPv4 addresses, has a length of 32 bits.

Connections between two end points are established between HIs instead of IP addresses. So, the HIP socket is bound to an HIT (in IPv6 hosts) or LSI (in IPv4 hosts). The new protocol layer introduced with HIP is located between the IP and transport layers. It provides translation of HIs to IP addresses (for sending packets via an application) and vice versa (for receiving packets). So, HIP is a kind of new waist in the traditional TCP/IP (and UDP/IP) protocol stack, but it also provides interoperability between IPv4 and IPv6 due to its HI-to-IP address (and vice versa) translation functions.

Regarding the structure of the HIP packet, it is almost identical to an IPv6 header, but instead of IPv6 addresses, the sender's HIT and receiver's HIT are used. Further, HIP requires connection establishment via a so-called four-way handshake. This creates a HIP association between the two end points by providing mandatory authentication of hosts. For security purposes HIP uses IPsec's encapsulated security payload (ESP) security associations (SAs), one in each direction. However, if the HI is unknown to the host (this is called opportunistic mode) then the destination IP address is used until the host learns the other party's HI. The HIP uses a new IPsec mode called BEET (bound end-to-end tunnel). BEET enhances the existing ESP tunnel as well as transport modes and provides end-to-end tunnels. It is intended to support new uses of ESP such as for mobility and multihoming (e.g., when the host has multiple locators simultaneously rather than sequentially as in the case of mobility) [10]. In regular transport mode the IP header is not changed, whereas in regular tunnel mode an outer IP header is created at the sending side and discarded at the receiving side of the connection (i.e., association). In the BEET mode the original IP header is replaced with another one on both input and output.

How does HIP supports mobility? When a host moves to another IP network, it obtains a new IP address (i.e., a new locator), but the HI (as a connection end point) remains the same. So, with HIP each connection is bound to constant HIs that do not change during the host movement, and therefore connections do not break (even when the IP address is changed). However, a mobile host notifies its peer of the new IP address by sending a HIP UPDATE packet containing a locator parameter, which is then acknowledged by the peer [10]. To provide reliability, HIP defines the retransmission mechanism to be used in case an UPDATE packet is lost. On the other side of the connection, the peer authenticates the content of the UPDATE packet based on its signature and the keyed hash of the packet (although the host may decide to rekey its security association, e.g., when using the ESP transport format [11]). However, with the use of HIs, mobility between IPv4 and IPv6 is also supported.

Additionally, HIP also has a standardized HIP mobile router that allows networks to move with HIP hosts and use private IP address space in the moving network. Such a router can be used in heterogeneous wireless and mobile environments, such as 3G, 4G, and Wi-Fi networks, with different address families (IPv4 and IPv6). So, mobile and wireless public networks are one potential market for

HIP deployment (if it is used sometime in the future). With its standardized multihoming features, it can provide high reliability in all-IP environments.

What benefits are possible from the use of HIP? Network operators may benefit from a more controlled network, because HIP requires four-way handshaking before data transfer, which may prevent certain attacks (e.g., DoS). Also, the multihoming features of HIP may contribute to higher resilience (e.g., no single point of failure). In enterprises HIP may be used to provide easier integration of IPv4 and IPv6 environments, as well as additional security. For residential users (i.e., individual users) HIP is suitable for home servers (e.g., for certain smart home or home cloud solutions).

What are HIP's obstacles; that is, why is HIP just an experimental protocol and architecture? Well, HIP is a well-standardized protocol (with RFCs 5201 to 5207) and architecture [9]. However, all RFCs belong to the experimental track, so HIP is just experimental. Why? Well, the Internet has been based on just two name spaces (IP addresses and domain names) from the time of ARPANET, and it has grown significantly during the past couple of decades in terms of the number of nodes and hosts attached to it. With a continuously changing architecture (IP networks come and go, hosts appear and move or disappear), adding a new name space (the HI space, which is needed for HIP) might add uncertainty to the global Internet network—which no one wants to risk. On the other hand, most of the functions provided by HIP are also being provided by other standardized protocols and architectures, such as MIP for mobility management and SCTP on the transport layer for multihoming and multistreaming. So, HIP remains experimental. Its approach may become useful in the future, but not for the present day. When speaking about the current time—the second decade of the 21st century—and all-IP mobility, we should focus on an all-IP packet core in 4G mobile networks as being the most representative.

6.3 All-IP Packet Core for 4G Mobile Networks

3GPP has been dominant in the mobile world on a global scale since the 1990s. Therefore, the standardization of an all-IP concept is an important step toward an all-IP telecom world. Note that there is a worldwide agreement for 4G to be all-IP, as stated in the IMT-Advanced umbrella specification from ITU-R [12].

The introduction to system architecture evolution (SAE) is given in Chapter 5 of this book. The focus in this section is on QoS and mobility management functionalities in 4G (from 3GPP), which are influenced by the architecture of the mobile networks operated by mobile operators. The SAE introduces a so-called "flat architecture" in mobile networks that consists of a terrestrial radio access network and core network, the E-UTRAN and EPC, respectively. The main characteristic of SAE is its simplified mobile network architecture, which consists of only two tiers: (1) base stations (e.g., eNodeBs), and (2) centralized gateways (in the core network).

The SAE architecture suits well the all-IP nature of LTE/LTE-Advanced. Additionally, SAE makes use of separated control and user traffic (i.e., control and user plane); that is, different planes are serviced by different gateways in the core network.

In summary, the main part of the SAE is the EPC, which is the core network for LTE and LTE-Advanced radio access networks.

6.3.1 Evolved Packet Core (EPC)

EPC consists of several types of gateways that are targeted to carry either control or user data traffic, or both. The main functionalities in EPC are in fact regrouped from initial SGSNs and GGSNs as central and edge gateways in 2.5G (GPRS) and 3G (UMTS/HSPA) mobile networks from 3GPP. The main functions are mobility management, QoS support, and AAA functions. To support those functions, there are three main gateway nodes in the EPC:

- Mobility management entity (MME): The MME is the main control element in the EPC, which is responsible for signaling-related mobility management including user tracking, paging procedures, and bearers' activation and deactivation. Because mobility is related to roaming between different mobile networks, MME has functionality for authentication of the user, and therefore it uses a signaling interface for the users' database home subscriber server (HSS). Also, the MME provides interfaces for interconnection with previous 3GPP core networks from 2G and 3G, for handling the mobility between the heterogeneous mobile networks. It is a control-only network node (no user data passes through the MME).

- Serving gateway (S-GW): The S-GW is the main gateway for user traffic (i.e., user plane) within the core network, which establishes bearers for user traffic with eNodeBs (base stations in LTE/LTE-Advanced) on one side, and with the packet data network gateway on the other side (toward the global Internet). Since MME is controlling the mobility, S-GW has a control interface with MME, which is important for bearer switching at handovers (e.g., a mobile terminal makes a handover from one eNodeB to another, thus resulting in a change of the bearer between the S-GW and the eNodeBs).

- Packet data network gateway [i.e., PDN Gateway (P-GW)]: The P-GW provides connectivity from mobile terminals to external packet data networks. P-GW also acts as an anchor point between the 3GPP core network and non-3GPP networks (e.g., Wi-Fi, WiMAX). Also, it performs typical functionalities for the edge routers of a given core network, such as packet filtering, policies enforcement, charging support, and so forth. The P-GW and S-GW are connected via an S5/S8 interface, and both nodes are parts of the so-called SAE gateway (SAE GW).

The EPC architecture shown in Figure 6.4. Besides the three main nodes (MME, S-GW, and P-GW), EPC has two other important control nodes:

- Home subscriber server (HSS): The HSS is a centralized database in the mobile network that contains user-related information, such as location of the user (i.e., the MME to which it is connected to) and the subscriber profile including available services to the user, allowed PDN connections or roaming

6.3 All-IP Packet Core for 4G Mobile Networks

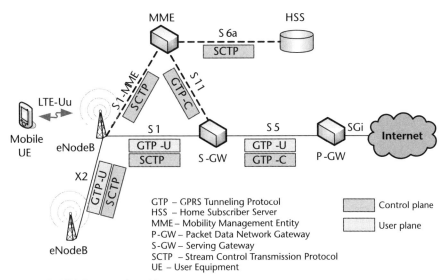

Figure 6.4 SAE/EPC network architecture.

in visited networks, and so forth. User authentication is performed via the Authentication Center (AuC), which is usually a part of the HSS.

- Policy and charging rules function (PCRF): The PCRF is a software-based network node in the EPC that is responsible for policy and charging control (PCC). The PCRF detects the service flow and determines policy rules for it in real time. Also, it enforces the charging policy in the mobile network. In the SAE architecture the PCRF provides PCC information for bearer setup to the enforcement function located in the P-GW, which is named the policy and charging enforcement function (PCEF). When user applications communicate directly with the IMS, the application function (AF) in the service domain (i.e., service stratum according to the NGN terminology) requires the PCRF to apply PCC rules for dynamic policy or charging control. PCRF is also a central node in the general NGN architecture connecting the service stratum and transport stratum.

The main EPC nodes (MME, S-GW, and S-GW) are interconnected with eNodeBs and to each other via so-called interfaces (standardized by 3GPP). Because 4G mobile networks are all-IP (i.e., EPC and E-UTRAN are all-IP networks), all interfaces are based on Internet technologies, that is, standardized IETF protocols (from the network layer up to the application layer).

6.3.2 Protocol Stacks for 4G Interfaces

Although 4G mobile networks by 3GPP are referred to as all-IP, most of the QoS provision and mobility management is based on a different protocol stack being used between different pair of nodes. In general, there are two planes: the user plane, which is used for data originating from or terminating at mobile terminals, and the control plane, which is used to carry signaling traffic, such as AAA signaling, mobility and location management signaling, and service-related signaling. For example, eNodeBs, the base stations for LTE/LTE-Advanced, are connected to each other (to neighboring ones that can hand over connections from one to another) and additionally have interfaces with central nodes MME and S-GW. Each of the interfaces is in fact based on standardized (and sometimes specific) protocols and mechanisms that belong to the Internet technologies. For example, the X2 interface provides connections between eNodeBs and supports mobility for user equipment (UE) in the connected mode. It is a many-to-many interface, so each eNodeB can connect with many eNodeBs. The signaling protocol on the X2 control plane interface is called the X2 Application Protocol (X2AP) and is carried over SCTP/IP to provide reliable transmission, where SCTP is a TCP-based multihoming and multistreaming protocol that provides for the use of parallel streams through different paths between two end points of a communication, thus increasing the reliability needed for signaling traffic.

If one considers the interface between the UE (i.e., mobile terminals) and eNodeB, then it may conclude that it is working on layer 2 (with PDCP over RLC over MAC over the physical layer); then, the eNodeB relays the IP packets into the GPRS Tunneling Protocol (GTP) tunnels between eNodeB and all other nodes in the mobile network to which it is connected to (other eNodeBs, S-GW, P-GW, and MME). Note that it is referred to as GTP-U for transport of user plane traffic and GTP-C for control plane traffic (Figure 6.5). GTP uses the UDP/IP protocol stack for user traffic and SCTP/IP for control traffic on all interfaces between each pair of nodes (except the radio access links). So, the protocol layering in the network nodes of 4G mobile networks is based on tunneling protocols over UDP/IP or SCTP/IP protocol stacks, with an application part (e.g., X2AP is the application part on the X2 interface, which connects two adjacent eNodeBs).

So, to provide more complete coverage of Internet technologies used in 4G mobile networks, one needs to have more details regarding the SCTP and GTP.

6.3.2.1 Stream Transmission Control Protocol (SCTP)

SCTP is used as a transport protocol for the control plane in E-UTRAN and EPC of 4G mobile networks from 3GPP. Why is this needed at all when TCP or UDP can be used? Well, SCTP is primarily needed for its built-in reliability. Although there are application-level control protocols (e.g., DNS, RADIUS) that cope with reliability on the application layer by having at least two different servers (in case one is down) that is not enough for signaling in mobile networks, where handovers occur frequently and there should be no additional delay or unexpected packet losses or stalling of the throughput for signaling that refers to mobility management (e.g., handovers) and QoS support (e.g., reservation of the required resources in the new cell and on the new path to the new eNodeB). Also, the TCP version standardized by the IETF at the end of the 1990s had several other drawbacks that are "fixed" with SCTP. The TCP disadvantages included the following [13]:

- Strict byte-order delivery from one to other peer application that causes so-called head-of-the-line (HOL) blocking when TCP segments are lost and should be retransmitted by the sender.

6.3 All-IP Packet Core for 4G Mobile Networks

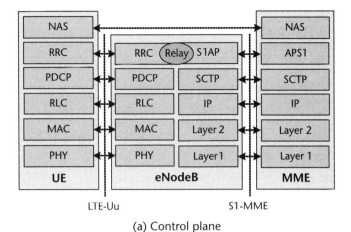

Figure 6.5 Internet protocol stack in 4G mobile networks: (a) control plane; (b) user plane.

- Stream orientation of legacy TCP, without any message boundaries, which makes the transport protocol layer unaware of anything that comes from the application (at the sender's side) or goes to the application (at the receiver's side).
- Multihoming is not supported by either TCP or UDP, and it is required to provide high reliability for transport of signaling traffic across IP networks.

The SCTP was standardized by the IETF in the 2000s [13], initially designed as a transport protocol to be used for carrying signaling traffic over IP networks with the "five nines" reliability (i.e., 99.999% reliability) that is accepted as a standard in the telecom sector. It is a reliable protocol due to several of its features:

- Error-free nonduplicated transfer of user data (which is also provided by TCP) is possible as is data fragmentation according to the discovered MTU size.
- Multistreaming capability provides sequenced delivery of messages over multiple streams, to reduce the impact of HOL blocking to a single stream instead of an entire connection.
- Multihoming capability allows two SCTP end points to set up a so-called association with multiple IP addresses on each end, thus providing the possibility of using an alternative path if the primary one fails. This increases the reliability of the end-to-end communication with SCTP.
- Congestion control is provided by using TCP SACK (selective acknowledgment) mechanisms. Congestion control is mandatory in SCTP, whereas it is optional in standard TCP.
- SCTP has enhanced security at the connection establishment phase, based on a four-way handshake for establishing an association, instead of a three-way handshake in the case of TCP for connection establishment. That prevents blind DoS attacks, such as SYN attacks (based on the three-way connection establishment of TCP with a set SYN field in the header).

SCTP communication is shown in Figure 6.6. The capabilities of SCTP can become important features of the transport layer on longer timescales, such as its extension in future to provide multihoming capabilities to a mobile host attached to several different access networks at the same time (e.g., SCTP could be used to

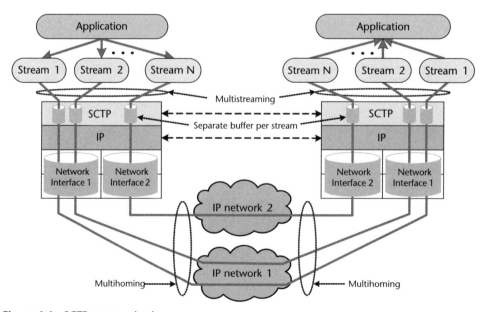

Figure 6.6 SCTP communication.

provide more efficient mobility in heterogeneous environments, as well as aggregation of bit rates from different radio access networks).

However, the use of SCTP is well established for the first time on a global scale in 3GPP radio access and core networks, in parallel with fundamental Internet protocols (e.g., IP, UDP) and certain tunneling protocols (e.g., GTP, GRE). On the other side the tunneling of IP traffic in 3GPP mobile networks is being standardized by the 3GPP (not the IETF) and it is closely related to mobility management and QoS framework.

6.3.2.2 GPRS Tunneling Protocol (GTP)

GTP is standardized by 3GPP [14], and it is used in packet-switched parts of the GSM/GPRS (2.5G) and UMTS (3G), as well as LTE/LTE-Advanced mobile networks with SAE/EPC (4G) [15]. The main purpose of GTP is encapsulation of data for its transport across terrestrial radio access networks and core networks. The three main types of GTP are as follows:

- GTP-C: This is GTP specified for transfer of signaling traffic, such as bearer activation and modification or deletion in GPRS, UMTS, and LTE/LTE-Advanced with SAE/EPC. There are three standardized versions: version 0, version 1, and version 2. In LTE and EPC/SAE the GTP-C uses the SCTP/IP protocol stack.
- GTP-U: This is used for carrying user traffic in the user plane. 3GPP has standardized two versions of GTP-U: version 0 and version 1. It is used in GPRS, UMTS, and LTE/LTE-Advanced mobile networks.
- GTP: (GTP for charging data): This is based on the same messages as GTP-U and GTP-C, but it is targeted to carry charging data from charging data functions to charging gateway functions in GSM and UMTS.

Generally, the GTP performs IP packet encapsulation from one node to another via so-called interfaces. The creation of GTP messages is shown in Figure 6.7.

In practice, GTP is a very important protocol that connects legacy 3GPP mobile network approaches and the Internet protocol stack in network nodes (base stations, gateways), which is needed for both mobility and QoS management. For interconnection between different mobile networks, GTP is used on interfaces that connect two 3GPP mobile networks (e.g., two SAE/EPC, or UMTS and SAE/EPC), while for interconnection of 3GPP mobile networks and non-3GPP wireless and mobile networks (e.g., Wi-Fi, WiMAX) PMIP is used. In the latter case, PMIP can be also used on the interface between P-GW and S-GW (although GTP is the default choice).

So, one goal of GTP is mobility support. The IP address of the mobile node in the connected mode remains unchanged and packets are being forwarded to/from mobile terminals via tunneling between P-GW and the serving eNodeB (to which the UE is attached) via S-GW. The UE's IP address remains hidden (regardless of whether it is a public or private IP address), which provides additional security for the mobile users as well as the mobile operator. LTE/LTE-Advanced mobile networks, which include E-UTRAN and SAE/EPC, together referred to as the Evolved

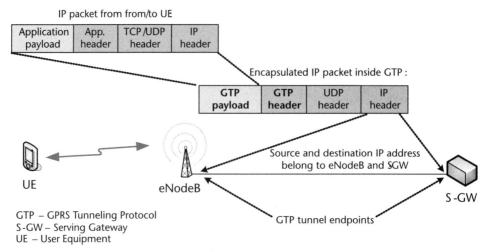

Figure 6.7 GTP messages in E-UTRAN and EPC.

Packet System (EPS), use the latest GTP versions, that is, version 1 for GTP-U and version 2 for GTP-C.

Finally, the creation of GTP tunnels (in either the user or control plane) is triggered by creation of so-called bearers, which define the basis for the QoS framework in 4G mobile networks by 3GPP, as discussed in the following section.

6.4 QoS Framework for 4G Mobile Internet

A pure IP network has no QoS support. However, the IETF has standardized several QoS mechanisms, such as DiffServ, IntServ, and MPLS (these mechanisms are covered in Chapter 5). But, all of them do not provide enough of the functionality that is required for real-time support (i.e., short delay) of many handovers going in parallel from many users connected to the mobile network. For that reason, the QoS framework is specified by 3GPP and based on a bearer philosophy (which is directly related to tunneling with GTP). On the other hand, QoS is tightly coupled with mobility management in 3GPP mobile networks and its AAA functions.

What is a bearer? A bearer is defined as a service that provides transmission of information between given entities in the network, with certain QoS attributes, capacity (bit rate), and traffic flow characteristics. The concept of bearers is present in a similar manner in 3G and 4G mobile networks that are standardized by 3GPP.

6.4.1 QoS in UMTS

In UMTS the QoS is defined with 4 QoS classes that are further used for definition of QoS categories. The fours classes defined in UMTS are (1) conversational, for voice and video telephony; (2) streaming, for real-time video services; (3) interactive, for web and other non–real-time data services; and (4) background, for messaging and other delay-tolerant data transfers.

Different QoS classes in UMTS can be further differentiated in QoS categories by using allocation and retention priority (ARP) and traffic handling priority

(THP) [16]. However, the definition of QoS classes and categories does not mean that they are all used in deployed UMTS mobile networks. In practice, most of the classes are not used. The main QoS differentiation is typically between VoIP and Internet access as a service, but UMTS typically uses the circuit-switched part for voice service (which is based on GSM technology); hence, there is no real QoS differentiation in practice. On the other side, SMS is provided via a separate bearer that is not IP based, whereas MMS can be provided over the Wireless Application Protocol (WAP) or IP protocol stacks, and it is typically provided via a separate bearer.

When there is no QoS differentiation, all services are multiplexed into the same bearer, for example, the radio access bearer (RAB) in the radio access network. But, if QoS differentiation is applied, then services are mapped into separate bearers according to the supported QoS class, ARP, and THP (in the case of UMTS). For example, VoIP (if present) shall be mapped into guaranteed bit rate conversational RAB, streaming into guaranteed bit rate streaming RAB, push-to-talk into interactive RAB, THP = ARP = 1, browsing into interactive RAB, THP = ARP = 3, and MMS and Internet access as a service to background RAB. However, this is only an example. Most UMTS networks do not use VoIP (provided by the mobile operator) or push-to-talk. Also, browsing is typically not QoS differentiated from other Internet services such as email. However, mobile operators may have contractual relations with certain third-party providers to offer so-called value-added services to their subscribers (e.g., games, songs, pictures).

6.4.1.1 PDP Context

To get the "big picture" on QoS in 3GPP mobile networks, we need to define the so-called Packet Data Protocol (PDP) context and its role in UMTS (and GPRS) mobile networks. What is PDP context? It is a data structure that contains a subscriber's session information when the given mobile terminal has an active (i.e., established) session, which is present in two main nodes in UMTS, as well as in GPRS: SGSN and GGSN. Each PDP context includes a subscriber's international mobile subscriber identity (IMSI), allocated IP address, as well as tunnel end point identifiers at the gateway nodes (SGSN and GGSN). The PDP context is not used in LTE/LTE-Advanced mobile networks.

6.4.1.2 Access Point Name (APN)

When a mobile terminal sets a PDP context, a so-called access point is selected. It is identified by its access point name (APN).

What is an APN? The APN is an identifier stored in DNS that identifies the packet data network (PDN) and the network operator. The APN is used in GPRS, UMTS, and LTE/LTE-Advanced mobile networks. It has two parts [17]:

- Network identifier: This defines to which external network the GGSN (in GPRS or UMTS) or P-GW (in EPS) is connected and optionally identifies the service requested by the mobile terminal. The network identifier is a mandatory part of the APN. It contains at least one label (e.g., "internet," "mms")

and has a maximum length of 63 bytes. However, if it contains more than one label, then network ID should correspond to Internet domain name.
- Operator identifier: This defines a particular operator's packet domain in which the GGSN (in GPRS or UMTS) or P-GW (in EPS) is located. This part of the APN is optional. The default operator ID (i.e., domain name) has the format "mnc<MNC>.mcc<MCC>.gprs," where MNC is the mobile network code and MCC is the mobile country code. Both are taken from the IMSI, which differ from the country code for mobile numbers according to E.164 (i.e., MSISDN numbers). Because the MCC is globally unique, the operator ID is also unique.

An APN that consists of both the network ID and operator ID corresponds in practice to the DNS name of GGSN or P-GW (it has a maximum length of 100 bytes). However, besides network identification, APN can be used (and it is used in many deployed GPRS and UMTS networks) for identification of a specific service, such as a WAP-based connection to the Internet, MMS, and so forth. Also, there is a so-called wildcard APN, which is coded as APN with and asterisk (*) as its single label (the asterisk is coded using its 8-bit ASCII code). If a mobile operator allows wildcard APN, then it allows the subscriber to access any network of the given PDP type. A wildcard APN is received by the SGSN from the mobile terminal when activating a PDP context.

6.4.2 QoS in LTE/LTE-Advanced Mobile Networks

Although different so-called QoS classes were defined for 3G mobile networks (e.g., UMTS/HSPA), the main approach in 3G was provisioning of the voice service via the circuit-switched domain and delivery of Internet network-neutral services (e.g., WWW, email) via the packet-switched domain of the mobile network. However, in EPS all services are based on the IP protocol stack including VoIP. So, efficient QoS support in the IP-based mobile network becomes a necessity, especially for real-time services that are sensitive to delay (e.g., VoIP, mobile IPTV, online gaming). The QoS concept in SAE is directly related to bearers, similar to the QoS framework in UMTS. Several EPS bearers are shown in Figure 6.8.

EPS consists of the E-UTRAN and EPC. Both nodes used for user traffic in the EPC, S-GW and P-GW, perform QoS enforcement acting as policy enforcement points (PEPs). The S-GW performs policy and QoS control at the packet level, whereas the P-GW provides policy enforcement at the service level. Similar, S-GW performs charging functionalities at the packet level and P-GW generates charging records at the service level. S-GW has direct interfaces with previous RANs from 3GPP, such as UTRAN (3G RAN) and GERAN (GSM EDGE Radio Access Network, i.e., 2.5G RAN). For the non-3GPP IP Connectivity Access Network (IP-CAN), the mobility anchor point is the P-GW. So, in the heterogeneous service environment of an all-IP network in SAE, the crucial part becomes the QoS support.

In the cases of 3GPP radio access networks, the S-GW terminates the GTP-U tunnels and starts the GTP-U tunnels toward the P-GW. As discussed in the GTP subsection, the GTP tunnels are used between S-GW and P-GW, where GTP-U and GTP-C are used for transfer of user traffic and control traffic, respectively. This

6.4 QoS Framework for 4G Mobile Internet

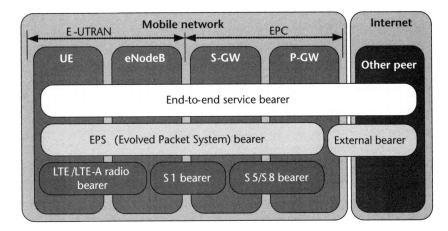

E-UTRAN – Evolved UMTS Terrestrial Access Networks
EPC – Evolved Packet Core
P-GW – Packet data network Gateway
S-GW – Serving Gateway
UE – User Equipment

Figure 6.8 EPS bearers for LTE/LTE-Advanced mobile networks.

is typical for 3GPP radio access networks where S-GW is, in fact, a GTP-U relay between the mobile terminals in the access network and P-GW. Another option is to use PMIPv6 in the control plane for interconnection with non-3GPP networks, such as Wi-Fi and WiMAX (from IEEE).

Both central gateways in the SAE, the S-GW and P-GW, may be physically located in a single unit (i.e., SAE gateway). The S-GW and P-GW are connected with the PCRF via control interfaces by using the Diameter protocol. Such interfaces control the PCEF located in the S-GW and P-GW. For access to non-3GPP networks, both gateways must implement Diameter interfaces toward external AAA nodes.

Every mobile terminal that is connected to the SAE/EPS-based mobile network has at least one established bearer for IP connectivity, called the default bearer. Further additional bearers may be set up for the mobile terminal. Also, different flows with similar QoS requirements can be mapped on the same bearer.

A bearer is set up when an application in a mobile terminal initiates connections (e.g., web client connects to a web server) or when an incoming connection (e.g., voice call) is received. When QoS support is needed, the connection is established between the mobile terminal and a certain application server on the operator's side (e.g., via the IMS). The EPS has three main bearers:

- Radio bearer (on the LTE/LTE-Advanced radio interface);
- S1 bearer (on the S1 interface, between the eNodeB and S-GW); and
- S5/S8 bearer (on the S5/S8 interface, between the S-GW and P-GW).

So, when the application server requests an EPS bearer, it is established with signaling for lower layer bearers, that is, the radio bearer, S1 bearer, and S5/S8 bearer. The main goal of bearer setup and existence is to minimize QoS knowledge and configuration in mobile terminals [16].

Regarding the QoS identification SAE uses QoS class identifiers (QCIs), which are based on three QoS parameters: priority, loss probability, and delay. An EPS bearer is characterized by the following parameters:

- Allocation retention priority (ARP): This parameter refers to the priority used for the allocation and retention mechanisms.
- Guaranteed bit rate (GBR): This parameters is only applicable to bearers that require guaranteed QoS for services such as voice or streaming.
- Maximum bit rate (MBR): The MBR parameters help to set a limit on the data rate expected for the related service.
- QoS class identifier (QCI): This is a scalar value that is used as a reference to a set of access network-related QoS parameters, for the transmission between the terminal and the eNodeB.

Two resource types are defined in SAE/EPS for QoS: GBR services and non-GBR services. Further, nine QCI values with nine priorities are defined, as shown in Table 6.1. Highest priority is given to IMS signaling, which therefore can be served with a non-GBR resource type (without signaling no connections can be established via SAE, so its "number one" position on the priority list is with a purpose). Further, priority two is given to VoIP, but the lowest delay constraint has real-time gaming due to higher interaction between machines (to keep the action synchronized) than between humans. Video calls have lower settings regarding the delay budget and loss rate than voice only due to the need for synchronization of voice streams with bursty and more bandwidth-demanding video streams. Finally, the GBR group is completed with streaming services that can accommodate longer delays due to the unidirectional character of the service and the possibility for longer playout buffering at the receiving side. There is less QoS differentiation for non–real-time traffic, due to the current network-neutral characteristics of such traffic (i.e., all services are served in a best-effort manner, based on network neutrality for OTT services).

Although a QoS framework in defined in 4G mobile networks (with LTE in the radio part and EPS in the terrestrial part), the last service to typically transfer

Table 6.1 QoS Class Identifiers (QCI) for LTE/LTE-Advanced

QCI	Resource Type	Priority	Delay Budget (ms)	Loss Rate	Example Application
1	GBR	2	100	10^{-2}	VoIP
2		4	150	10^{-3}	Video call
3		5	300	10^{-6}	Video streaming
4		3	50	10^{-3}	Real-time gaming
5	Non-GBR	1	100	10^{-6}	IMS signaling
6		7	100	10^{-3}	Voice, live video, interactive gaming
7		6	300	10^{-6}	TCP applications (web, email, P2P file sharing, http video, chat, buffered video streaming)
8		8			
9		9			

to all-IP is voice. Most of the deployments of LTE mobile networks in the period from 2010 to 2015 were targeted to provision of Internet access with higher bit rates (than UMTS/HSPA), whereas voice has continued to be delivered via GSM or the CS domain in UMTS. However, we can expect that full transition of mobile telephony to VoIP (with QoS support), similar to transition of fixed telephony to VoIP according to the NGN framework, will be implemented in the second half of the 2010s. We refer to this technology here as 4G mobile VoIP.

6.4.3 QoS in 4G Mobile WiMAX

QoS support is provided in both Mobile WiMAX release 1 (3G) and release 2 (4G) by the IEEE 802.16e and IEEE 802.16m standards, respectively. The QoS in WiMAX is based on service flows [18]. It may be based on contention among mobile stations (similar to Wi-Fi) or it may be contention free.

Contention-based service flow type is the best-effort (BE) service, whereas contention-free services include the unsolicited grant service (UGS) and real-time polling service (rtPS). Further, a non–real-time polling service (nrtPS) can be used in contention-based or contention-free mode. For the contention-free service type, the base station (BS) polls the mobile station (MS) periodically. In contention-based service types, the BS allocates bandwidth to the MS upon receiving a bandwidth request (BR) message.

In all versions of WiMAX, the UGS is targeted for VoIP service (also rtPS can be used for VoIP with silence suppressions), while rtPS and extended rtPS (ertPS) are designed to be used for video and multimedia streaming (e.g., IPTV). Further, nrtPS and BE are targeted to non–real-time services (e.g., file transfer, WWW, email).

QoS service types in Mobile WiMAX are listed in Table 6.2. The IEEE 802.16m advanced air interface provides a more flexible and efficient QoS framework by adding one new service type, adaptive granting and polling (aGP) service, which is targeted to services that demand adaptation and flexibility, such as online gaming, VoIP with adaptive multirate (AMR) codecs, as well as TCP-based delay-sensitive services.

Table 6.2 Service Flow Types in 4G Mobile WiMAX

Service Flow Type	Traffic Type
Unsolicited grant service (UGS)	Real-time traffic with fixed-size packets, for TDM-like services including VoIP
Real-time polling service (rtPS)	Real-time traffic with variable-size packets; for VoIP with silence suppression, IPTV
Extended rtPS (ertPS)	Real-time traffic with variable-size packets and active/silence intervals, targeted to video and multimedia streaming
Non–real-time polling service (nrtPS)	Delay-tolerant traffic with minimum reserved bit rate, for file transfers, web services
Best-effort (BE)	No guarantees (best-effort Internet concept), for web, email, peer-to-peer services, and so forth
Adaptive granting and polling (aGP)	Flexible QoS support for both allocation size and interarrival time; for online games, VoIP with AMR codecs, and delay-sensitive TCP-based services

6.5 4G Mobile VoIP

Although voice has been IP based in Mobile WiMAX and Wi-Fi networks from their inception, the use of VoIP in mobile 3GPP networks is lagging behind the transition of the networks toward an all-IP core and the terrestrial access part. The main reason is the performance of CS telephony for different users' mobility requirements (with different velocities, from nomadic users at home, office, or public places, to highly mobile users in cars and trains). CS telephony has high standards with regard to the required QoS, including smooth handovers without noticeable losses, delays, and jitter. So, mobile voice (with QoS support, specified in the SLA between the subscriber and mobile operator) will transit to mobile VoIP when it provides at least the same performance as CS voice.

Several prerequisites were needed for mobile VoIP. VoIP needs to be a real-time (i.e., requires short delays) conversational service in that it should provide the "feeling" to two people conversing on mobile phones that they are talking to each other face to face. It requires QoS provisioning all the time (in a given cell, at handover), which requires a well-specified and well-working signal during call establishment (e.g., AAA) and good mobility and QoS management during the call (e.g., dropping a call at a handover is considered to be more disturbing for the end user than rejecting a new call attempt due to certain admission control in the mobile network).

Why mobile VoIP? The main target for transition to mobile VoIP in 3GPP mobile networks is convergence of all services to a single all-IP network infrastructure. However, mobile WiMAX (as well as Wi-Fi) has used VoIP from the beginning because it is an Internet-native technology. VoIP uses the RTP/UDP/IP protocol stack, so each VoIP packet has an IP header, UDP header, and RTP header, which additionally increase the required bit rate for voice connections. So, the needed throughput is higher with VoIP than with CS voice for the same voice codec (i.e., coder-decoder), as shown in Table 6.3. Also, mobile VoIP requires a small delay budget because end-to-end delay for telephony (which refers also to VoIP) is recommended to be kept below 150 ms, but in any case it should not be higher than 400 ms (according to ITU-T G.114 [19], with the goal of preventing people from interrupting each other while talking). Mobile VoIP (as well as fixed VoIP) is also not tolerant to jitter (i.e., delay variation). Such requirements become more complex with added mobility events such as handovers. So, we can assume that mobile

Table 6.3 Mobile VoIP with Different Packet Sizes (Frames)

Frames/Packet	Packets/sec	Payload	GSM Codec Full Rate (Kbps)	Packet Size	VoIP (Kbps)*	VoIP Higher Bit Rate (%)	Frame Delay (ms)
1	50.00	33	13.2	87	34.80	264	20
2	25.00	66	13.2	120	24.00	182	40
3	16.67	99	13.2	153	20.40	155	60
4	12.50	132	13.2	186	18.60	141	80
5	10.00	165	13.2	219	17.52	133	100

* IPv4 header = 20 bytes + 14 bytes options; UDP header = 8 bytes; RTP header = 12 bytes.

VoIP with guaranteed QoS parameters (network QoS parameters are described in Chapter 5) is complex; therefore, the transition of mobile telephony in the PLMN (e.g., GSM) to mobile VoIP is happening later than in PSTN (the late 2010s).

Mobile telephony in 3GPP has evolved from 2G (i.e., GSM) to 4G (i.e., LTE/LTE-Advanced). In contrast, the 3GPP mobile network architecture for 4G (i.e., EPS) is a flat one, without controllers of base stations (or NodeBs), with the goal being to reduce the delay budget as much as possible and accommodate the all-IP mobile network architecture to delay-sensitive services, such as VoIP. (Note that every service benefits from smaller delays and higher bit rates; taken together, these lead to an overall better mobile experience for end users.)

Overall, for operator-provided mobile VoIP, there are several requirements in the mobile network:

- Location management (for delivery of terminating calls to the mobile users) and mobility management (for handling the handovers of established calls);
- A well-defined and applied QoS framework for VoIP connections; and
- Signaling defined through control plane protocols (e.g., SIP, Diameter) and standardized interfaces between each pair of mobile network nodes.

6.5.1 Carrier-Grade Mobile VoIP Implementation

In 4G mobile networks (and we can expect that it will continue in the same manner in 5G mobile networks, after 2020), carrier-grade VoIP (as a replacement for PLMN) is standardized to be deployed by using common IMS functions and the IETF protocols SIP and Diameter. SIP is used mainly for call establishment and control, whereas Diameter is used on interfaces toward the HSS. According to the discussions in Chapter 5, the 4G mobile networks are in fact implementations of NGN specifications in terms of VoIP services. For the signaling purposes of mobile VoIP, the main "player" in the service stratum is IMS, with its CSCFs. For example, Proxy CSCF is the first entity on the way from mobile terminals toward the core network, which intercepts the SIP signaling messages and provides AAA functionalities via communication to the CSCF-S and finally HSS (as the standardized users' database in the network). Mobility management and admission control is further provided by communication of service control functions (SCF), that is, CSCF entities of IMS, with the following NGN functional entities: mobility management and control functions (MMCFs), network attachment control functions (NACFs), and resource and admission control functions (RACFs).

The end user has to be authenticated and authorized for VoIP services in each of the mobile and wireless networks during established voice connections; this is done to provide seamless handovers. In practice, UEs indicate the possibility for handovers by measuring the received signal strength (RSS) in a given radio access network (RAN) and use layer 2 signaling to a given base station (e.g., eNodeB for LTE) or access point (for Wi-Fi). The goal is to provide a smooth transfer of the voice session. This is accomplished in the typical handover scenario by "make-before-break," which means establishing a new layer 2/layer 3 path (to the new location of the mobile terminal in the network) before handover, and then release of the previous path for the given voice session (e.g., late path switching in LTE networks [16]).

Figure 6.9 shows VoIP service establishment from a user connected to an LTE/LTE-Advanced mobile network (mobile operator A) to a destination user B, connected to mobile operator B, which uses Mobile WiMAX. The VoIP service is established by using SIP signaling messages. Both NGN strata are involved in the process. The originating user terminal, terminal A, sends a service request to establish a VoIP call with user B to the P-CSCF, which then forwards it to the S-CSCF until it finally reaches the I-CSCF of mobile operator A (since the domain of user B belongs to another operator). Then, the I-CSCF performs DNS resolving of the domain name of the destination mobile operator to obtain the IP address of the I-CSCF in the target network. The receiving I-CSCF (in mobile operator B) forwards it to S-CSCF entity in the destination network, with the goal of obtaining the P-CSCF to which the called user is registered in the mobile network. Finally, a service request from user A is forwarded to the called user B (i.e., terminating call for mobile operator B), which provides a response that is carried using the reverse path back to the originating user terminal. On such a path, each P-CSCF (in both networks) communicates with the RACF for the purpose of admission control (for the given VoIP service request) and also does the QoS setup in the transport IP networks of each mobile operator. For a positive admission control decision and

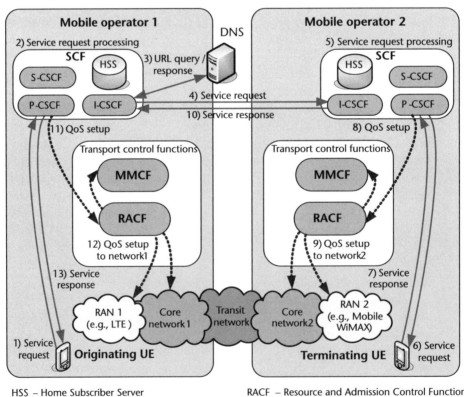

Figure 6.9 Establishing VoIP service between different NGN mobile operators.

resource allocation by RACF (which communicates with MMCF for handling the mobility when it is necessary), the positive response is forwarded back to the VoIP call originator.

After the VoIP session has been established (which is done simultaneously in both directions, as is typical for telephony in all environments), the voice data are being transferred between end users in both directions. The voice data transfer is based on a voice application (with voice codecs, such as G.723, G, 729, GSM full or half rate, AMR) over RTP/UDP/IP for data transfer, through the transport stratum (i.e., user plane in the mobile network) as given in Figure 6.10. Each RACF (in the operator's network on each side) sets up the preestablished path (with signaling, prior to VoIP service establishment) with the goal of maintaining the required QoS level. The MMCF maintains QoS support during mobility of the users with established VoIP connections (e.g., EPS bearer setup/release for LTE/LTE-Advanced).

From a practical point of view, carrier-grade mobile VoIP is replacing the PLMN (e.g., GSM), but such a process is going more slowly than in fixed networks for several reasons, with mobile terminals' capabilities (e.g., SIP agents in mobile terminals for signaling) being a highly important one.

6.5.2 Over-the-Top Mobile VoIP Consideration

In addition to carrier-grade mobile VoIP, OTT VoIP services are frequently used nowadays in mobile and wireless environments, such as Skype and Viber. However, such OTT VoIP services are provided via the Internet as a service access to mobile users (or Wi-Fi users in the home, office, or public places), which means that they are served in a best-effort manner (i.e., without any differentiation from other Internet services that the user may be using in parallel, such as web browsing, video streaming, file download, chat, email). If there is no congestion in the network (i.e., no bottlenecks end to end), such services can be used with a satisfactory QoE, especially for voice only. Best-effort (i.e., OTT) VoIP implementations are covered in more detail in Chapter 7 of this book.

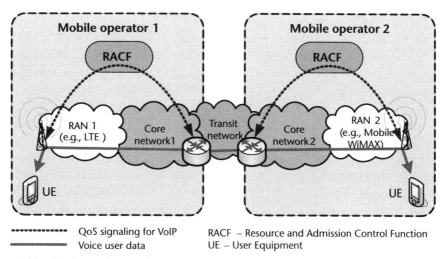

Figure 6.10 Carrier-grade VoIP data transfer.

6.6 Mobile TV

According to the ITU-T [20], Internet Protocol Television (IPTV) is defined as multimedia services such as TV, video, audio, pictures/graphics, and data, delivered over IP-based networks that are designed to support the needed level of QoS, QoE, interactivity, reliability, and security. Providing IPTV for viewing on mobile (handheld) devices is referred to as mobile IPTV.

In general, IPTV is the successor to digital TV broadcast networks (e.g., terrestrial TV broadcast, cable TV) in a converged IP environment. Because digital TV requires higher bit rates for delivering the media (i.e., video synchronized with audio channels), the IPTV is becoming possible with deployment of broadband access networks.

For example, with existing codecs, standard definition TV content requires 2 to 3 Mbps when it is delivered as IPTV to end users. Due to the smaller screen sizes of mobile devices, typical bit rates for mobile TV in mobile networks are approximately 10 times smaller for the same content, for example, 250 to 300 Kbps per TV stream.

In practice, several attempts have been made to standardize TV for handheld devices such as mobile terminals. The most significant standards were made by Digital Video Broadcasting (DVB) with the DVB Handheld (DVB-H) standard, as well as Multimedia Broadcast Multicast Service (MBMS) standardized by 3GPP for UMTS (3G) and LTE (4G) mobile networks. However, the DVB-H has faced commercial failure (although there were several network launches during the 2000s), while MBMS is looking forward to certain success. MBMS was introduced in 3GPP release 6, for the UMTS, but with no major deployments during the 3G era. ITU-T has done ongoing work for the standardization of IPTV for both environments, fixed and mobile, based on the NGN framework.

6.6.1 Mobile IPTV Standardization by ITU-T

Mobile IPTV is defined as an application [21] and has the same functional architecture (standardized by ITU-T [22]) for fixed and mobile environments. The basic IPTV architecture is shown in Figure 6.11. It consists of four IPTV domains [21]: (1) content provider, (2) service provider, (3) network provider, and (4) end user. However, a single provider can be involved in one or more IPTV domains.

Most of the architecture for mobile IPTV is the same as that for fixed IPTV. The main difference is the requirement to have wireless connectivity between the terminal device (e.g., smartphone, laptop) and the network provider. Considering the NGN terminology (refer to Chapter 5), a mobile IPTV terminal is connected to the transport functions of the mobile operator's network. However, one of the main features of IPTV is being able to communicate and access the established IPTV services through different (heterogeneous) wireless and mobile access networks (e.g., 3G, 4G, Wi-Fi). On the other hand, differences in various wireless/mobile access networks may derive changes in terms of available throughput (i.e., bit rates), network characteristics (e.g., during movement of the user), mobile network policies and configurations, and differences among home network and visited networks at roaming (in terms of the mobility) [23]. Such differences will certainly affect the QoS of the assessed IPTV services, and finally the QoE for the mobile users.

6.6 Mobile TV

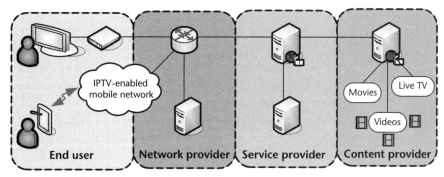

Figure 6.11 IPTV domains.

Figure 6.12 shows the general mobile IPTV architecture. Each mobile network consists of cells. Group of cells defines a service region, which can include more than one base station (e.g., eNodeB in LTE networks). The IPTV service controller is located in the root of the tree (from the IPTV stream source toward BSs and MTs). It chooses some base stations to create a service region by using location information provided by mobile terminals that are requesting a particular IPTV service. Each service region can be differentiated as a broadcast or multicast service region. However, the delivery of IPTV service to mobile devices (e.g., IPTV stream of a certain TV program) can be performed by using unicast, multicast, and broadcast approaches.

So, mobile IPTV services require support for broadcast and multicast capabilities in the mobile network. In practice, such standardized systems already exist and include the following:

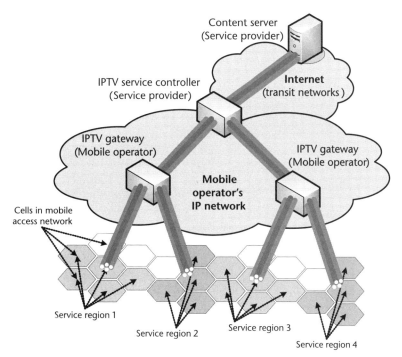

Figure 6.12 Mobile IPTV architecture.

- MBMS by 3GPP [24] and
- Multicast and Broadcast Service (MBS) by the WiMAX Forum [25].

Several other requirements are specified for architectures used for delivery of mobile IPTV services by ITU-T [21]. These include changing of the service region (due to the mobility of mobile devices), transmitting via all BSs or APs in the service region, support of multicarrier operations (for higher bit rates for IPTV delivery), switching between different transmission modes (unicast, multicast, and broadcast), seamless horizontal and vertical handovers, monitoring of mobile devices receiving IPTV service, collecting and managing environmental information (e.g., location, status of the mobile device, channel usage information), and so forth. There are also specific functional requirements in terms of QoS and QoE, such as resource allocation in the mobile or wireless network, managing the priority of traffic scheduling that is associated with the mobile IPTV service, and the capability of a mobile architecture to provide information about the available resources for IPTV services to reconfigure service regions, to release and reuse radio resources dedicated to the service, and so forth. Finally, mobile IPTV terminals have certain requirements, including selection of a specific radio interface among multiple network interfaces, identifying the service region with which the device is associated, measuring the quality of radio resources, as well as capabilities to select and receive IPTV contents.

Special-use cases for mobile IPTV are specified in ITU-T Supplement 20 to Y-series [26], which defines content delivery methods such as these:

- Content delivery using the multicast service zone: In this case the service zone is created only when and where a mobile user requests a certain mobile IPTV service for the first time.
- Content delivery using the dynamic service zone: This continues the IPTV service when a mobile device leaves the multicast service zone by extending the zone to the service zone boundary.
- Content delivery using multiple access networks: This provides IPTV service delivery simultaneously through multiple available radio interfaces (e.g., 3G, 4G, Wi-Fi, future 5G) with the goal of increasing the efficiency of radio resource usage as well as quality (e.g., higher aggregate bit rates will be available in this case than in the case of IPTV delivery through a single radio interface).
- P2P-based IPTV content delivery to mobile terminals: This approach is targeted to provide efficient cooperation between the mobile devices (i.e., end users) for delivery of real-time IPTV services through the mobile network.

In the practical implementation of any different delivery methods, there can be different so-called use cases (i.e., mobile IPTV scenarios). Some of them may be dedicated to mobile devices only, while others may also be targeted to fixed devices (e.g., a TV device in the home). The following use cases are foreseen for mobile IPTV [26]:

- N-screen service: This allows an IPTV service subscriber to use its contents on PC, TV, and mobile terminal screens. One approach is IPTV contents sharing among different devices, while another is to provide service mobility (e.g., change of the device used for delivery of IPTV service, such as moving a live event from a mobile device to a TV or laptop while continuing to watch it). Also, content sharing may be possible by using cloud computing technologies (e.g., for watching VoD content and changing the receiving device).

- Location-based service (LBS): In this case IPTV service is provided to mobile users based on location information obtained via GPS (locally in the terminal) or other positioning techniques (e.g., assisted by mobile operators, "fingerprint" databases of Wi-Fi networks). Such information is given to the IPTV service provider for filtering the IPTV contents delivered to users based on their location. Advertising IPTV services are also one of the possible use cases, although they may be annoying to mobile users and face the unsuccessful story of broadcast SMS services (as an example).

- Hybrid IPTV service: This case refers to a combination of traditional non-IP mobile TV distribution techniques (e.g., satellite TV, terrestrial TV) and IPTV. However, mobile TV (which is non-IP) is a unidirectional service, whereas mobile IPTV is bidirectional. Two scenarios are possible: (1) service transition between mobile TV and mobile IPTV (e.g., changing the viewing contents from a TV interface to an IPTV interface on the same device), and (2) service cooperation between the mobile TV and mobile IPTV (e.g., mobile IPTV can provide the uplink path to mobile TV for interactivity purposes).

The practical deployments of mobile IPTV services are lagging behind their standardization (e.g., MBMS has not been commercially deployed although it has been introduced in UMTS with 3GPP Release 6, while enhanced MBMS, i.e., eMBMS, is following with similar destiny in 4G mobile networks). Besides mobile IPTV, other standards had limited deployments. One was DVB-H [27], which has not been a successful story like its "twin" technology DVB Terrestrial (DVB-T), which is currently the main terrestrial technology for digital TV distribution. The main drawback is that DVB-H uses a dedicated interface for mobile TV, and that is completely at odds with the main philosophy of telecommunications regarding the convergence of networks and services, which means that every network interface (wireless or wired) should be capable of supporting various IP services. Regarding the mobile environments, several different approaches have been suggested for deployment of mobile TV and mobile IPTV [26]:

- Mobile TV plus IP: This approach uses traditional broadcast networks for delivery of IP-based audio, video, graphics, and other broadband services to the mobile terminals, which also have a bidirectional connection through the mobile network to which they are attached (e.g., 2.5G, 3G, 4G).

- IPTV plus mobile approach: This is the same IPTV service that is delivered to fixed hosts, but in this case it is provided to mobile hosts by means of IMS functionalities.

- Cellular-based IPTV approach: This refers to an end-to-end framework for IP-based broadcasting networks, which are bearers independent (any transport technology can be adopted), and dedicated to mobile networks.
- Mobile Internet-based IPTV approach: This is a highly diversified approach in which anybody can be a content provider, a service provider, or a customer. Although it has a global reach (e.g., similar to the Web), the QoS is not guaranteed and it is provided in a best-effort manner. On the other hand, since it is not related to any particular access technology, mobile users can access IPTV services through different wireless and mobile networks (e.g., Wi-Fi, 3G, and 4G).
- 3D-image delivery: When images are delivered via multiple access networks (e.g., 4G, Wi-Fi) then the content is 3D, otherwise it is 2D.
- Web-based mobile IPTV: In this case IPTV service is discovered by using HTTP as the IPTV service discovery function, which further provides the metadata to end users. Web-based IPTV (also called WebTV) is provided in a best-effort manner and it is based on Internet network neutrality.
- Mobile IPTV with cost-effective peering: In this case a given mobile user receives IPTV content that is stored in some other user terminal that has requested the same content and is accessible through a more cost-effective network (e.g., Wi-Fi resources are cheaper than LTE or Mobile WiMAX radio resources). It requires a P2P control platform in the mobile operator's core network for locating and managing communication with appropriate peers.
- Mobile IPTV service with cloud computing technologies: This approach uses third-party IPTV services in a cloud computing manner, thus saving mobile operators the costs of maintaining an IPTV overlay service. In this case the mobile IPTV platform provides customer-oriented service features based on NGN functions (for IPTV service provisioning).

The reader can see that various mobile IPTV service deployment scenarios and use cases have been devised. However, mobile users currently prefer to access VoD by using third-party video-sharing services (e.g., YouTube), which is not considered to be IPTV (although IPTV also covers web-based IPTV contents delivered as VoD). However, the leading mobile technologies nowadays, such as LTE and Mobile WiMAX, have standardized multicast and broadcast systems. The MBMS from 3GPP is discussed in the following section as an example.

6.6.2 Multimedia Broadcast Multicast Service (MBMS)

The 3GPP has standardized MBMS for UMTS/HSPA mobile networks, and has continued that development with eMBMS in LTE/LTE-Advanced mobile networks. The main idea was to exploit the multicast bandwidth savings for linear TV (i.e., a TV program with a given schedule, so every user watches the same TV show when tuned to the same TV program) in the mobile access network (in a cell or group of cells).

In the LTE system, the eMBMS either uses a single-cell transmission or a multicell transmission. The multicast transport channels in the LTE radio interface are mapped to shared physical channels for multicast data transmission (point to multipoint), and scheduling is done by the eNodeBs. To avoid unnecessary eMBMS transmissions in a cell where there is no eMBMS user, the network can detect the presence of users interested in the eMBMS service by polling or through a UE service request.

The eMBMS reuses the LTE network infrastructure (unlike DVB-H, as an example) and provides flexibility in the use of network resources. It includes synchronous broadcast transmission on a same frequency in multiple cells, that is, multicast broadcast over a single frequency network (MBSFN). The eMBMS architecture has a multicell/multicast coordination entity (MCE) as a logical entity. The MCE can be a separate entity or part of another network element. It performs functions for admission control and allocation of the radio resources used by all eNodeBs in the MBSFN area and also determines the radio configuration (e.g., modulation and coding scheme). A single eNodeB may belong to multiple MBSFN areas. While MBMS for UTRAN was specified in release 6, the eMBMS was developed in release 9 and additional features were specified in the releases beyond it. For example, release 10 introduced priorities between eMBMS sessions and dynamic adaptive streaming over HTTP (DASH), while release 11 added support for service continuity.

Overall, the QoE of services provided through eMBMS depends heavily on the available bit rate and size and resolution of the mobile screens. For example, UTRAN offered up to 256 Kbps per MBMS bearer service (e.g., linear TV stream) with up to 1.7 Mbps per cell (e.g., up to six different linear TV programs). Such limitations have caused MBMS to be unsuccessful. Although LTE/LTE-Advanced offers higher aggregate bit rates (e.g., over 3 Gbps aggregate bit rates), the success of eMBMS in the future will depend on the development of successful business models that will be adopted to the deployment models presented for mobile IPTV, which cover a diverse set of use cases.

MBS for Mobile WiMAX is a combination of MBMS (by 3GPP) and DVB-H. Both technologies, mobile networks by 3GPP and mobile WiMAX, may be able to make possible multicast and broadcast systems in the near future even for terrestrial TV broadcast, due to their much higher aggregate throughputs per cell in their 4G versions, which will make them competitive with DVB-T providers, satellite TV, and even cable TV providers, especially for mobile IPTV combined with service mobility and cloud computing platforms (for recorded contents). Although such contents are available through many video-sharing sites on the Internet that are served in a best-effort manner (due to Internet network neutrality), the main differentiation for mobile IPTV services supported by a mobile operator is the guaranteed QoS and QoE, which is specified in the SLA with the mobile user.

6.7 Mobile Web

Web services for mobile hosts are provided in manner similar to that used for fixed hosts. Looking back at history, at the end of the 1990s and beginning of the 2000s, the then-existing mobile handheld devices had limited processing and memory capabilities (as well as battery support) that were not strong enough to support the HTTP/TCP/IP protocol stack needed for "normal" web communication between a mobile client and web servers. So, the appearance of the first packet-switched mobile systems as 2.5G (e.g., GPRS by 3GPP) faced low capabilities on mobile devices, which led to the standardization and use of the WAP protocol stack as a replacement for the normal web protocol stack for mobile devices. The WAP protocol architecture is shown in Figure 6.13, with corresponding mapping of related WAP and HTTP protocol stacks. In such cases, typically a WAP-HTTP proxy (usually referred to as WAP gateway) was placed between the mobile terminals and the Internet, in the mobile operator's core network. For the same reasons that contributed to WAP's introduction, the first MMS systems (by 3GPP) were based on the WAP protocol stack. For completeness in looking back at the first years of the mobile web, we should also mention iMode (which was successful in Japan around 2000, developed by Japan's DoCoMo company) and XHTML Basic (a simplified version of HTTP 1.1, which excludes certain elements such as frames, image maps, nested tables, bidirectional text, and text editing).

However, mobile devices have enjoyed a continuous increase in their processing power and memory. That led to introduction of the HTTP protocol stack in mobile devices (for web services) in the second half of the 2000s, thus making the WAP protocol stack obsolete for smartphones. (However, it is still being used for MMS, in parallel with HTTP, due to backward compatibility with "older" mobile devices.)

The web is generally covered in Chapter 4 of this book. However, mobile devices always have certain limitations when compared to computers and laptops, because the smaller size always results in them having lower capabilities (simply due to parallel development of desktop computers and smartphones, although the "gap" between their capabilities has become smaller).

Smaller screen sizes and limited capabilities influence creation of so-called mobile web services (e.g., mobile web portals), which are accommodated to mobile devices. Also, the personal character of mobile communications and hence location information (when it is available) influence the design of mobile web services. For example, penetration of mobile users is higher when compared to other types of communications that involve a human on one or both ends, such as the number of users who have fixed telephony or fixed Internet access. On the other side, web traffic contributes with the most transferred bytes and established connection on a global scale and web traffic is continuously increasing due to the increase of penetration of mobile (and fixed) broadband access to the Internet, and hence richer

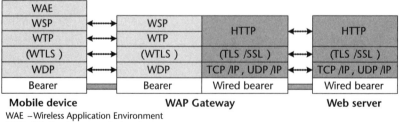

Figure 6.13 WAP protocol stack.

web contents. That leads to continuously increasing Web usage through mobile devices, which are referred to as mobile web. So, mobile devices and mobility influence the web design and application.

Note that mobile web design is "the art" of web communication within mobile environments. With smartphones having implemented the complete HTTP/TCP/IP protocol stack in the second decade of the 21st century (and beyond), there is no need for the WAP protocol stack, but there are several methods for implementation and provision of mobile web, including the following:

- Do nothing approach: This means that same design for fixed websites is used for mobile hosts, so there are no differences.
- Content adaptation (strip images and styling): Content adaptation provides a better fit of web pages to the smaller screens of mobile device. This adaptation requires detection of the screen size through the use of mobile agents for given mobile device types. A proxy server, as shown in Figure 6.14, automatically adapts web contents to fit the existing limitations of a mobile web client. Note, however, that they may not be able to handle effectively all web content requested by the mobile web clients.
- Content differentiation: In this case the HTTP User-Agent header can be used to provide customized content to the browser in the mobile device.
- Mobile specific site/application: In this case a separate website is customized for access by handheld devices through a mobile or Wi-Fi network. However, this can lead to separate and not synchronized desktop and mobile websites.

According to the W3C [28], when the mobile web first began to be used, the primary use was for choosing an area (e.g., city, which can be further used for LBS), conducting a web search (e.g., Google, Yahoo), and accessing an event calendar. Nowadays, however, primary uses also include access to LBS services (e.g., weather checking, finding an object at a given location), watching streaming video from video-sharing sites (e.g., YouTube), social networking via mobile, and cloud access (e.g., Dropbox and Google Drive for files storage, Steam for gaming).

Regarding the web browsers, mobile versions are available for all popular desktop PC browsers in the 2010s: Chrome, Firefox, Opera, and so forth. However, there are also several mobile browsers that have attracted users with their well-designed user interfaces, such as the Dolphin browser for the Android operating

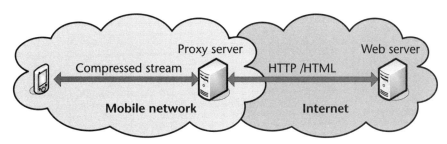

HTML – HyperText Markup Language
HTTP – HyperText Transfer Protocol

Figure 6.14 Content adaptation for mobile web.

system. All browsers, however, have their versions accommodated to the OS in mobile terminals (e.g., Android from Google, iOS from Apple).

Regardless of the type of mobile browser, the best practice when creating mobile websites is to provide a consistent user experience, so the user can have almost the same experience regardless of whether he or she is accessing web contents through a desktop computer or a mobile device. For that purpose, web designers should be aware of device limitations, which can be obtained in the first instance from the header information (when it is available) in the request that is received from the mobile user. Also, web designers should not use unsupported features, such as pop-up windows, image maps, or nested tables. Additionally, certain contents, such as graphics, style sheets, tables, and cookies, should be adapted to the size limitations of mobile browsers. General best practices include not sending contents that mobile browsers do not support and designing web applications to fail gracefully when certain features are not supported. Another best practice is to adapt the layout, which means dividing large pages into smaller sections (if possible) and using relative measures (not pixels).

User interaction also comes into the focus for the mobile web. To provide the user with a good experience, designers should keeps URIs short where possible (to reduce the need for typing), use a consistent and balanced means of navigation through the web pages (i.e., balance the breadth and depth of the web page hierarchy), and provide more effective page titles (e.g., shorter) for display and bookmarking. Scrolling in one dimension on the mobile browser also is beneficial for a better user experience. Because Internet access bit rates are typically smaller than in fixed-access networks and also variable (e.g., they depend heavily on the location of the user and received signal strength, the error ratio, and any congestion in the radio access network due to traffic from other mobile users in the same cell or cells) mobile web content should be limited to only what the user needs, and it should be kept short and simple (e.g., minimize external images). Because there are many different websites worldwide, it is hard to optimize mobile browsers for all different design approaches to web pages. Therefore, the most successful websites [29] , such as google.com, facebook.com, and youtube.com, have developed their own HTTP User Agents optimized for the specific site, which are called applications. They are in fact browsers accommodated to the web technologies that are used on the specific website, such as the YouTube application for Android, to be used for access to youtube.com in a customized manner. (Note, however, that the given website is also accessible through general-purpose mobile browsers, such as Chrome, Firefox, and Opera).

6.8 5G Mobile Broadband Developments

Mobile broadband Internet after 4G will continue with the next generation, called 5G, which is currently in the research phase. The first applications of these technologies should appear around 2020. The current question is whether 5G will be just an evolution of 4G, or if the new mobile broadband technologies will cause such a significant disruption that full rethinking of mobile principles will be required.

From an Internet technology viewpoint, we might expect that 5G will be primarily IPv6 networks, which may bring new possibilities regarding seamless mobility, and user-centric orientation of service provisioning. At the time of 5G (2020 and beyond), the IPv6 address space is expected to become similar in size to the IPv4 address space, and to prevail further during the 2020s.

Overall, the 5G mobile network concept is seen as a user-centric concept [30], instead of the operator-centric concept of 3G or the service-centric concept of 4G. We might expect that the terminals will have access to different wireless technologies at the same time and that the mobile terminals should be able to combine different flows from different access technologies (i.e., heterogeneous networks) [31] to obtain very high bit rates (e.g., 10 times higher for individual users than those in 4G, and 1,000 times higher aggregate bit rates). So, 5G might involve a fundamental shift in the mobile networking philosophy toward user-centric mobile broadband networks and services.

6.8.1 5G Architectures

The 5G networks will need more diverse cell types than the standard macrocells used in earlier generations. Small cells boost a network's capacity, which is almost impossible to do with just the macrocell coverage. The radio access network is not the only important aspect of 5G development. Fronthauling (for interconnection of radio units, for different cell types) and backhauling (which is the transport network toward the Internet) are also very important. When very high bit rates are needed in the radio access part, the first choice for the backhaul is certainly fiber.

Practically, how can 5G services be implemented over a heterogeneous access infrastructure? Well, 5G is expected to support the massive growth in the number of connected devices, as well as massive growth in traffic (for example, 1,000 times more than LTE). Also, 5G will support low latencies on one side and gigabit-per-second bit rates for rich multimedia on the other side. We can conclude that 5G can be realized by using a combination of additional spectrum and more flexible resource usage (e.g., via spectrum sharing in the radio part, and infrastructure sharing in the fronthaul and backhaul) and by using numerous small cells via heterogeneous radio networks (e.g., HetNets). An example of a possible 5G architecture is shown in Figure 6.15.

However, the mobile fronthaul and backhaul become a major challenge in the development of 5G (Figure 6.15), so how should we proceed? There are two possible approaches, one is Centralized RAN, called cloud RAN (C-RAN) [32], and the other one is Distributed RAN (D-RAN).

D-RAN has baseband units (BBUs) that are colocated with antennas. In this approach the entire transport is via the backhaul. Exchange of data is only partly local at macrocell sites (more is transferred over the backhaul network). In contrast, C-RAN is based on centralized (as in a "pool") BBUs and many remote radio access units (RAUs). In this case all transport is so-called fronthaul (in the RAN). The concept of a trade-off between C-RAN (full centralization) and D-RAN (full decentralization) in 5G is referred to as RAN as a Service (RANaaS), which only partially centralizes functionalities of the RAN depending on the actual needs, as well as network characteristics [33].

In the 5G access part, densified small cell deployment with overlay coverage is emerging as a viable solution [34]. In such a heterogeneous and dense environment, one possibility is also to apply software-defined networking (SDN) and

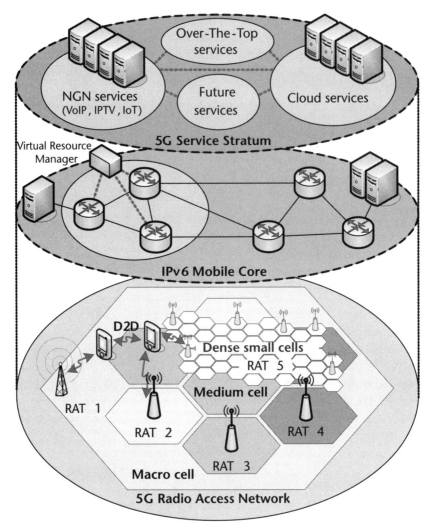

D2D – Device-to-Device communication
IoT – Internet of Things
IPTV – IP Television
RAT – Radio Access Technology
VoIP – Voice over IP

Figure 6.15 5G architecture.

provide centralized control via a standardized OpenFlow protocol for allocation of capacities, processing power, memory usage, and so forth, in network nodes in the RAN. SDN provides separate control and data paths by taking control logic from the underlying switches and routers (e.g., base stations in 5G) by centralized (or distributed) SDN controllers. Thus SDN may provide reconfigurable 5G HetNets, as well as easier implementation of new protocols and functions in the network.

What other technologies might be used by future 5G mobile operators? Well, besides high bit rates in the HetNets, the 5G mobile operator will also require much higher data rates for backhaul and fronthaul that can support the capacity provided in the radio access part. Such an architecture requires so-called aggregation points in the fronthaul (which can be colocated with some BSs). This means that the 5G architecture will require an optical network (e.g., Metro Ethernet) to the radio units, which is sometimes called Fiber-Wireless (Fi-Wi). Because the "Fi" will always have higher capacity than the "Wi," future 5G networks will need to "mix" them to provide the desired traffic volume.

6.8.2 5G Services and Business Aspects

When considering the future services provided by 5G, we should take into account different viewpoints, because many options are on the table, including operator-provided services, as well as OTT services that will continue to expand further with the development of mobile broadband access. Mobile users are expected to continue to use multimedia services (e.g., multimedia streaming) with richer content (e.g., higher resolutions, 3D), which require more bandwidth. Mobile cloud services are expected to continue the increasing trend and most likely to become the "platform" for all mobile services in near future; that is, smartphones will be used primarily as a gateway to access services performed in the cloud [33]. We can expect further expansion of social networking services (which are also cloud based) with the addition of richer content (e.g., videos, rich multimedia).

Further, machine-to-machine (M2M) and device-to-device (D2D) services are foreseeable in future 5G mobile environments. In fact, D2D communications enable the exchange of data traffic directly between mobile devices without the use of BSs. In addition, coupling of M2M and 5G will provide a practical implementation of the Internet of Things (IoT) concept [32]. We might also expect more LBS and user-centric services to be available to end users and businesses.

What is the user-centric mobile Internet? This term refers to an existing or new service that is used in a given mobile user context (e.g., when the user wants to be contacted or to receive certain information, or how to use information or communicate from home, the office, or while in a car). However, that is just one viewpoint. Other user-centric services may refer to giving control to the user regarding the setup of preferences for mobile and wireless network connections (e.g., in heterogeneous wireless environments) [31]. We could go further by mentioning the possibility of mobile users virtually personalizing their mobile devices, for example, users might have the option to control the bandwidth and resources allocated to different OTT services (since mobile operators are typically not allowed to do so, due to network neutrality principle).

Speaking about the next decade, we could say that cloud computing is the future "killer" application/service [33]. Why? Well, first, because cloud computing provides an excellent back-end for applications on mobile devices by providing access to valuable resources (for handheld mobile devices) such as computing power, memory storage, and so on. Further, in terms of services and contents, clouds provide the possibility for end users to access cloud-based services at any time, from any device, and from any location. Also, cloud services can be accessed from both fixed and mobile environments (since both are converging onto the Internet).

In parallel with the growth of mobile data traffic (which doubles almost every year, so in 10 years it will increase about $2^{10} \sim 1{,}000$ times), everyday life is becoming more and more based on running many different applications that are specific for mobile devices. Such applications are used for entertainment, business, conversational communication, shopping, social networking, traveling, information, and news. But, is this exponential growth in mobile and wireless traffic matched to parallel improvement on mobile terminals (e.g., smartphones)? Well, the fact is that the processing power of mobile devices follows Moore's law, more or less; that is, it doubles every 1.5 to 2 years. Batteries are improved with the same pace (according to the Koomey's law [35]), enabling many more mobile computing applications to be feasible, but is it enough? If the 5G target is set to much higher data rates, then processing power and batteries cannot improve at the same pace, as different services may demand. This determines a kind of widening gap between the energy required to run demanding applications (existing and new ones) and the available energy in mobile devices. How do we solve this situation? One way is to create energy-efficient applications and services. The other possible way is to enable mobile devices (where possible and needed for the application) to offload some energy-consuming tasks to some dedicated servers (in a cloud) or to other devices.

Then, cloud + mobile = mobile cloud. So, when a mobile user accesses cloud services through a mobile handset, it is called mobile cloud computing (MCC). However, one of the major limitations of today's mobile cloud services is the energy consumption associated with the radio access technology as well as the delay experienced in reaching the cloud provider through a mobile network.

Then, with the estimated capacities of 5G mobile networks and MCC-based services, people's work patterns and habits can be dramatically changed. We are transiting into a point in time where most Internet users will work primarily through Internet-based applications in clouds, accessed through various networked devices including fixed and mobile ones. Future MCC applications in 5G are expected to have a major impact on most of the activities in our personal and business lives. Humans will become connected to the network and dependent on it more than ever before.

Will be there something else in 5G with regard to mobile devices? Well, although smartphones and laptops will remain the largest market segments, we can expect increased connectivity in other business sectors such as mobile health, connected smart homes, connected cars, and connected transportation and logistics (which are using LBS services, as an example). At the same time the whole environment around the user can become smart and controlled via the 5G devices. Of course, that user should be authenticated and authorized to do so, which raises concerns about the security issues in 5G that should be solved on the run, especially in heterogeneous wireless/mobile environments.

Mobile networks such as 5G are expected to be highly reliable (considering heterogeneous infrastructure), so they will be able to support new applications related to the control of critical infrastructure, such as electrical grids, or to essential social functions such as traffic and smart-city management (i.e., public services). Also, with the very high data rates of 5G, it will be possible to deploy remote video monitoring and surveillance applications for certain businesses or homes. However, space must be allocated for some unforeseen applications that will appear in 5G mobile Internet.

Finally, let's look at the timeline for 5G. Industry developments for 5G mobile are focused on enabling a seamlessly connected society in the 2020 timeframe and beyond. That brings together people along with things, data, applications, transport systems, and cities in a smart networked communications environment. For this purpose ITU and other standardization bodies have recognized the relationship between the IMT and the 5G system, and they are all working and planning together for realization of the future vision of mobile broadband communications. In that manner, the ITU-R started a program in 2012 titled "IMT for 2020 and Beyond." This program marked the start of 5G research activities on a global scale. As a continuation of ITU's work in the development of radio interface standards for mobile communications, including IMT-2000 (for 3G) and IMT-Advanced (for 4G), work is being conducted toward standardization of IMT-2020 as the umbrella for 5G technologies. It is expected that the timeframe for 5G proposals will be focused around 2018 [36]. Afterwards, during the 2018–2020 time period, proposals will be evaluated for definition of the new 5G radio interfaces to be included in IMT-2020. Finally, the whole process should be completed in 2020 with a new ITU-R recommendation that will contain detailed specifications for the 5G radio interfaces. In conclusion, the path is already leading toward the future 5G mobile networks and services.

6.9 Regulation and Business Aspects for Mobile Broadband Internet

The many emerging mobile services can contribute to a better society in general. However, that process involves interrelated regulation and business activities that should provide efficient mobile broadband environments, which are accessible by all users and foster ICT innovations toward new services and new ways to use the mobile technologies.

6.9.1 Regulation Aspects for Mobile Broadband

In mobile environments the main targets for regulation are QoS, pricing, spectrum management, and addressing issues, such as mobile telephone numbers, domain names, and IP addressing.

6.9.1.1 Regulation of Mobile QoS

When applying QoS regulations, chosen QoS parameters should be easily understood by the public and be useful to them. Such parameters should be capable of verification by independent organizations, verification that can be performed by direct measurements or by audit of the operator's measurements. However, QoS parameters for regulation purposes should be limited in number and focus to hot topics (e.g., bit rates for Internet access). Overall, the QoS model for mobile services has four layers [37] as shown in Figure 6.16:

- The first layer is network availability, which defines QoS from the viewpoint of the service provider rather than the service user.
- The second layer is network access, which is defined from a user's point of view regarding basic requirement for all the other QoS aspects and parameters.
- The third layer contains other QoS aspects: service access, service integrity, and service retainability.

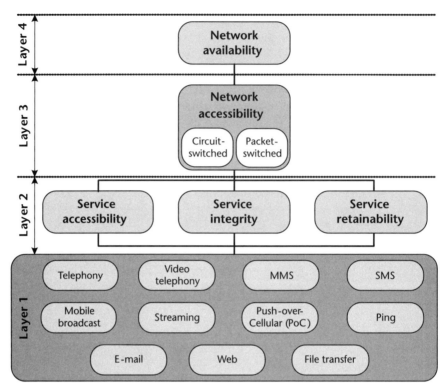

Figure 6.16 QoS model for mobile services.

- The fourth layer includes all different services, and the outcome is the QoS as perceived by the mobile user.

The QoS parameters (Figure 6.16) in layers 1 and 2 are common for all services, whereas QoS parameters in layers 3 and 4 are specific for a given service (do not confuse this layering with OSI protocol layers). Hence, QoS parameters for layer 1 and layer 2 in mobile networks are as follows: layer 1, radio network unavailability (%), and layer 2, network selection and registration failure ratio (%). Table 6.4 defines service-specific QoS parameters for what are currently the most relevant services in mobile markets, on layer 3 and layer 4 according to the layered mobile QoS model. However, regarding QoS regulation for mobile services, the most relevant parameters are the user-perceived end-to-end parameters. So, the QoS parameters (on all layers in the layered mobile QoS model) are possible targets for regulation regarding the QoS in mobile networks.

Mobile networks are specific because there are different network conditions on different geographic locations due to different capacities, different distances between BSs and MTs, different mobility speeds of the terminals, and so forth To measure the QoS parameters as perceived by users, drive tests are typical tools that are used by both regulators and mobile operators.

Mobile QoS regulation is harder than fixed QoS regulation due to constantly changing conditions in mobile environments. However, with selection of appropriate key performance and quality indicators, limited in number and important to end users (for which the QoS is important overall), it is possible to monitor and

Table 6.4 Mobile QoS Parameters

Service	Layer 3 QoS parameter	Layer 4 QoS parameter
Telephony	Telephony service nonaccessibility (%)	Telephony setup time (s)
		Telephony cut-off call ratio (%)
SMS	SMS service nonaccessibility (%)	SMS end-to-end delivery time (s)
		SMS completion failure ratio (%)
MMS	MMS send failure ratio (%)	End-to-end delivery time (s)
		MMS end-to-end failure ratio (%)
Video telephony	VT service nonaccessibility (%)	VT audio/video setup time (s)
		VT cut-off call ratio (%)
Streaming video	Streaming service nonaccessibility (%)	Streaming service access time (s)
		Streaming reproduction cut-off ratio (%)
Web browsing (HTTP)	HTTP service nonaccessibility (%)	HTTP setup time (s)
		HTTP session failure ratio (%)
		HTTP data transfer cut-off ratio (%)

enforce QoS in mobile networks. The contracted QoS for mobile services (e.g., via SLAs) should be monitored and measured by mobile operators and regulators, where regulators publish such information publicly (e.g., on a public website, in public magazine). Also, nonconformant operators may become subject to certain penalties or fines by the regulator.

The future development of mobile QoS regulation will lead toward creation of public databases taken from measurements of users' terminals.

6.1.1.2 Spectrum Management for Mobile Broadband

Spectrum is a limited resource and must be managed to ensure efficient and equitable access for services that use radio communications. However, radio waves do not respect national borders as people do, thus coordination of access to spectrum depends on international cooperation.

The goals and objectives of the spectrum management system are to facilitate the use of the radio spectrum within the ITU's radio regulations and in the national interest [38]. Spectrum management is dependent on regulation in a given country. Although no two countries are likely to manage spectrum in exactly the same manner, for spectrum that is important on international level, such as mobile broadband spectrum, there is a need for regional and global cooperation in terms of spectrum usage. Spectrum management on a global scale is managed and led by the ITU-R.

With the transition of analog TV to digital TV (i.e., to DVB-T), a new portion of spectrum with excellent radio propagation characteristics (around 700 to 800 MHz), called the digital dividend, has appeared. In addition to establishing the conditions for using the 800-MHz band (i.e., 790 to 862 MHz) in some regions (the "first" digital dividend), further spectrum allocations to the mobile service are in the 700-MHz band (i.e., 694 to 790 MHz), which is called the "second" digital dividend [39].

In the near future the ITU-R will consider additional spectrum allocations to mobile services on a primary basis and identification of additional frequency bands (e.g., 470 to 698 MHz and 4000 to 4999 MHz) for IMT, including IMT-2000, IMT-Advanced, and IMT-2020, to facilitate further the development of terrestrial mobile broadband services.

6.9.2 Business Aspects for Mobile Broadband

The convergence of telecommunications on Internet networks and services influences the mobile business sector. With the introduction of mobile Internet access, soon all mobile users will become at the same time Internet users. Since the penetration of mobile broadband (i.e., mobile users with smartphones) is higher in developed countries at the beginning, we can expect that most of the smartphone sales during the next years will originate in the developing countries, especially due to the growth of low-cost mobile smart devices [40].

Overall, mobile broadband is recognized as the fastest growing technology in human history. As shown in the Figure 6.17, much of the growth in mobile broadband has occurred in the developing world [40].

Mobile broadband connectivity is not a substitute or universal remedy, but when integrated with existing systems, it can enhance certain service delivery (in a country or a region) or facilitate development of new services with the goal of delivering the best results. For example, mobile health could potentially save many lives in Africa, or it could generate several hundreds of billions of dollars of savings in developed countries [40]. Also, mobile learning may significantly contribute to educational development, including the creation of globally accessible learning environments.

Business processes and business models have been transformed by mobile broadband development, which increases the overall efficiency of workers and their capabilities due to the availability of information on the run.

The expanding broadband ecosystem (e.g., application ecosystems) becomes an engine of society transformation in all parts of the world and provides sustainable

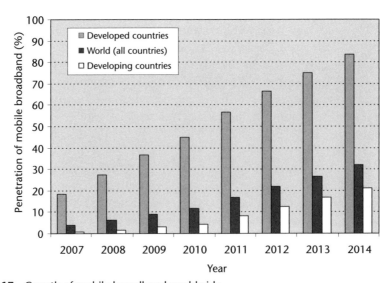

Figure 6.17 Growth of mobile broadband worldwide.

development. The aim of broadband is not only to allow businesses to generate more revenues, but also to provide humanity with access to information and services as a fundamental human right worldwide.

6.9.3 Discussion

The mobile industry is continuously developing and introducing IP-based services such as Voice-over-LTE, mobile IPTV, and rich communication services. The established competition in all segments (including mobile operators and OTT service providers) is generating innovations that provide benefits to individual mobile users and to society. In practice, however, the mobile sector is among the most intensively regulated industry sectors, subject to many common rules, including consumer protection and privacy, network interoperability, security, emergency calls (for carrier-grade voice), lawful interception of customer data, universal service, and more. Most of these rules are under the jurisdiction of national regulatory agencies (NRAs). Additionally, mobile operators use spectrum, for which they pay heavy levies. All requirements as well as many national taxes, levies, and fees directly influence the business of mobile operators.

In contrast, Internet OTT players, which offer equivalent voice, messaging, and streaming services, are subject to none of these requirements or the cost of compliance [40]. For example, a Skype call is not bound by the same rules as a mobile phone call provided by a mobile operator. Further, there are many antidiscrimination and data protection rules that apply only to telecom operator companies, and not to the communication taking place via unregulated platforms of OTT services. There is a certain view by operators' associations (e.g., GSMA [40]) that the same service should be subject to the same rules, that is, there should not be any regulatory double standards. Also, the responsibility for protection of privacy and personal data is often not widely understood by end users. Many users locate such responsibility with the mobile operator, while in reality the mobile operator has no control over which data a third-party application captures from a given user's device (e.g., smartphone). So, this current asymmetric regulation has resulted in uneven competition for services, which is heavily in favor of the OTT players. But, responsibility for the regulation of services is with the governments and NRAs on a national level and with international organizations that provide harmonization on a global scale (e.g., ITU).

However, a sustainable and stable mobile industry (with healthy competition and established market forces) is fundamental to the spread of mobile broadband Internet access around the world. Mobile services cannot be provided without broadband mobile networks. On the other hand, OTT services are the main drivers for the introduction of mobile and fixed broadband access, due to the heterogeneity of such services and their fast innovation (short time to the market), which is also driven by the network neutrality principle. Overall, fair rules for mobile business (including mobile operators, OTT service providers, and vendors) create an environment that leads to the best experiences for citizens everywhere in the world, through continuing competition and innovation.

References

[1] C. Perkins, "IP Mobility Support," RFC 2002, October 1996.
[2] T. Janevski, *Traffic Analysis and Design of Wireless IP Networks*, Norwood, MA: Artech House, May 2003.
[3] C. Perkins, "IP Mobility Support for IPv4, Revised," November 2010.
[4] C. Perkins, D. Johnson, J. Arkko, "Mobility Support in IPv6," July 2011.
[5] R. Koodli, "Mobile IPv6 Fast Handovers," RFC 5568, July 2009.
[6] H. Soliman et al., "Hierarchical Mobile IPv6 (HMIPv6) Mobility Management," RFC 5380, October 2008.
[7] S. Gundavelli et al., "Proxy Mobile IPv6," RFC 5213, August 2008.
[8] R. Moskowitz et al., "Host Identity Protocol," RFC 5201, April 2008.
[9] R. Moskowitz, P. Nikander, "Host Identity Protocol (HIP) Architecture," RFC 4423, May 2006.
[10] P. Nikander et al., "End-Host Mobility and Multihoming with the Host Identity Protocol," RFC 5206, April 2008.
[11] P. Jokela, R. Moskowitz, P. Nikander, "Using the Encapsulating Security Payload (ESP) Transport Format with the Host Identity Protocol (HIP)," RFC 5202, April 2008.
[12] ITU-R Report M.2134, "Requirements Related to Technical Performance for IMT-Advanced Radio Interface(s)," 2008.
[13] R. Stewart, "Stream Control Transmission Protocol," RFC 4960, September 2007.
[14] 3GPP TS 29.060, "GPRS Tunnelling Protocol (GTP) Across the Gn and Gp Interface (Release 13)," March 2015.
[15] 3GPP TS 29.274, "Evolved General Packet Radio Service (GPRS) Tunneling Protocol for Control Plane (GTPv2-C); Stage 3 (Release 13)," March 2015.
[16] H. Holma, A. Toskala, *LTE for UMTS: Evolution to LTE-Advanced*, 2nd ed., New York: John Wiley, March 2011.
[17] 3GPP, "Numbering, Addressing and Identification (Release 13)," TS 23.003, March 2015.
[18] M. Alasti et al., "Quality of Service in WiMAX and LTE Networks," *IEEE Communications Magazine*, May 2010.
[19] ITU-T Recommendation G.114, "One-Way Transmission Time," May 2003.
[20] ITU-T Recommendation Y.1901, "Terms and Definitions for IPTV," March 2010.
[21] ITU-T Recommendation Y.1903, "Functional Requirements of Mobile IPTV," January 2014.
[22] ITU-T Recommendation Y.1910, "IPTV Functional Architecture," September 2008.
[23] ITU-T Recommendation Y.1911, "IPTV Services and Nomadism: Scenarios and Functional Architecture for Unicast Delivery," April 2010.
[24] 3GPP TS 23.246, "Multimedia Broadcast/Multicast Service (MBMS); Architecture and Functional Description (Release 13)," March 2015.
[25] WiMAX Forum, "Mobile WiMAX: The Best Personal Broadband Experience!," June 2006.
[26] ITU-T Y.1900-Series Supplement 20, "Supplement on Scenarios and Use Cases of Mobile IPTV," May 2008.
[27] ETSI EN 302 304, "Digital Video Broadcasting (DVB); Transmission System for Handheld Terminals (DVB-H)," November 2004.
[28] World Wide Web Consortium, "Mobile Web Design," July 2006.
[29] "Actionable Analytics for the Web," www.alexa.com, accessed April 2015.
[30] T. Janevski, "5G Mobile Phone Design," IEEE Consumer Communications and Networking Conference 2009, Las Vegas, NV, January 10–13, 2009.

[31] T. Shuminoski, T. Janevski, "Radio Network Aggregation for 5G Mobile Terminals in Heterogeneous Wireless and Mobile Networks," *Wireless Personal Communications*, Vol. 78, pp. 1211–1229, 2014.

[32] B. Bangerter et al., "Networks and Devices for the 5G Era," *IEEE Communications Magazine*, February 2014.

[33] R. Peter et al., "Cloud Technologies for Flexible 5G Radio Access Networks," *IEEE Communications Magazine*, May 2014.

[34] X. Duan, X. Wang, "Authentication Handover and Privacy Protection in 5G HetNets Using Software-Defined Networking," *IEEE Communications Magazine*, April 2015.

[35] J. G. Koomey et al., "Implications of Historical Trends in the Electrical Efficiency of Computing," *IEEE Annals of the History of Computing*, pp. 46–54, 2011.

[36] E. Dahlman et al., "5G Wireless Access: Requirements and Realization," *IEEE Communications Magazine*, December 2014.

[37] ITU-T Supplement 9 to ITU-T E.800-Series Recommendations, "Guidelines on Regulatory Aspects of QoS," December 2013.

[38] ITU-R Report SM.2012-3, "Economic Aspects of Spectrum Management," 2010.

[39] ITU-R Recommendation M.1036-4, "Frequency Arrangements for Implementation of the Terrestrial Component of International Mobile Telecommunications (IMT) in the Bands Identified for IMT in the Radio Regulations (RR)," March 2013.

[40] ITU, "The State of Broadband 2014: Broadband for all," Broadband Commission Report, September 2014.

CHAPTER 7

Broadband Internet services

The development of fixed and mobile broadband Internet is driven by broadband Internet services, and vice versa. We can define broadband Internet services in general as services that require high bit rates in one or both directions (to/from end users). However, several other classifications of services are possible. One typical classification of Internet services (regardless of their bandwidth requirements) that are provided via broadband access is based on their delay requirements:

- Real-time services: These services have strict end-to-end delay (and jitter) requirements, and typically include conversational services (e.g., telephony and video telephony, Skype, Viber) and video streaming (e.g., IPTV, VoD).
- Non-real-time services: These services have less strict requirements on delay and jitter, and typically include all best-effort services that are based on Internet network neutrality (e.g., Web, BitTorrent).

Another classification of services can be made based on the particular services delivered by a provider:

- Carrier-grade services: These services are provided by fixed or mobile telecom operators based on SLAs between the operator and the subscriber. In this case the network provider (i.e., telecom operator) is also a service provider. Typical services that belong to this group are VoIP and IPTV, both with QoS support end to end.
- OTT services: These services are based on best-effort and network neutrality principles on the public Internet. This type covers all services that are provided by service providers other than telecom operators through which the users (of the service) connect to the Internet.

This chapter covers the main broadband Internet services that are provided via broadband Internet access, including services offered by telecom operators (e.g., QoS-enabled VoIP and IPTV) and OTT services (e.g., multimedia streaming, peer-to-peer services, social networks), as well as emerging ones (e.g., Internet of Things).

7.1 VoIP as the PSTN/PLMN Replacement

With the transition to all-IP networks, telephony will transit to VoIP. This transition to operator-provided VoIP is standardized within the NGN release 1 framework. In fact, the first target of NGN standardization in the 2000s was standardization of carrier-grade VoIP as a replacement for PSTN/ISDN digital telephony. NGN-based, QoS-enabled VoIP requires backward compatibility with its predecessor PSTN/ISDN, so such a transition is transparent to the end users of voice services through telecom operators (either fixed or mobile ones).

In general, NGN has two implementations for the provision of VoIP as PSTN/ISDN replacement (Figure 7.1) [1]:

- Emulation: This refers to provision of most of the existing PSTN/ISDN service capabilities and interfaces by adapting them to the IP environment in NGN. In this case the user receives standard PSTN/ISDN services (i.e., the end-user experience remains unchanged), although the provider's network is changed to NGN (i.e., changed core network). The adaptation function (ADF) is typically implemented via access gateways that are located at the NGN provider's premises.

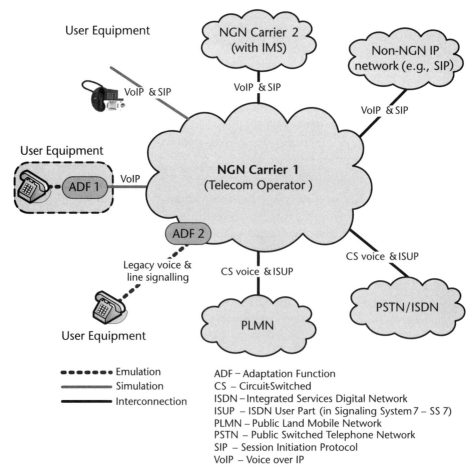

Figure 7.1 PSTN/ISDN emulation and simulation for carrier-grade VoIP.

- Simulation: This refers to the same service provision in the NGN as the emulation approach, but there is no guarantee that all PSTN/ISDN features are available to end customers. However, this approach may provide additional new features and capabilities that have not been available previously in PSTN/ISDN. The typical approach for PSTN/ISDN simulation in the NGN is an IMS-based approach. In this case ADF is implemented in residential gateways (e.g., in user homes).

Generally, the ADFs may be implemented in access or residential gateways. The ADF on one side interfaces with the IP multimedia component by using SIP as a signaling protocol, whereas on the other side it interfaces with PSTN/ISDN terminals and provides simulation and emulation services (Figure 7.1).

PSTN/ISDN emulation [2] is more complex and requires several different network nodes for its implementation. The emulation approach requires different signaling protocols, such as H.248 and SIP (in IMS-based networks). However, the communication between the signaling gateways and the external PSTN/ISDN networks is implemented by using ISUP signaling as part of the SS7 signaling, which is carried by use of SIGTRAN (i.e., ISUP/SS7 over IP-based networks) in all-IP core networks.

The PSTN/ISDN simulation in the NGN is based on IMS, which is also used for different services in the NGN (e.g., IPTV, VoIP, web-based services). Also, IMS uses unified signaling between all signaling entities in the operator's signaling network, including SIP communication between IMS entities and IP phones (at end-user premises) and access gateways (in cases when legacy user equipment is used).

Additionally, for PSTN/ISDN-to-NGN migration (or evolution) the practical approach is to install multiservice nodes [3], which can provide QoS-enabled VoIP service through different broadband access networks to the Internet. So, migration from legacy digital telephony networks to QoS-enabled VoIP requires changes only to the underlying transport technologies for voice (e.g., Internet technologies are used for carrying voice traffic end to end instead of circuit switching in digital networks) without the need for changes in the end user's equipment (e.g., dummy fixed telephones). It also provides end-to-end QoS guarantees that are specified in the SLA between the telecom operator and the subscriber.

However, another type of voice applications offered by third-party service providers has appeared on the Internet scene by using the network neutrality aspect of broadband Internet access. These voice services are provided independently from fixed and mobile network providers (i.e., telecom operators) while using the broadband access to the Internet provided by the telecom operators. They are referred to as OTT VoIP.

7.2 Over-the-Top (OTT) Voice-over-IP

After the standardization of SIP at the end of the 1990s and beginning of the 2000s (RFC 2543 in 1999 was the first SIP standard, and it was made obsolete by RFC 3261 in 2002), Skype became the most successful peer-to-peer VoIP services provided as a third-party OTT service by using the network neutrality of Internet access.

In general, all OTT VoIP services are proprietary, which means that they are not based on open protocols and standards such as RFCs from the IETF or ITU-T recommendations. OTT VoIP design, however, has the advantage of being different than other similar solutions; for example, it has a user-friendly interface and allows for specific user identities. At the same time, that is a disadvantage because different OTT services have different addressing schemes for end users and proprietary signaling and voice data transfer mechanisms that are not compatible or interoperable like legacy telephony is.

7.2.1 Skype

Skype is the best-known OTT VoIP service in fixed and mobile environments. It first appeared as an application to be used in fixed hosts, but it moved to the mobile playground with the spread of mobile broadband networks. In this subsection we discuss the most important technical characteristics of Skype that are known as a result of analysis, because Skype is a closed technology solution (i.e., not standardized), and hence it is not based on open standards.

Skype is referred to as being a peer-to-peer VoIP client developed by KaZaa in 2003 [4]. It is a type of overlay P2P network across the global Internet, consisting of two main types of nodes:

- Ordinary node: This hosts a Skype application that can be used to place voice calls (optionally with video) and send text messages.
- Supernode (SN): Any ordinary node with an assigned public IP address and sufficient processing power, memory capacity, and network bandwidth is a candidate to become a Skype supernode.

The Skype network architecture is shown in Figure 7.2. An ordinary host must connect to a supernode and must register itself with the Skype login server to accomplish a successful login. User names and passwords are stored at the login server, so user authentication at login is also done at this server. This server also ensures that Skype login names are unique across the Skype name space. Additionally, Skype encrypts communication end to end and for that purpose it uses the Advanced Encryption Standard (AES), and for that purpose the users' public keys are certified by the Skype server at login. Apart from the login server, the Skype network has no central server. Online and offline user information is stored and propagated in a decentralized fashion and so are user search queries.

Skype provides end-to-end communications (after discovery of the peers between each other using the decentralized login servers), so it must work across NATs and firewalls. It is believed that each Skype node uses a variant of the STUN protocol [5] to determine the type of NAT and firewall it is behind.

The Skype network is a virtual network from the viewpoint of telecom operators, or an overlay network from the viewpoint of the public Internet. To maintain connectivity, each Skype client should build and refresh its own table (called a host cache) of reachable Skype nodes. Such a host cache also contains IP addresses and the port number of supernodes.

7.2 Over-the-Top (OTT) Voice-over-IP

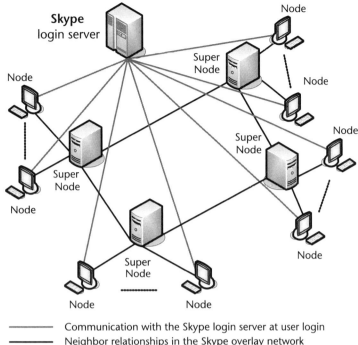

Figure 7.2 Skype network architecture.

Skype probably has implemented a so-called "global index" technology, which is guaranteed to find a given user if that user has logged into the Skype network in the past 72 hours [4]. For voice data traffic Skype uses wideband codecs (50 to 8,000 Hz), which provide reasonable call quality with bandwidth of about 32 Kbps.

Regarding the transport layer protocols, Skype uses TCP for signaling, and both UDP and TCP for transporting voice media traffic. However, signaling and media traffic are not sent through the same ports. A given Skype client opens a TCP and a UDP listening port at the port number configured in its connection dialog box, which is randomly chosen upon installation. In addition to the random port numbers, which are not well-known ports assigned by IANA, a Skype client also listens on TCP port 80 (i.e., an HTTP port) and TCP port 443 (i.e., an HTTPS port). Because it is a proprietary application/service (not a standard), there is no default TCP or UDP listening port, something that is typical for many OTT services.

As a contact list Skype uses a so-called "buddy list," which is digitally signed and encrypted. However, the buddy list is local to one machine and is not stored on a central server (which is different from some other OTT VoIP applications, such as Viber), so when a given user uses Skype on a different machine to log onto the Skype network, he has to reconstruct the buddy list.

Finally, Skype's presence on the global market for more than a decade shows that it has long-lasting business models, which is based on the use of prepaid cards for cheaper calls to public telephone networks, deployment of so-called Wi-Fi Skype hotspots (e.g., in hotels), and so forth. However, note that Skype has changed the "big picture" for voice traffic globally, especially regarding international calls (from one country to another), which are free for Skype-to-Skype users (of course, Internet access is needed and usually it is not free).

7.2.2 Other OTT VoIP Applications

Besides Skype, several other OTT VoIP applications have had notable success, including Viber, Google Talk, ICQ, Nimbuzz, Yahoo, Fring, Vonage, WeChat, and Tango, which are typically targeted for use in smartphones [6]. There are, however, several differences among them that influence their global popularity:

- Authentication and numbering: Two main methods are used for authentication, although both are based on a username and optionally a password:
    - Username/password authentication: This type is used by OTT VoIP applications that were primarily designed for fixed hosts (e.g., computers) or equally targeted to fixed and mobile hosts. Skype belongs in this group because it uses typical username/password authentication, as does Google Talk (requires a valid Gmail account), Yahoo (requires a valid Yahoo account), and others.
    - Number of the mobile user: This is typically used by OTT VoIP applications developed for mobile devices that have E.164 numbers assigned to them via SIM or USIM cards. Such VoIP applications include Viber, Fring, Vonage, WeChat, and Tango.
- Encryption: All VoIP applications encrypt messages, but voice communication is encrypted only in a few OTT VoIP applications (Skype, Viber, Nimbuzz, Tango, and so on).
- Contact list: Some OTT VoIP applications, such as Skype, do not upload a contact list obtained from a user's devices (typically a mobile phone), whereas most of the others (e.g., WhatsApp and Viber) upload the address book from the mobile device.

Almost all OTT VoIP applications, especially for mobile devices, show certain security threats, especially from man-in-the-middle attacks at the authentication phase (Viber, WhatsApp, and others) [6]. Because OTT applications use the network neutral approach of Internet access with best-effort service, no additional security guarantees are deployed by telecom operators. Most of the mobile OTT applications for VoIP (including messaging) use mobile phone numbers for authentication [7], where verification is done via a SMS sent to the given mobile number (Figure 7.3).

One possible problem with OTT VoIP applications for mobile devices is the privacy concern regarding the automatic import of user's contacts. Most of them, including Viber and WhatsApp, upload the entire address book from the mobile device to the system's servers and also compare the phone numbers from the contact list to already registered phone numbers stored on servers for the given OTT VoIP provider. Several security problems could arise in such a situation, including enumeration (finding active mobile phone numbers), identification of the operating system of a given user, and privacy concerns, as well as control regarding access to the Internet (e.g., Wi-Fi or mobile access network).

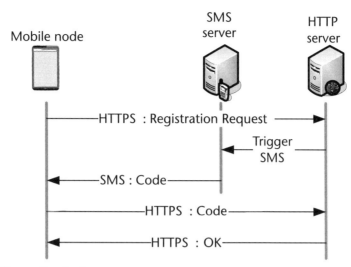

Figure 7.3 Viber authentication.

Finally, regarding QoS, the OTT VoIP applications have no performance guarantees at all; that is, they are provided end to end in a best-effort manner and using the network neutrality of the public Internet. So, a lack of QoS support and therefore a lack of real-time mobility support are some of disadvantages of OTT VoIP applications. Also, they are not obliged to conform to policies applied to telephony on a national level in each country (e.g., emergency numbers, privacy guarantees, numbering, number portability). However, their use through fixed broadband access (which also includes Wi-Fi access) makes them available for free in cases when both users use the same application (e.g., Skype-to-Skype, Viber-to-Viber), which is especially attractive for international calls. Also, OTT VoIP services provide for the higher possibility for innovation of services, which may be more attractive to users at a given time.

Overall, OTT VoIP services will continue to exist in parallel with QoS-enabled VoIP as a PSTN/ISDN replacement, because they have different features and characteristics, as summarized in Table 7.1, that can be used by different user groups or by the same users for different purposes (e.g., communication, collaboration, entertainment, gaming) in different situations (e.g., at home, at the office, abroad).

7.3 IPTV Services

Besides telephony, the television is also transferring to an all-IP environment. To assist with this transition, NGN release 2 was dedicated to standardization of IPTV (note that NGN release 1 was dedicated to voice).

IPTV has different implementations based on four IPTV domains: the end user, network provider, service provider, and content provider [8] (discussed in Chapter 6). However, the first IPTV implementations were based on proprietary platforms for signaling with use of standardized Internet protocols, such as the RTP/UDP/IP protocol stack for transfer of IPTV data, as well as DNS and HTTP for discovery of IPTV channels and finding the locations (i.e., IP addresses) of IPTV media servers.

Table 7.1 Comparison of QoS-Enabled VoIP and OTT VoIP

	QoS-Enabled VoIP	OTT VoIP
VoIP providers	Telecom operators (fixed and mobile)	Global third-party service providers
QoS guarantees	End-to-end QoS	Best effort (no QoS)
Mobility	Designed for high-speed seamless mobility for voice	Mobility support as provided for all best-effort Internet services (the lowest level)
Innovation	Standardization of all architectures, protocols, and service types	Proprietary applications over standardized Internet protocol stack
Market	National	Global
Main business target	National-level calls	International calls
Regulation	Strict regulation of voice (QoS, legal sniffing, phone numbering)	No regulation (network neutrality principle in public Internet)
Additional services	Different services provided via different bearers and/or applications	Additional services typically embedded in the VoIP application (e.g., messaging, file sending)

In practice, several terms are typically used to mean TV delivery over IP networks: IPTV, Internet TV, or Web TV. What is the difference between them? IPTV is delivered to TV sets through the Internet architecture and networking methods that provide QoS guarantees (regarding bit rates, packet delay, losses, and jitter) as well as AAA functions. Internet TV and Web TV, on the other hand, refer to delivery of television to a PC or other devices connected to the Internet (e.g., smartphones) in a best-effort manner (i.e., without QoS support) and typically without any AAA mechanisms (e.g., free to watch on a national, regional, or global level). Overall, note that IPTV provides better QoS than Internet streaming video (e.g., YouTube) and it is defined and standardized by ITU-T within NGN as a broadcast TV replacement in all-IP networks.

In general, IPTV services can be grouped into two main groups:

- Broadcast/multicast IPTV service: In this case linear TV programs (i.e., scheduled programs) are sent to a large number of subscribers. Each subscriber individually tunes to a given channel and typically cannot influence the playback.
- VoD: In this case stored videos are streamed to subscribers on request, so each subscriber can control the video playback (typically using VCR commands, such as "play," "pause," "stop," "forward").

According to the NGN framework, the IPTV functional architecture provides the necessary network and service functions to deliver TV content via an all-IP network to end users. There are three different types of IPTV functional architectures [9]: non-NGN IPTV, NGN-based non-IMS IPTV, and NGN-based IMS IPTV functional architecture.

7.3.1 IPTV Content Delivery

IPTV delivery includes signaling for discovery of media servers as well as resource reservation, followed by data transfer (i.e., IPTV channel stream). A given IPTV functional architecture can be used with different delivery mechanisms, including multicast delivery [10], unicast delivery [11], or a combination.

7.3.1.1 Multicast IPTV Delivery

Multicast delivery of IPTV content is based on standardized Internet protocols and mechanisms for multicast (e.g., IGMP for IPv4, MLD for IPv6, multicast routing protocols), which are used in several defined models. In that manner, ITU-T defines four different functional models for multicast-based delivery of IPTV content in NGN environments [10]:

- Network multicast model: In this case multicast capabilities are provided by the network provider domain via multicast transport functions (MTFs). The end-user IPTV function registers to a specific multicast group by using multicast protocol messages for joining to the group, such as IGMP in IPv4 networks [12], and Multicast Listener Discovery (MLD) [13] in IPv6 environments. In this model the network provider manages the multicast capabilities in its network, and the obtained performances of the IPTV services are dependent on the network conditions such as traffic load (therefore, QoS mechanisms should be implemented).
- Cluster model: This model uses multicast capabilities provided by the service provider domain, and multicast is supported by a content delivery function (CDF). The clusters for IPTV delivery (within the CDF) are placed in certain chosen locations.
- Peer-to-peer model: The P2P model is based on multicast capabilities supported and controlled by EU. In this case an end user can also receive IPTV content from other users and simultaneously distribute the IPTV content to other peer users. However, end users distribute contents to other end users by using unicast transport since they cannot create multicast paths over the transport networks. Also, this model lacks efficient management mechanisms (e.g., traffic management, AAA).
- Hybrid model: The hybrid model is a combination of the cluster and P2P models. It uses multicast capabilities from the CDF in the service provider's domain (provided in telecommunication equipment, and operated by a service provider) and at the same time assists with P2P delivery in the end-user domain. This model has the same characteristics as the cluster model and higher scalability due to possible P2P content distribution in the end-user domain. It provides better management possibilities than the "pure" P2P IPTV delivery model.

Although multicast routing has been defined in the Internet from the beginning of the 1970s, such protocols were not deployed practically until IPTV deployment began to be driven by development of broadband Internet access. As concluded in Chapter 3, of the many multicast protocols standardized by the IETF, the most used one is PIM-SM [14], which is used for intra-AS multicast routing, which is typically used in the network provider model for IPTV delivery (Figure 7.4). However, besides multicast, all IPTV delivery models also use unicast-based IPTV delivery in some parts of the end-to-end connection between media servers (i.e., servers with CDF) and end-user equipment, for example, set-top boxes (STBs).

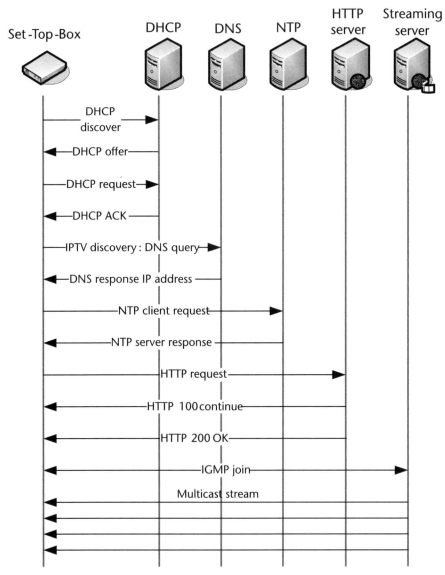

DHCP – Dynamic Host Configuration Protocol
DNS – Domain Name System
HTTP – Hypertext Transfer Protocol
NTP – Network Time Protocol

Figure 7.4 IPTV network architecture and networking protocols.

7.3.1.2 Unicast IPTV Delivery

Unicast-based delivery of IPTV is a typical scenario in cases where users are nomadic within the network, and when the IPTV service provider is not located in the network provider domain of the end user. There are two scenarios for unicast-based IPTV content delivery:

- Nonroaming scenario: This scenario assumes that IPTV terminal functions (ITFs) can use standard mechanisms (e.g., signaling and control protocols for IP networks) for communication with service control functions (SCFs), including IMS-based and non-IMS implementations.
- NGN roaming scenario: In the roaming NGN IPTV scenario, both SCFs of both NGN operators should be capable of communicating with each other (e.g., both SCF to be IMS based).

The IPTV content in both cases (unicast and multicast) is distributed by using video and audio codecs (e.g., MPEG-2, MPEG-4) over the RTP/UDP/IP protocol stack on the end hosts (i.e., IPTV server for content distribution, and IPTV client in the end-user equipment for IPTV, i.e., STBs).

Regarding the signaling related to IPTV service provision, non-IMS implementations appeared in the second half of the 2000s at many telecom operators that started to penetrate into the TV services market by provision of bundled services (e.g., combination of telephony, broadband Internet access, and/or IPTV, in fixed and mobile networks). However, in the long term, we could expect that NGN IMS-based IPTV deployments will have the dominant market share, although they will likely coexist with non-IMS IPTV implementations in parallel.

7.3.2 Internet Technologies Used for IPTV Service

Generally, the Internet technologies used in non-NGN IPTV implementations (i.e., the initially deployed overlay IPTV networks) and NGN-based IPTV implementations with IMS functionalities have similarities and differences.

A typical non-NGN IPTV communication between the end-user IPTV equipment (e.g., STB) and nodes/servers in the provider's network (either network or service provider) is shown in Figure 7.5. In the general case, the STB on the end user's side requires an IP address, which can be allocated via DHCP. The IPTV service discovery is typically based on HTTP URLs (e.g., iptvservice.telecomoperator.example.com), which are resolved by the DNS. IPTV is a real-time service and hence synchronization of the STB and the network side equipment is needed in scenarios with a network provider IPTV delivery model (which includes a QoS provision), which is also sometimes referred to as "in-house" IPTV service delivery (i.e., it is not accessible via Internet access through some other operator's network). For purposes of synchronization the Network Time Protocol (NTP) is typically used. NTP was initially standardized in 1985 with RFC 958; the current standardized version is NTP version 4 (RFC 5905 [15]). Finally, access to the media streaming servers is through HTTP and web servers, which contain metadata to access the contents. Hence, each STB also contains an adapted HTTP client. After discovery of the media servers, the RTP/UDP media stream (based on MPEG-2/MPEG-4,

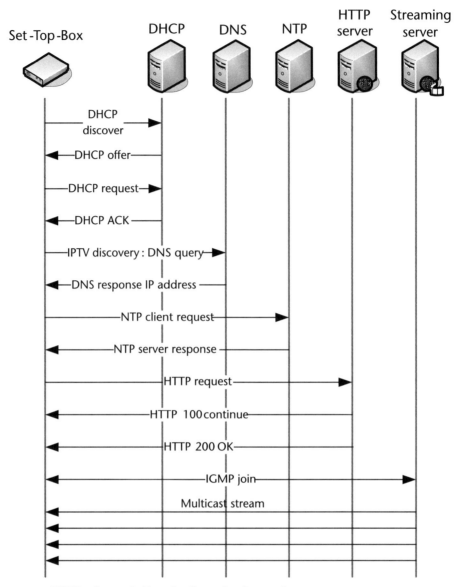

Figure 7.5 Internet technologies used for general non-NGN IPTV provision.

and H.264) is delivered to the STB and finally to the TV device. Note that several proprietary reliable UDP implementations for IPTV content delivery (not standardized by IETF) were available that provided guaranteed-order packet delivery and simple flow control (less complex and faster than TCP). Finally, the user control for display of the IPTV stream typically is based on the RTSP as an out-of-band control protocol.

In the NGN IMS-based IPTV scenario, the ITFs use standard mechanisms, such as signaling and control protocols for IP networks, for communication with the

7.3 IPTV Services

core IMS (based on the SIP and Diameter protocols for communication with other entities in the service stratum), as shown in Figure 7.6. In the nonroaming scenario, end-user functions interact with the network functions located in the home NGN network, where the NACF performs access authentication. In the case of successful authentication, the end-user functions perform IPTV application and service discovery and selection by using application-based functions. After service discovery and selection have been completed, the end-user functions set up an IPTV session via interaction of the core IMS with the CDF in the NGN. For QoS provision to IPTV (i.e., admission control and resource reservation for the IPTV session), the core IMS exchanges information with the RACF of the home NGN. Finally, the CDF starts sending IPTV content to end-user functions. Similar to the non-NGN IPTV case, the content is delivered by using video/audio codecs (e.g., MPEG-2, MPEG-4) over the RTP/UDP/IP protocol stack.

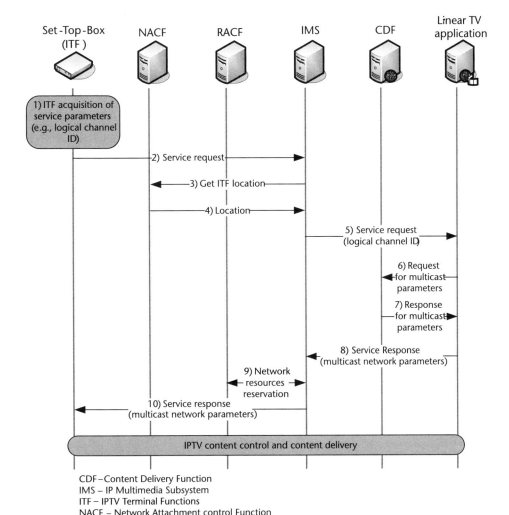

Figure 7.6 NGN IMS-based IPTV delivery.

7.3.3 Traffic Management, QoS, and QoE for IPTV

IPTV services require QoS support. For that purpose there is needed traffic management in different parts of the IPTV network architecture, including the core network, access network, and home network [16]. However, all three parts of the network are usually shared between a number of applications such as data (e.g., OTT applications) and voice (i.e., operator-provided VoIP).

The core network of the telecom operator is expected to be well engineered, which means deployment of Internet QoS mechanisms (covered in Chapter 5), such as DiffServ or IntServ, and MPLS with traffic engineering (with applied network policies regarding traffic from different services and for different types of users). Regarding the current time and near future, we can expect that the core network will be equipped with DiffServ functionalities due to their scalability to a large number of flows. For example, if IPTV is provided with DiffServ by using expedited forwarding (EF), and a token bucket with parameters (B, R) is used for traffic shaping, then the delay of any packet departing an interface (of a network node) is limited by the following value:

$$D \leq B/R + E_a \tag{7.1}$$

In this equation R is the token bucket rate (in bytes per second), B is the token bucket size (in bytes), and E_a typically represents the value of non–rate-dependent delay (a combination of fixed minimum delay due to propagation and processing, and variable delay due to buffering) [16]. However, the QoS for IPTV can also be guaranteed with the Internet QoS solutions, where DiffServ is currently the most used approach.

In contrast, the access networks for IPTV delivery should have the means to control per-flow (for each IP packet) and per-class connection admission control (CAC). For example, if a subscriber has access to the Internet through an ADSL line with 8 Mbps in the downlink direction and one SD IPTV stream requires a 2.5- to 3-Mbps dedicated bit stream, then the maximum number of TV channels that can be simultaneously streamed through such Internet access is two. However, the remaining throughput is used typically for best-effort Internet access, thus the number of streamed TV channels in such a case influences the experience for the OTT services used in parallel with IPTV services (in the given example with ADSL access). If, however, access is provided via 50 to 60 Mbps in the downlink direction, then it can receive multiple streams in parallel (e.g., on multiple TV sets). Further, the access network should provide per-user and per-service (e.g., linear TV, VoD) QoS provisioning. Since the last meters in each access network (for IPTV delivery) are based on Ethernet access, per-packet/flow mapping between IP DSCP markings and Ethernet priorities (IEEE 802.1Q) should also be provided.

The access networks carry traffic in both the upstream and downstream directions. In the upstream direction, network nodes should be able to guarantee a minimum bandwidth (in bits per second) and QoS per service (e.g., VoIP, IPTV, video conference). For such purposes, the following are typically used: 5-tuple classification and policing, priority queuing (typically VoIP has the highest priority, followed by IPTV priority over the best-effort traffic), and traffic shaping (e.g., token bucket, which limits the maximum bit rate per flow with the token bucket size).

In the downstream direction, for QoS support the network nodes are requested to guarantee the minimum bandwidth per service (e.g., per IPTV flow/stream). For that purpose, access networks should support per-flow and per-class CAC, traffic shaping and policing per subscriber, classification of IP packets based on the differentiated services code point (DSCP) field in the IP headers (if used, this approach is used only in the downlink due to network control of that direction), and finally mapping between DSCP mappings and IEEE 802.1Q Ethernet priorities.

Finally, each distribution of IPTV traffic ends in the home network of the subscriber, which is typically a LAN (Ethernet) or wireless LAN (Wi-Fi). The home gateway (HG) is the node placed in a user's home that acts as an interface with the home network to provide network access. Typically, one HG is used for different service types (e.g., operator-provided VoIP, IPTV, and best-effort Internet access) and it is connected through an Ethernet switch to the STB devices (there is typically one STB per TV set). Such home Ethernet switches may support user priority bits (known as p-bits), which define up to eight priority levels that can be used to identify traffic classes such as VoIP, IPTV, and best-effort services. In the case of Wi-Fi home networks for IPTV delivery, QoS support can be provided only with access points (APs) that have included IEEE 802.11e (in such a case, different access categories are used for different services).

The QoS provides an objective measure for IPTV delivery. However, QoE for IPTV is also influenced by subjective factors, which include human components (e.g., emotions, experience in usage of IPTV service, billing aspects) [17]. QoE consists of subjective quality as experienced by the end user, known as the mean opinion score (MOS). Typically, for IPTV services the media delivery index (MDI) [18] is used for monitoring and troubleshooting networks carrying any IPTV traffic. The MDI measurement gives an indication of expected video quality, that is, QoE based on network-level measurements. The MDI predicts the expected video quality based on the IP network layer (independently of the encoder type), and it is defined as a combination of media delay factor (DF) and media loss rate (MLR). The DF refers to the time for which the IPTV flow is buffered on the receiving side at a nominal bit rate when there are no packet losses. The MLR is defined as the number of lost MPEG packets in 1 sec, and its relation to MOS is given in Table 7.2.

Overall, the QoE in the case of IPTV depends on two main technical elements:

- IPTV application: This influences the QoE via quality of the source, resolution (e.g., SDTV, HDTV), bit rate, encoder output, group of pictures (GoP) structure in MPEG, and so forth.
- Network transmission parameters: IPTV is highly dependent on the type of data loss (especially for I and P frame losses from the MPEG stream). Also it is dependent on the used codec, MPEG transport stream packetization, loss distance and loss profile, outage due to multicast router table recovery, and so forth.

Finally, carrier-grade IPTV is typically provided with QoS end to end and it is not part of the public Internet. The contents are strictly regulated by national laws for media and monitored by national regulatory agencies. Also, the relationships

Table 7.2 MOS Dependence from MPEG Packet Losses for IPTV Service

MPEG Packet Loss per Second	Description Quality	MOS
<20	The best	5
20–100	Very good (periodical freezing)	4
100–160	Good (loss of video frames)	3
160–230	Bad (not clear picture)	2
>230	Worst (frozen picture or black screen)	1

and responsibilities among service providers (including network providers) and subscribers are regulated via SLAs. Hence, IPTV services are typically billed per TV package, per IPTV services (e.g., flat rate for linear TV, pay per view for VoD). On the other hand, broadband access to the Internet provided ground for development of many video-sharing or video-streaming services offered through websites as OTT services.

7.4 Over-the-Top Multimedia Streaming

Multimedia streaming is based on a similar protocol stack in both cases: (1) for streaming provided by a telecom operator with guaranteed QoS support (e.g., IPTV, VoD) and (2) for OTT multimedia streaming without QoS. The main protocols used for multimedia streaming are covered in Chapter 4. Typically, nowadays MPEG-4 (MPEG-2 in the past) over RTP/UDP/IP protocol stacks is used for transmission of video and audio data (as parts of the multimedia streaming process) in real time.

Internet OTT multimedia streaming includes two main types (excluding VoIP and video telephony, which are primarily conversational services):

- Streaming stored audio/video: This refers to on-demand requests for multimedia files (including audio, video, and related data such as subtitle, description). The best-known OTT examples are YouTube (globally) and Netflix (in some regions).
- Streaming live content: This refers to live broadcasting of TV/radio programs over the Internet (e.g., web-based TV).

In the case of OTT multimedia streaming, because there is no QoS support end to end, the delay is variable as is the jitter. So, smooth and synchronized playing of the video and audio (and possibly other data) at the receiving side should be provided by the application (e.g., multimedia player) by using buffering on the receiving side (since multimedia streaming is typically unidirectional) and using a playback point. So, to eliminate the jitter (i.e., to shape the video and audio streams), the arrival times of IP packets carrying the multimedia data are separated from the playback times. This requires a playback buffer, which stores packets that are received until it is time to play them back. The relation between the actual sending time and playback time of the multimedia content is shown in Figure 7.7.

7.4 Over-the-Top Multimedia Streaming

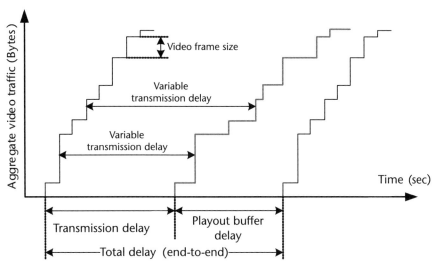

Figure 7.7 Playback point for multimedia streaming over Internet.

However, interactive OTT streaming (i.e., conversational multimedia) is based on UDP due to its sensitivity to delay. In such a case the media streams use an application codec over the RTP/UDP/IP stack (e.g., Skype video telephony). In case of on-demand multimedia streaming, the HTTP/TCP/IP protocol stack is typically used because OTT VoD is located through web browsing (on video-sharing sites such as YouTube). For live streaming, which is encoded in real time and sent from the sender to the receivers, both implementations—video streaming over RTP/UDP/IP or HTTP/TCP/IP (e.g., adaptive HTTP streaming [19])—are possible.

7.4.1 YouTube Technologies

YouTube is a video-sharing website on which users can watch various videos on demand, including short clips and full movies. It appeared in 2005, and was bought by Google in 2006. Currently it is a part of Google's services portfolio. A high-level description of the video retrieval process from the YouTube system is shown in Figure 7.8.

The user accesses a given video from YouTube at www.youtube.com either by browsing the web portal looking for the desired content or directly accessing the video web page by clicking on a URL. So, before the video streaming actually starts, mostly static web-based information and small pictures (thumbnails) of suggested videos are presented to the end user.

When an actual video is selected, the front-end replies with an HTML page in which the video is embedded using an Adobe Flash Player plug-in, which then takes over the download and playback of the video [18]. However, the name of the server that provides the video is encoded using a static URL. Then, the content server name is resolved to an IP address (by the client on the end user's side) via a DNS query to the local DNS server. Finally, the client queries the content server by using the HTTP to get the requested video. However, the actual selection of the IP address by the local DNS server is not arbitrary, because the DNS resolution is exploited by YouTube to route clients to appropriate content servers according to

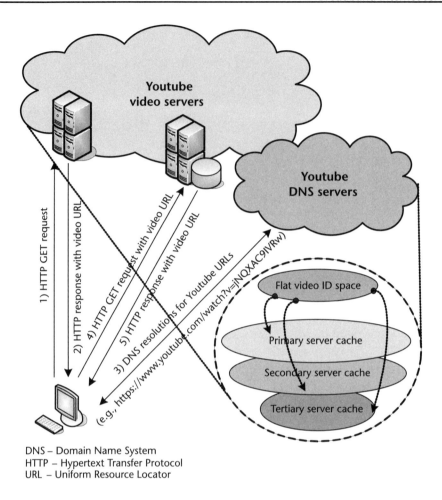

Figure 7.8 YouTube video delivery process.

established YouTube policies (e.g., geographically closest cache, which contains the requested video). Also, often, the preferred YouTube server cannot provide the content and therefore the client is redirected by that server to another one in, for example, a different data center.

The design of the YouTube video delivery system faces important and enormous challenges because of its constantly increasing user base (e.g., mobile users with smartphones), especially when keeping in mind that it is the largest and most visited video-sharing website in the world (according to the trusted websites ranking at www.alexa.com). The design of the YouTube video delivery system is based on three main components [20]:

- Flat video ID space: This is based on unique identifiers with a fixed length (string of 11 characters, given in the form https://www.youtube.com/watch?v=jNQXAC9IVRw) and employs fixed hashing with the goal of mapping the videos (and their IDs) to the logical namespaces. Because the character sets used for video IDs are the lowercase and capital letters from A to Z as well as digits (0–9), there are 64 different characters and hence 64^{11} IDs in the YouTube video ID space. YouTube creates URLs for the videos without informing the end users who request them where the videos are actually stored or cached (e.g., location of servers). The mapping between the video ID space and logical video servers (called anycast DNS namespaces) is fixed.

- Multilayered logical server network: In practice, YouTube uses multiple DNS namespaces (which are anycast based), where each represents a given collection of logical video servers. The anycast in this case refers to the capability of logical video servers to be mapped to multiple physical servers (each with a different public IP address), which may be at the same location or different locations (e.g., for load balancing between servers). YouTube uses HTTP redirections for transfer of user requests from one logical video server layer to another one. All logical video servers (i.e., DNS namespaces) have DNS mappings to the physical cache hierarchy [20].

- Tiered physical cache hierarchy: YouTube uses three tiers of caches: a primary cache (most of the cache locations belong to this tier), a secondary cache, and a tertiary cache (the least number of locations belong to this tier). Each location contains a variable number of IP addresses that locate (i.e., identify) the physical video servers. The mapping between the IP address of a video server and the DNS name is one to one.

Regarding the selection of the data center (where the YouTube video servers are physically located), the main role plays the RTT between the end users and the data centers [21]. However, in some cases a user's request may be redirected to a nonpreferred video server (e.g., for load balancing).

In one decade YouTube has paved the way for OTT VoD services, introducing user-generated contents besides content produced by professionals (e.g., movie studios, TV channels). That was a big shift in a new era of video services, where ordinary end users from different ages have become content generators in addition to being content users.

However, another technology also has a significant impact on file sharing among ordinary Internet users, and that is OTT peer-to-peer technology.

7.5 Over-the-Top Peer-to-Peer Services

File-sharing systems began to appear on the Internet at the end of the 1990s and beginning of the 2000s, driven by the introduction of higher access bit rates for individual Internet users at that time (i.e., the start of broadband access deployments).

The initial P2P file-sharing systems were based on centralized architectures, in which all content was obtained from a single server. Typically, a user downloads and installs a P2P file-sharing application, for example, from websites that host such applications). The application has the ability to connect to centralized servers to locate a peer that has a copy of the requested file, for example, an *.mp3 (song), *.avi (movie), or *.pdf (book) file. When peers that host the file are located and listed by the P2P file-sharing application, the user chooses one of the peers (through the GUI of that application, which is proprietary) and the application copies the file from the remote peer typically using a nonstandardized application layer protocol embedded in the P2P application, which is typically based on certain FTP and HTTP features. If HTTP is used as the application layer transfer protocol, then the user uses a web client to download the file from a web server running on the other peer's machine. So, in P2P applications each peer simultaneously runs a client component (to access contents from other peers) and a server component (to provide its own content to other peer hosts, by using defined policies and rules in the P2P applications). Simply said, all peers are also servers, which makes P2P network highly scalable.

Regarding the known P2P file-sharing implementations, one of the first "famous" OTT P2P applications was Napster, while the most successful one is BitTorrent.

7.5.1 Napster Technology

Napster was originally based on a centralized P2P network architecture. Centralized directory servers stored information about peers (their IP addresses) and content (e.g., music files in *.mp3 format). Then, upon a request-query from a peer host for a certain file (e.g., a song from given artist), the centralized directory server would send a response with information, including the IP address of the peer hosts, as well as the file's name and type matching the query. However, due to copyright infringements as well as the absence of a legal basis for content sharing over the Internet, Napster was eventually shut down. Napster, however, paved the way for further development and optimization of P2P file-sharing systems (e.g., Gnuttela, eDonkey, eMule). In the first decade of the 21st century, P2P file-sharing systems significantly contributed to the Internet traffic volume on a global scale, being comparable to the WWW share. However, with the appearance of web-based video-sharing sites (since videos as media type generate the highest traffic volume), such as YouTube in 2005 and many others later, the traffic share of P2P file-sharing systems decreased significantly over a decade. Nowadays P2P file sharing constitutes around 6% of the total Internet traffic globally, from which about half of such traffic is due to the BitTorrent application, which has established itself (from its introduction more than decade ago) as a major P2P file-sharing system on the Internet, constituting nearly 3.5% of the total Internet traffic. Hence, BitTorrent has become one of the most important OTT applications.

7.5.2 BitTorrent

BitTorrent was created by Bram Cohen in 2001. One of the reasons for the success of BitTorrent lies in its design, which provides robustness of the service [22]. With BitTorrent many users can download a given file and also upload pieces (called "chunks") of that file to other users downloading the same file. Each downloader reports the files it has to all of its download peers. The integrity of the chunks is verified by so-called torrent files. A torrent file is a hash of all file chunks. A hash is an algorithm that generates a fixed-length string (called a "message digest" or merely "digest") from given input data. The hash function is not reversible, so a given message digest can be compared only with another message digest for verification. BitTorrent includes trackers (working on the top of HTTP servers) that are designed to help downloading peers find each other. In summary, the BitTorrent network architecture consists of three types of nodes: peers, torrent trackers, and web servers for hosting torrent files.

BitTorrent results in a cost and performance redistribution for downloading of a given file. However, for downloading of a given file with BitTorrent, the total upload bit rate of all peers in the network (the Internet) must be equal to total download bit rate of all peers. So, each peer must allow a certain bandwidth in the uplink to be used for uploading the downloaded pieces (chunks) to other peers, when it is downloading a certain file in the downlink. For example, the BitTorrent application provides the possibility that the end user will be able to set a maximum upload rate (in kilobytes per second) from 1 Kbps (as the minimum) up to an unlimited rate (i.e., it is only limited by the bit rate in the upstream direction of its own access to the Internet). In that manner, P2P file sharing provides higher scalability in the cases when more peers are hosting a given content. However, that is also related to the popularity of the content—the more popular contents (i.e., files) have more peers that provide such content, and vice versa.

Let's compare the P2P technology of BitTorrent with client–server service architectures, as shown in Figure 7.9, in terms of their scalability and file distribution time:

- Client–server distribution time: In this case the server sequentially sends N copies. So, we denote the file size by F, and us is the upstream bit rate of the server. A given client i requires a download time of F/di where di is the downstream bit rate for that client. The time to distribute the file with size F to N clients is equal to:

$$t_{client-server} = \max\{N*F/u_s, F/\min(d_1, d_2, \ldots d_N)\} \quad (7.2)$$

- Peer-to-peer distribution time: In this case a server must send one copy that is uploaded for time period F/us for a file of size F and server upstream bit rate of us. The client i takes time F/di to download the file. Then, the fastest possible upload bit rate (assuming theoretically that all other peers are sending chunks to the same peer host) equals:

$$u_{fastest} = u_s + \sum_{i=1}^{N} u_i \quad (7.3)$$

Then, the P2P file distribution time can be calculated as:

$$d_{p2p} = \max\left\{F/u_s, F/\min(d_1, d_2, \ldots, d_N), NF/(u_s + \sum_{i=1}^{N} u_i)\right\} \quad (7.4)$$

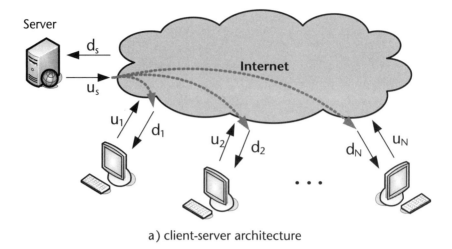

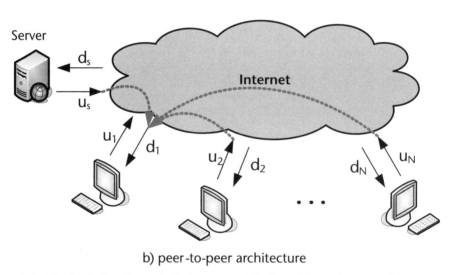

Figure 7.9 File distribution time: (a) client–server architecture; (b) peer-to-peer architecture.

Using the above equations for file distribution time with both the client–server and P2P (such as BitTorrent) architectures, the result points to the benefit of P2P architectures with an increasing number of N hosts that send the chunks of the same file to a single requesting peer. In both cases, the limiting parameter is the downstream capacity of the receiving host. In an ideal case, when the number of peers N is high enough, then the file distribution time is lower with the P2P architecture than with the client-server one because the client-server architecture has a bottleneck on the server's side; the bottleneck is the available upload bit rate from the server to the given client, as well as the server's processing capabilities in some cases. In contrast, the P2P architecture of BitTorrent is not efficient when the number of peers is limited (or absent) or their upstream bit rate (allowed for BitTorrent use) is limited by their owners. BitTorrent's file distribution is shown in Figure 7.10.

Generally, P2P file-sharing systems have proved that self-sustaining overlay networks are possible on the Internet, based on OTT P2P applications. Due to its

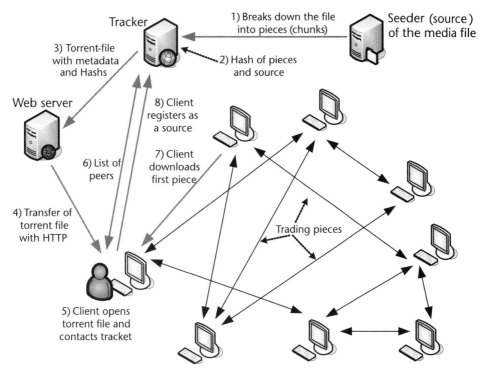

Figure 7.10 BitTorrent file distribution.

high scalability and end-user participation in the file distribution (or sharing) process, BitTorrent has established itself as a fundamental OTT P2P application that is not restricted by the technical capabilities of dedicated servers or copyrights (since each end user provides a chunk of a file, which is typically not usable without other chunks).

Other technologies that have grown almost in parallel with P2P file-sharing systems in the 2000s and have impacted the OTT Internet world are the Internet social networks.

7.6 Over-the-Top Social Networks

Social networks are present in the everyday life of people and relate to their connections and mutual interests. When such relations are mapped on the Internet, using the well-established Internet technologies, then they are referred to as online social networks or social networking. The most popular social networks at the present time are Facebook, Twitter, Google+, LinkedIn, and Foursquare [23]. The more general-purpose social networks (e.g., Facebook, Twitter) attract a bigger user base than the more specialized ones.

Social networks can be classified as an interdisciplinary science, because they include the input of knowledge from ICT people, sociologists, physicists, and mathematicians [24]. In general, social networking is based on the assumption of independent links (i.e., relationships) between social actors (i.e., users of the online social network).

Regarding the technology side of the story, online social networks are websites that provide content sharing (e.g., text, pictures, video files, audio files) among users, according to personalized settings for the availability/unavailability of such content to different groups of users with which the given user has connections.

Let's create a more formal definition of a social network. A social network consists of actors and the relationships among them (Figure 7.11). We can therefore represent a social network with points (i.e., nodes) that represent the actors, and lines that interconnect a pair of points, thus representing the ties. Further, we can weight the lines, with the goal of strengthening the tie. For example, people who exchange emails on a regular daily basis have stronger ties than people who only occasionally exchange emails. In the simplest form of a social network, an actor is an individual user. In a more complex form, an actor may represent a group of individuals who share the same interest on a given subject. Examples of targeted social networks are collaborations (e.g., co-authorship) and decision making (e.g., in a company).

7.6.1 Social Network Concepts and Technology

Online social networks imitate the behavior of real-life social networking, which can be extended outside the real-life neighborhood at home (e.g., family, neighbors) or in the office (e.g., colleagues). So, in most social networks the important feature is balance between the actors in terms of exchange of information (e.g., messages, pictures, videos). Also, we can expect that exchange of "support" on longer terms (e.g., weeks, months) between actors with ties is balanced. Further, the "role" of

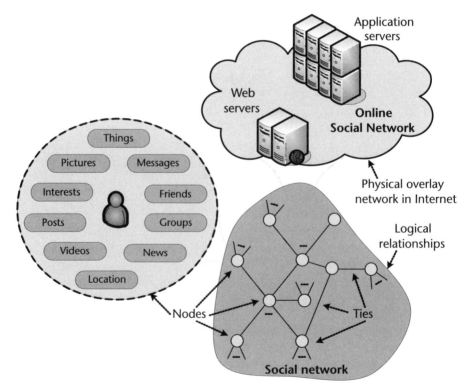

Figure 7.11 Social network graph.

an actor is very important. For example, in Twitter there are actors that follow a given actor (they are called followers). The popularity of the actor (e.g., a celebrity, a politician) in real life typically has an impact on the number of followers (actors in the social network that receive messages from the followed actor). So, the importance of an actor in a social network is also referred to as "prestige," which is typically expressed through the number of followers (e.g., in the case of Twitter) or number of ties with other actors (i.e., individual users) in the social network (e.g., Facebook, LinkedIn).

Current social networks are completely web based. Each such social network is formed by set of web pages that interconnect to other web pages. So, a social network is a part of the WWW. Two major tasks should be accomplished: (1) social network extraction from the web (e.g., to provide decentralized social networks via the so-called semantic web) and (2) social network analysis (e.g., for finding relevant targets for ads and other types of context-based marketing, as the main source of revenues for social network providers).

The semantic web is a group of extensions to web standards by W3C. It is also referred to as a web of data [25]. In fact, the semantic web relates to a web that is defined and linked in such a way that it can be used by people (and processed by machines) in a wide variety of applications. In practice, it enables online social information to be explicitly represented.

While the front-ends of online social networks are web servers, the back-end application servers perform social network analysis (SNA) with the goal of providing a sustainable business model by targeting certain offers (e.g., ads, information about stores, books, hotels, games, courses) to certain individual users or groups according to their interests and/or relationships in the social network.

7.6.2 Comparison of Existing Social Networks and the WWW

Existing social networks target equally users who use a fixed Internet host (e.g., PCs) and mobile users with smartphones [26]. If location information is not used in a certain context, then there is no difference between social networks for fixed and mobile Internet hosts.

We can compare the main targets of the most popular social networks in the second decade of the 21st century. For example, Facebook users typically search for users with whom they have or had real-life (i.e., offline) connections, whereas Twitter concentrates on microblogging services and helps people to communicate with other users who have an interest in similar topics. So, regarding Twitter, we could ask if it should be classified as a social network (which is more balanced in both directions between a pair of users) or news media [27], since the majority of the tweets are posted headlines or news on the web portals from different media. As a compromise, Twitter may be classified as a combination of both, news media and social network.

Besides Facebook and Twitter, which currently have the largest user bases, there is also FourSquare, a location-based social network website targeted mainly for use by mobile devices. FourSquare encourages its users to share location-based information (e.g., information about coffee shops, shopping malls, monuments, hotels) by giving certain rewards. However, the users of this social network can link their accounts to other social networks such as Facebook and Twitter, and share location-based information through them [26]. So, there is also vertical interaction between the different social networks.

There are more than 100 social networks in the world [28]. However, only those with a large user base have an impact on a global scale; therefore, focus is given to existing popular networks at the present time. Further, different social networks also have developed APIs to provide programmers with the ability to develop new functions and services for a given social network, or to be used by third-party service providers. Most of the popular social networks have developed mobile applications for the most popular mobile operating systems (e.g., Android, iOS, Windows Mobile), which are adopted to smaller screens and limited processing capabilities of mobile terminals.

Finally, the early standard WWW consisted of web pages that were hosted on web servers. It is referred to as Web 1.0 and users could only read the web content; no other interaction was possible. The web technologies that provided higher interaction between the end users and web servers, such as social networks, are referred to as Web 2.0. Web 2.0 technologies allow users to read web content and generate web content through social networks or content-sharing sites. Finally, in such a classification, the future semantic Web (i.e., web of data) is also referred to as Web 3.0 (data structures are reusable on the web by using queries through standardized formats such as RDF, XML, and microformats). However, such a classification of WWW may be considered as jargon in the literature.

7.6.3 The Future of Social Networks: Decentralization

Most large online social networks are centralized (e.g., Facebook, Twitter, LinkedIn), that is, centrally operated and controlled [29]. This means that the details of each actor, social graphs, and ties (i.e., relationships among actors) are held exclusively by the online social network service provider. In such a situation, with centralized social network management, it is not a trivial task for a given user to reuse his or her data from the social network, including messages to friends, shared photos, and so forth. Also, there are no efficient mechanisms to port such data from one social networking platform to another one [30], because social networks try to attract users to their own platform and most of the other platforms can be considered to be competition.

The limitations of centralized online networks lead to their decentralization, as shown in Figure 7.12. However, there is also a significant difference between the existing closed data and centralized control and open data and decentralized control. For example, the largest online social network uses the fundamental web technologies, such as HTTP, HTML, CSS, and JavaScript [30], and it does not output them in a structured way such as via the resource description framework (RDF).

One possible future development of social networks is decentralization. A decentralized setting may give back to the users the control of their own data, which is controlled by the social network service provider in the centralized case. In particular, three user aspects will benefit from decentralization [30]:

- Privacy: In decentralized social networks this refers to a user's ability to decide who to show information to and what kind of restrictions to put on their data.

7.6 Over-the-Top Social Networks

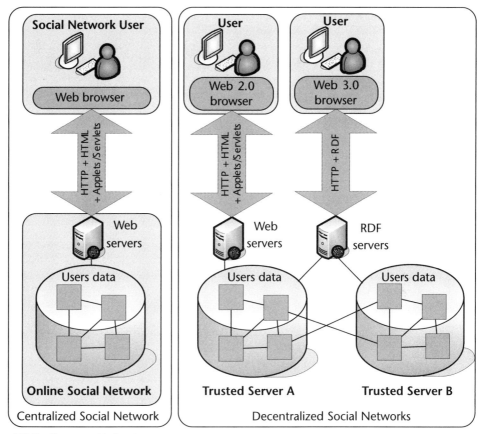

Figure 7.12 Centralized versus decentralized social network architectures.

HTML – Hypertext Markup Language
HTTP – Hypertext Transfer Protocol
RDF – Resource Description Framework

- Ownership: This refers to storage of personal data on a trusted server (e.g., provided by a telecom operator) or on the local computer (e.g., in a home network), so that users have complete ownership of their data. This way users are protected even in cases when the proprietary service hosting their data decides to shut down without prior notice.
- Dissemination: This allows users to disseminate their information according to personally defined users preferences and their friendship relationships.

Implementation of such decentralized social networks is a challenge. However, it may be provided not only by global social network service providers, but also by telecom operators (fixed and mobile ones), which may offer such services through cloud computing and semantic web, and to guarantee privacy, security, QoS, and so forth, based on SLAs and national legislation in a given country. Another possibility is to have OTT decentralized social networks, which may provide vertical aggregation of users' information on different centralized social networks (e.g., Facebook, LinkedIn). In that manner, another massive "decentralization" of the Internet is expected to happen through the practical development and deployment of the Internet of Things.

7.7 Internet of Things (IoT)

Initially, the vision for the IoT was described in detail in an ITU report [31], which covered the potential technologies, market potentials, challenges and implications, as well as benefits to all countries including developing ones.

According to the ITU-T definition [32], the IoT is a global infrastructure for the information society that enables advanced services via interconnection of different physical and virtual things, based on existing (standardized) and evolving information and communication technologies. The merging of the IoT and WWW results in a Web-of-Things (WoT), which is defined as a concept for making use of IoT where the things (either physical or virtual) are connected and controlled via the WWW [32]. The ITU-T has taken a role in defining a unified approach for the development of technical standards for enabling the IoT on a global scale through its Global Standard Initiative on IoT (GSI-IoT).

The IoT paradigm is expected to have a long-term influence on technology as well as society. In that respect, the IoT can be considered to be the standardization of an information architecture for enabling information society in practice. The IoT adds another dimension, referred to as "any-thing communication" to the ICT world [33], besides the other two dimensions, "any-time communication" and "any-where communication," as shown in Figure 7.13.

The "things" can have associated information, which can be either static or dynamic. Also, the "things" can be physical or virtual objects. The virtual things (e.g., multimedia content, application software) are present in the information world, and such things (i.e., objects) can be stored, processed, and accessed. In the IoT system physical objects can be represented via certain virtual objects, but virtual objects can also exist without any association with physical objects. Physical objects are all types of devices that can provide certain information to other devices or to humans (via applications or services). The devices (i.e., physical objects) can communicate through a gateway toward the network (e.g., the NGN) or without a gateway (which is possible if they have IP connectivity capabilities), or directly with other devices via a local network (e.g., Ethernet, sensor network, ad hoc network).

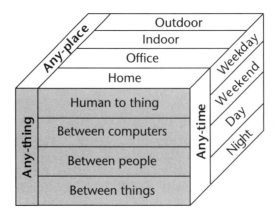

Figure 7.13 IoT space.

In the IoT framework devices are categorized into several categories, including the following:

- Data-carrying devices: These are used to connect physical things to the network.
- Data-capturing devices: These refer to reader/writer and sensing and actuating devices (to detect or measure certain parameters and convert them into digital signals; and in the opposite direction to convert digital signals from the network into certain device operations).
- General devices: Devices not covered by the preceding two categories belong in this category, including industrial machines, smartphones, electric appliances, and all such devices that have the ability to communicate through wired or wireless access networks.

The IoT applications are very similar to Ubiquitous Sensor Network (USN) applications and services [34]; in fact, USN services can be considered a subset of IoT services and applications. Emerging examples of IoT applications are e-health (i.e., electronic health systems for online health monitoring and diagnostics), smart grids, smart homes, and intelligent transportation systems. In the beginning such applications are typically based on proprietary application platforms. However, NGN provides common functionalities for AAA, devices and resource management, and common application and service support systems.

The IoT is characterized by a huge scale of applications and devices that are connected to the global information infrastructure (e.g., via the NGN). With "things" becoming Internet hosts, the number of Internet hosts (as unique devices connected to the Internet) will go well beyond the number of individual human users and their devices.

The reference model for IoT includes four layers as follows:

- Application layer: This layer contains the IoT applications (e.g., e-health, smart home, smart city).
- Service support and application support layer: This layer may include general support capabilities (e.g., data processing and storage in databases) as well as specific support capabilities.
- Network layer: This layer provides access and transport resource control functions, AAA, mobility management, and connectivity for the transport of IoT services.
- Device layer: This layer consists of device capabilities and gateway capabilities:
  - The device capabilities are used for direct or indirect (e.g., via a gateway) interaction with the network, as well as ad hoc networking.
  - The gateway capabilities include multiple-interface support, which can be provided on the device layer (e.g., Wi-Fi, Bluetooth, Zigbee) or on the network layer (e.g., 3G/4G mobile networks, Ethernet).

The IoT builds a huge ecosystem around itself that is composed of different business players. There are device providers, network providers, software platform providers, application providers, and finally IoT application users. The central role in the IoT ecosystem belongs to the network providers (e.g., telecom operators), which connect all other device providers on one side with service and application providers on the other side. Also, end users are connected to network providers to access IoT applications. A single player in the IoT ecosystem may play one or more provider roles (e.g., a single telecom operator might own devices, networks, service platforms, and applications). Hence, different business models are possible for the IoT as shown in Table 7.3, depending on the providers' different types of ownership in the IoT ecosystem [33]:

- Business model 1: A single business player (Provider 1) operates all parts, including devices, network, platform, and applications and also provides services to the end user.
- Business model 2: One provider (Provider 1) operates devices, network, and the IoT platform (e.g., a telecom operator), while another business player (Provider 2), for example, a third-party IoT service provider, provides the application.
- Business model 3: One provider (Provider 1) operates the network and the IoT platform, while another provider (Provider 2) operates the devices on the user's side and provides IoT applications/services.
- Business model 4: One provider (Provider 1) operates the network, while another provider (Provider 2) is at the same time the device provider on one side and the platform and application provider on the other side.
- Business model 5: One provider (Provider 1) operates the network, a second provider (Provider 2) provides the platform, and a third provider (Provider 3) operates devices and provides IoT applications to the customers.

In general, the IoT is an ongoing evolution of the information and communication technologies that is targeted to connect every physical or virtual object to the Internet, to provide any form of benefit to people and society. In the longer term, the IoT paradigm's goal is to include almost everything, even the smallest objects such as molecules in the Internet of Bio-NanoThings paradigm [35]. The IoT must not represent only the advancement of technology, but must also be able to demonstrate value in terms of better living and better society.

Table 7.3 Business Models for IoT

Business Models for IoT	Device Provider	Network Provider	Platform Provider	Application Provider
Business model 1	Provider 1			
Business model 2	Provider 1			Provider 2
Business model 3	Provider 2	Provider 1		Provider 2
Business model 4	Provider 2	Provider 1	Provider 2	
Business model 5	Provider 3	Provider 1	Provider 2	Provider 3

7.8 Web of Things

Although the IoT is targeted to solutions for interconnecting things with Internet technologies, creation of an application that runs on the top of heterogeneous devices is a problem due to the many heterogeneous networks and technologies behind them. There is a general lack of interoperability across many proprietary platforms (e.g., hardware platforms, operating system, databases, middleware, and applications), as well as different types of data formats. The practical solution is to use a technology for the IoT that is applicable and already deployed worldwide in different types of devices, such as web technology. Web technology provides users with the possibility of interacting with devices (i.e., the things) in the IoT by using web interfaces. Such an approach is referred to as the Web-of-Things (WoT) [36].

A conceptual model for the WoT is shown in Figure 7.14. According to the model, end-user applications (e.g., web browsers) access the physical devices directly or through a so-called WoT broker, which has agents that provide an adaptation between interfaces of physical objects and web interfaces. Each agent within the WoT broker is dedicated to a specific interface (e.g., Wi-Fi, Bluetooth).

Considering device capabilities and web interfaces, in WoT we can distinguish among two general types of devices:

- Constrained device: This is a device that cannot connect directly to the Internet by using web technology, so it requires a WoT broker.
- Fully fledged device: This is a device that has all necessary web functionalities for connecting with services on the WWW without the need for a WoT broker, but it can also interact with it.

The WoT broker has functionality for communication between the WoT end user (e.g., web client) on one side, and devices on the other side (either constrained or fully fledged ones). So, the WoT broker is exposing devices on the web and thus provides its seamless integration onto the WWW. Each WoT broker contains agents that communicate with the physical devices and provide control functions for them. So, when a certain web application sends a request to access certain physical devices, the WoT broker performs adaptation of such a request to the interface toward the device (i.e., the thing).

Generally, three types of services can be distinguished within the WoT:

- Web service: This is a service that can be directly accessed on the web.
- WoT service: This is a service provided via an adaptor that provides 1:1 mapping with the services of the physical device.
- Mash-up service: This is a service that integrates the WoT with other services (e.g., IPTV, VoIP, cloud services) via the WoT broker.

In the WoT model all interfaces between each pair of WoT brokers, mash-up services, and web applications are based on HTTP. Fully fledged devices also use HTTP for communication with all other entities in the WoT model (i.e., WoT broker, mash-up service, and web application). It may also be possible to use a proprietary protocol between the fully fledged device and the WoT broker. On the other hand,

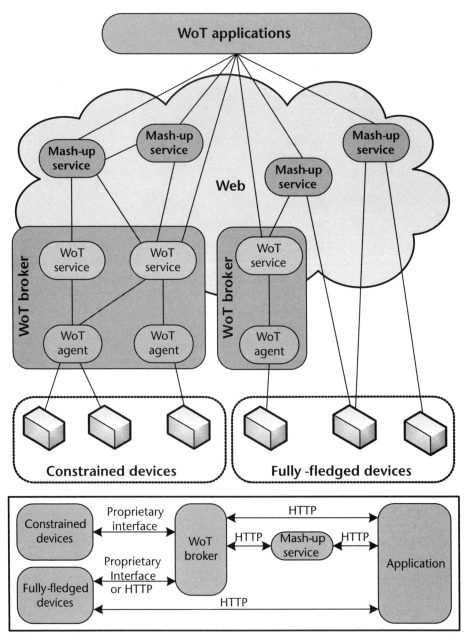

WoT – Web of Things

Figure 7.14 WoT model.

constraint devices communicate only through the WoT broker, and such communication is based on a device-dependent proprietary protocol. In this case all other communication between the WoT broker and mash-up service or web application is also performed with HTTP.

One typical WoT service is a smart home. An example of a smart home WoT service is shown in Figure 7.15. All devices in the home that need to be monitored or controlled are connected to a WoT broker, which has a separate agent for each

7.8 Web of Things

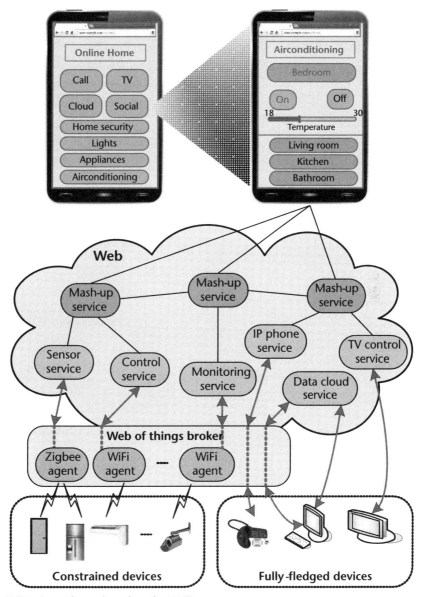

Figure 7.15 Smart home based on the WoT.

different technology that is used for connection to devices in the home (e.g., Wi-Fi, Zigbee). So, there will be a WoT agent for Wi-Fi, WoT agent for Zigbee, and so forth. The WoT devices in a smart home may include, but are not limited to, an air conditioner, heater, refrigerator, TV set, cooker, room lights, security cameras, front door, picture on the wall, and temperature sensor. Different WoT services are created for different "things" in the home, such as a cooking service for control of the cooker, heating service for control of the heater, cooling service for control of the air-conditioning system, light control service for control of the lights (e.g., on/off and adjust the light density), temperature monitoring service for control of the temperature sensor, picture display on the wall service, front door service for control of the front door, security camera service for control of the security camera (e.g., on/off, online home surveillance), and so forth.

Finally, the IoT paradigm still lacks practical mass deployments. So, to unlock the full potential of the IoT, there is a need for open ecosystems that will be based on open standards, including the identification, discovery, and interoperation of services from various vendors on a global scale. The WoT can be seen as the realization of the IoT in a form more suitable for human-to-machine interaction. The M2M solution (as part of the IoT) is covered in more detail in Chapter 9.

7.9 Regulation and Business Aspects for Broadband Internet Services

The main distinction between different broadband Internet services is the QoS support end to end. It influences the type of service provided and the commitments between the players in the value chain. In that respect, there are several business and regulation challenges for the services in the converged telecom/Internet worlds.

7.9.1 Business Aspects

Global challenges for Internet services are targeted toward a single packet-based transport infrastructure, with integrated (transport) services over IP-based networks, which can carry different types of application traffic. Then, predefined transmission planning of QoS has become a major challenge:

- Fixed allocation of resources is no longer possible, although planning resources per traffic type (e.g., voice, TV, data) are used.
- Quality requirements are not well defined end to end for OTT services, hence the responsibility for end-to-end QoS has been lost.
- Services must be considered as applications executed in terminal devices.
- IP networks cannot provide self-sustaining end-to-end QoS, but only transport classes that enable QoS differentiation.

These factors lead to several different aspects and challenges regarding network equipment, terminal devices, network providers, and service providers.

7.9.1.1 QoS Aspects for Network Equipment Manufacturers

Vendors of telecommunication network equipment have to rely on QoS-related performance requests from network operators and service providers. In practice, there are many proprietary solutions in the network equipment besides the standardized ones (e.g., routing protocols, AAA protocols).

In the ideal case, network equipment manufacturers would participate in the QoS standardization work of the different standards developing organizations, with the goal of standardizing the QoS and performance requirements between different parties involved in the networking business (users, network providers, and service providers). Many times there is no visible motivation for the short term, and long-term predictions are also driven by innovations in the application space that influence the telecommunications industry today more than ever.

7.9.1.2 QoS Aspects for Terminal Device Manufacturers

Terminal device manufacturers are playing in a global market, due to the harmonization of telecommunication/ICT standards on a global scale. This results in a larger market size, but also stronger international competition.

Manufacturers may move away from standardized minimum attachment requirements (including fixed and mobile terminals), which typically results in no harm to the network, but not necessarily a high QoS. Often OTT services are foreseen to be primarily used, so QoS provisioning is not a high-priority topic among device manufacturers.

Acceptance of terminal devices in the market is also based on other factors, such as price of the device, other functions of terminals (e.g., graphical user interface), applications available for that terminal (e.g., preinstalled applications, such as maps, social networking applications, cloud applications), and availability of an application ecosystem from the terminal vendor (this is especially important for mobile devices, where Android-based mobile terminals and Apple have ecosystems with over 1 million applications each). Additionally, the success of terminal equipment is influenced by the brand of the terminal, age group of the user, and the profession of the user, especially when the device is also used for business matters (e.g., collaboration, administration, access to documents).

7.9.1.3 Challenges for Network Operators and Service Providers

The main challenges for network and service operators are investments in the infrastructure (which mainly concerns network operators, although OTT service providers are also dependent on broadband access) and issues regarding the QoS end-to-end and related IP traffic management. However, network operators are likely to partially:

- Invest in new capacities: In the long term there is no other option for fixed access than PONs (e.g., G-PON, XG-PON, XLG-PON), as well as 4G mobile networks (and 5G in the future).
- Optimize the existing networks within their lifetime: Each technology and ICT equipment has a finite lifetime. For example, the processing power is doubled every 1.5 to 2 years, which also refers to networking equipment (e.g., routers) and operator's servers (they become "old" after several years, e.g., 4 to 5 years).

Another challenge to network and service providers is traffic management, which can increase the efficiency of managing existing network capacity and provide traffic balancing without changes in the network infrastructure. However, the appropriateness of different approaches to traffic management is one of the main parts of the network neutrality discussions (to continue with it or to change it). However, traffic management has beneficial aspects. For example, it is certainly needed and applied in existing all-IP networks for transfer of telephony traffic as QoS-enabled VoIP (i.e., emulation/simulation of PSTN/ISDN). So, each network that carries network neutral traffic differentiates such traffic (as best-effort traffic) from the QoS-enabled traffic such as telephony, IPTV, and leased lined (or VPNs dedicated to enterprises and organizations). Typically, traffic management is used to protect safety-critical traffic, such as VPN tunneling traffic, which has a significant part of the network traffic across the Internet (e.g., VPNs are typically used for IP interconnection between network operators).

7.9.1.4 Challenges for Customers

Telecommunication services are becoming user-centric; hence, the personal user experience is highly important and depends on, for example, discrepancies between advertised and actual delivery of QoS including speed of the network. Ordinary consumers may not be able to detect the actual applications of discriminating traffic management techniques and may find it difficult to distinguish between the effects of traffic management techniques on QoS from the effects of other quality-degrading factors such as congestion in the network or at the remote end. A consumer observing that traffic is routinely throttled may not know whether this is being done intentionally or is caused by other factors. In general, traffic management techniques and policies are difficult for consumers to understand.

7.9.1.5 Broadband Internet Service Markets

Broadband services can be divided into two main groups [37]:

- Broadband applications/services: These belong to the retail market because they are provided downstream from the individual end users.
- Broadband access: This can be classified into two markets: (1) retail markets (consisting of residential and business retail markets) and (2) wholesale markets (consisting of passive infrastructure and active services).

In the converged telecom/ICT world with all-IP networks and services, we should note the existence of the increasing trend toward service bundling (i.e., N-play economics in telecom/ICT sectors). That refers to telecom operators (e.g., bundled provision of voice, TV, Internet access as a service, mobile messaging, and so on, by a single operator) as well as OTT services providers (e.g., provision of voice, video, messaging, file exchange services, and so on, by a single OTT application/service).

In the case where a monopolistic company bundles the monopoly product (e.g., telephony monopoly by telecom operator with dominant market share, or global monopoly by OTT service provider) with a product that is sold in the competitive markets, the transfer of market power from one market to the other occurs. Also, bundling provides for the possibility of cross-selling. This way of putting together services A and B and selling them in packages as a service bundle makes it possible to transfer the possessing surplus from product B to A (and vice versa). Further, bundling provides for the possibility of service differentiation in comparison to rivals in the same market, on either a national or global level. A bundled service can have its price set at a higher level than in the case of a single-service offering, which may result in a better position compared to the competition.

Bundling has an effect on the telecom market structure, because it results in an increase in mergers and acquisitions. So, from the perspective of the telecom/ICT sector, the bundling strategy encourages the process of convergence, and vice versa. Overall, bundling reduces costs, increases the demand for different services, locks in the customers, and provides service differentiation to end users.

7.9.2 Regulatory Aspects

Regulation for broadband Internet services is mainly related to regulation of Internet access as a service (IAS), as well as QoS-enabled VoIP and IPTV. The main parameter for the IAS is the bit rate in the downlink and uplink directions, and it is typically specified in the SLA between the subscriber and network operator that provides IAS. Then, the subscriber uses IAS to access all OTT services, which are typically served in a best-effort manner by using the network neutrality principle.

Then, for IAS, regulations specify the need to monitor the IAS, either proactively or reactively. With the goal of preventing degradation of the IAS (e.g., variable bit rates in the downlink or uplink direction, different from those specified in the SLA), the NRA in a given country may choose between several regulatory tools [38], such as setting minimum QoS requirements on the operators that provide IAS (also referred to as ISPs, although they are in fact telecom operators since all telecommunication services are converging toward Internet technologies), a market mechanism (e.g., fostering competition in the country), proportionality (e.g., limiting the imposed requirements to operators to the adequate scope), and so forth.

In competitive markets, without barriers to entry, QoS regulation could be less stringent or not required at all. Where barriers to entry exist (such as spectrum availability), more stringent QoS regulation may be appropriate (e.g., mobile operators are required to obtain licenses on spectrum, as discussed in Chapter 6).

QoS regulation is considered part of customer protection, but it is broader than just QoS regulation; for example, it also covers sales activities, complaints, resolution procedures, and disconnection policies.

Regarding the jurisdiction of QoS regulation, regulators have to select parameters to be monitored, then measurements have to be published and targets set. However, parameters to be monitored should relate to the aspect of services that have the biggest impact on users (e.g., achievable bit rates in the downstream and upstream directions for the IAS, call success ratio for carrier-grade VoIP).

Regulation of IAS is important for network neutrality and OTT services, which foster innovation and development. ITU-T regulation guidelines [39] provide the fundamentals of QoS regulation, including QoS parameters and activities, recommended approaches, information gathering, and penalties and dialogue, and also specify parameters, levels, and measurement methods. Overall, QoS regulation aims to provide fairness and a high-quality user experience. That is equally important for QoS-enabled individual services (provided with support of the telecom operators), as well as IAS that is used for all OTT services that rely on high bit rates end to end.

7.10 Discussion

The openness of the Internet to new services and applications was a major reason why the Internet is becoming a single networking platform that is used to carry all legacy telecom services and native Internet services, while remaining open to development of new "killer" applications. The NGN standardization put certain Internet technologies within a given context (to enable end-to-end QoS provision, standardized AAA functions, and related signaling) for legacy services such as VoIP and IPTV. IAS is used by all OTT services that exploit the higher bit rates available in the access networks. Currently, there are many different OTT applications, which create the main Internet services "portfolio" for most users worldwide (e.g., YouTube, Facebook, BitTorrent, Skype, Viber). However, other variants of these OTT services can be found in different regions of the world due to regional and/or other constraints (e.g., Baidu offers many services based on the Chinese language, including a search engine for websites). Finally, IoT/WoT is an emerging area that is still looking ahead for its worldwide deployment. In that respect, the transition to IPv6 and Web 3.0 should foster the expansion of the Internet toward the IoT.

In practice, nobody can really predict exactly what the future "killer" applications will be. No one had predicted the Web before it appeared in the 1990s. No one had seen P2P file-sharing applications before they rolled out around 2000. Nobody could predict the success of social networks, as has happened since 2005. On the other hand, many people are working on many ideas for new innovative services, but only few of them will gain worldwide success. However, it is certain that cloud computing is one of the foreseen "winners" with regard to emerging broadband Internet services in the near future.

References

[1] ITU-T Recommendation Y.2262, "PSTN/ISDN Emulation and Simulation," December 2006.

[2] ITU-T Recommendation Y.2271, "Call Server-Based PSTN/ISDN Emulation," September 2006.

[3] J. S. Marcus, D. Elixmann, "The Future of IP Interconnection: Technical, Economic, and Public Policy Aspects," WIK-Consult, Final Report, Study for the European Commission, January 2008.

[4] S. A. Baset, H. Schulzrinne, "An Analysis of the Skype Peer-to-Peer Internet Telephony Protocol," IEEE Infocom 2006, Barcelona, Spain, April 2006.

[5] J. Rosenberg et al., "STUN—Simple Traversal of User Datagram Protocol (UDP) Through Network Address Translators (NATs)," RFC 3489, March 2003.

[6] A. Azfar, R. Choo, L. Liu, "A study of Ten Popular Android Mobile VoIP Applications: Are the Communications Encrypted?," 47th Hawaii International Conference on System Science, January 6–9, 2014.

[7] S. Schrittwieser et al., "Guess Who's Texting You? Evaluating the Security of Smartphone Messaging Applications," Internet Society Report, February 7, 2012.

[8] ITU-T Recommendation Y.1901, "Terms and Definitions for IPTV," March 2010.

[9] ITU-T Recommendation Y.1910, "IPTV Functional Architecture," September 2008.

[10] ITU-T Recommendation Y.1902, "Framework for Multicast-Based IPTV Content Delivery," April 2011.

[11] ITU-T Recommendation Y.1911, "IPTV Services and Nomadism: Scenarios and Functional Architecture for Unicast Delivery," April 2010.

[12] B. Cain et al., "Internet Group Management Protocol, Version 3," RFC 3376, October 2002.

[13] B. Fenner et al., "Internet Group Management Protocol (IGMP)/Multicast Listener Discovery (MLD)-Based Multicast Forwarding (IGMP/MLD Proxying)," RFC 4605, August 2006.

[14] B. Fenner et al., "Protocol Independent Multicast—Sparse Mode (PIM-SM): Protocol Specification (Revised)," RFC 4601, August 2006.

[15] D. Mills et al., "Network Time Protocol Version 4: Protocol and Algorithms Specification," RFC 5905, June 2010.

[16] ITU-T Recommendation Y.1920, "Guidelines for the Use of Traffic Management Mechanisms in Support of IPTV Services," July 2012.

[17] ITU-T Recommendation G.1080, "Quality of Experience Requirements for IPTV Services," December 2008.

[18] J. Welch, J. Clark, "A Proposed Media Delivery Index (MDI)," RFC 4445, April 2006.

[19] O. Oyman, S. Singh, "Quality of Experience for HTTP Adaptive Streaming Services," *IEEE Communications Magazine*, April 2012.

[20] V. K. Adhikari et al., "Vivisecting YouTube: An Active Measurement Study," IEEE Infocom 2012, March 2012.

[21] T. Torres et al., "Dissecting Video Server Selection Strategies in the YouTube CDN," *Proc. 2011 31st International Conference on Distributed Computing Systems*, June 2011.

[22] B. Cohen, "Incentives Build Robustness in BitTorrent," Workshop on Economics of Peer-to-Peer Systems, June 2003.

[23] L. Jin et al., "Understanding User Behavior in Online Social Networks: A Survey," *IEEE Communication Magazine*, September 2013.

[24] J. L. Moreno, "Who Shall Survive?," Beacon House Inc., New York, 1978.

[25] W3C, "Semantic Web," www.w3.org/standards/semanticweb, accessed May 2015.

[26] X. Hu, "A Survey on Mobile Social Networks: Applications, Platforms, System Architectures, and Future Research Directions," IEEE Communications Surveys & Tutorials, November 2014.

[27] H. Kwak et al., "What Is Twitter, a Social Network or a News Media?," WWW 2010, Raleigh, NC, April 2010.

[28] Wikipedia, "List of Social Networking Websites," http://en.wikipedia.org/wiki/List_of_social_networking_websites, accessed May 2015.

[29] B. Furht, *Handbook of Social Network Technologies and Applications*, New York: Springer, 2010.

[30] C. A. Yeung et al., "Decentralization: The Future of Online Social Networking," W3C Workshop on the Future of Social Networking, Barcelona, Spain, January 15–16, 2009.

[31] ITU Report, "The Internet of Things," November 2005.

[32] ITU-T Recommendation Y.2069, "Terms and Definitions for the Internet of Things," July 2012.

[33] ITU-T Recommendation Y.2060, "Overview of the Internet of Things," June 2012.

[34] ITU-T Recommendation Y.2221, "Requirements for Support of Ubiquitous Sensor Network (USN) Applications and Services in the NGN Environment," January 2010.

[35] I. F. Akyildiz et al., "The Internet of Bio-NanoThings," *Communications Standards* (Suppl. to *IEEE Communications Magazine*), March 2015.

[36] ITU-T Recommendation Y.2063, "Framework of the Web of Things," July 2012.

[37] ITU, Telecommunication Development Sector, "Regulatory & Market Environment—Regulating Broadband Prices," Broadband Series, 2012.

[38] "BEREC Guidelines for Quality of Service in the Scope of Net Neutrality," November 2012.

[39] ITU-T Supplement 9 to ITU-T E.800-Series Recommendations, "Guidelines on Regulatory Aspects of QoS," December 2013.

CHAPTER 8

Cloud Computing

The emerging trend of all Internet applications and services is toward the use of cloud computing, which is the current paradigm that is drawing much attention from standardization organizations, service providers, and end users. The idea of distance computing is not new (e.g., distributed computing), but recently it has taken a standardized framework under the name cloud computing. By definition [1], cloud computing is a paradigm for enabling network access to a scalable and elastic pool of shareable physical or virtual resources with self-service provisioning and administration on demand. Such standardization of cloud computing is accomplished by ITU-T as a continuing development after the NGNs, which were standardized to provide easy transition of legacy telecommunication services to the Internet environment and technologies. Cloud computing was initially based on Internet technologies, and there are many existing cloud services, including OTT cloud services (e.g., Dropbox, Google docs), as well as business cloud solutions offered by telecom operators with guaranteed QoS.

In that respect, this chapter focuses on different aspects of cloud computing, including the standardization framework by ITU-T, cloud architectures and different service models, and security and privacy issues in cloud environments. We also discuss mobile cloud computing as an emerging field, as well as implementation of cloud computing solutions as OTT and telecommunication services and their business and regulation aspects.

8.1 ITU's Framework for Cloud Computing

By definition, cloud computing is a paradigm that provides access to distant resources through an Internet access network. However, because the main idea of cloud computing is to use resources at a distance for storage, processing power, and son on, the key requirements for its deployment are access bit rates and delay. High bit rates, such as broadband Internet access, are required on the user's side, and this influences the delay budget. Higher bit rates provide lower delay, since the delay is inversely proportional to the sustainable bit rate end to end (e.g., from a cloud application on the user's side to the cloud servers on the cloud service provider's side). However, to provide lower delay, cloud servers are required to be closer to the users they serve, with the aim of lowering the RTT so that the end-user experience is similar to that supplied by locally available resources on the end-user equipment (e.g., PC, smartphone).

In general, cloud computing is an evolving paradigm at the present time. The key characteristics of this paradigm can be specified as follows:

- Broad network access: This refers to the ability of end users to access cloud computing resources from any Internet access network (broadband access network) by any device with which they are attached to the network.
- Measured service: This feature refers to metered delivery of cloud services, which means that they can be controlled, monitored, reported, and billed. The value provided by such cloud services is a switch from low-efficiency business models to high-efficiency ones.
- Multitenancy: In this context a group of cloud service users that form a given tenant shall belong to the same customer organization that uses cloud services. However, in public cloud deployments a group of cloud service users may consist of different cloud service customers. Also, a given cloud customer may have different tenancies with a single cloud (e.g., representing different groups in the same organization).
- On-demand self-service: This feature refers to automatic provision of cloud computing services, with minimal interaction by the end user, so that users have the ability to do what they need to do when they need to do it.
- Rapid elasticity and scalability: This refers to the ability of cloud resources to be quickly increased or decreased. To the end user they appear to be limitless resources; in practice, however, all resources are limited, so cloud service providers have to plan use of their cloud resources, which should be in line with offerings to the end user (e.g., storage capacity) and expected number of users.
- Resource pooling: This refers to aggregation of physical and virtual resources with the goal of serving one or more cloud customers. In general, customers do not have knowledge about where the resources are located or how cloud services are provided from a technology point of view—all the customer knows is that the cloud service works. However, even with such a level of abstraction, users might be able to identify the wider location of cloud resources (e.g., which data center, country, or region), especially when certain applicable legislation is required for stored data.

Depending on the availability of cloud resources to cloud customers, we can differentiate among several different cloud computing models [1, 2]:

- Public cloud: In this model the cloud services are potentially available to any customer. This cloud can be operated and managed by a business, governmental, or academic organization. Access to public clouds has limited or no restrictions. However, availability of the cloud to certain end users may be subject to jurisdictional regulations.

- Private cloud: In this model the cloud infrastructure is operated solely for a given customer organization that has control over its cloud resources. However, the cloud user may allow access to the cloud by third parties for its own benefit (e.g., from business aspects).
- Community cloud: In this model the cloud infrastructure and resources are shared among several organizations that have shared requirements (e.g., mission, compliance requirements, security requirements, policy) and a relationship with one another. The cloud resources are controlled by at least one member of such a community. So, a community cloud allows for limited participation by a shared group of cloud users (when compared to a public cloud), but it has broader participation than the private cloud.
- Hybrid cloud: This model is based on a combination of two or more cloud models (private, public, and community model), which remain as unique entities in the hybrid model, but are bound to each other with given technologies (either standardized or proprietary ones). A hybrid cloud can be owned, managed, and operated by the organization that owns it or by a third party, and it may exist on premises or off premises.

Each cloud service solution has certain capabilities, which may be grouped into three main capability types: infrastructure, platform, and application. The infrastructure capability type refers to a cloud service category that offers processing, storage, or networking resources. The platform capability type offers the ability to deploy, manage, and run customer-created or customer-acquired applications by means of programming languages as well as execution environments supported by the cloud service provider. The application capability type refers to the use of an application that is supported by the cloud service provider. The mapping between the cloud service categories and types is shown in Table 8.1. The main cloud service categories are Software as a Service (SaaS), Platform as a Service (PaaS), and Infrastructure as a Service (IaaS), which are mapped one to one to the application, platform, and infrastructure cloud capability types, respectively. However, other cloud categories can also be identified [1], such as Network as a Service (NaaS) and Data Storage as a Service (DSaS), that use all three capability types, as well as Communication as a Service (CaaS) and Compute as a Service (see Table 8.1).

Besides the cloud service categories listed in Table 8.1, there are also so-called informal cloud services categories that are currently used for OTT Internet applications and services, such as Email as a Service, Desktop as a Service, Security as a Service, Databases as a Service, Identity as a Service, and Management as a Service. Note that almost all OTT services nowadays use the cloud: email services such as Gmail and Yahoo mail, social networking services such as Facebook, video-sharing services such as YouTube, and gaming services such as Steam. Because the standardization of cloud computing is lagging behind its practical implementation (which is contrary to legacy telecommunication practice), it is hard to strictly standardize cloud services. So, we can expect them to remain in space between the standardized and proprietary solutions, allowing use of one or both approaches with the single goal of providing the desired services to end users.

Table 8.1 Mappings Between Cloud Service Categories and Types

Cloud Service Category	Cloud Capability Types		
	Application	Platform	Infrastructure
Software as a Service	X	—	—
Platform as a Service	—	X	—
Infrastructure as a Service	—	—	X
Network as a Service	X	X	X
Data Storage as a Service	X	X	X
Communications as a Service	X	X	X
Compute as a Service	X	—	—

8.2 Cloud Systems and Architectures

Meeting the goal of standardizing the cloud computing system and its architecture requires an architectural framework, which is given by ITU-T [3]. Such an architecture describes cloud computing roles and subroles, cloud computing activities, cross-cutting aspects, and the functional components of the cloud architecture.

A cloud computing system can be described by using a viewpoint approach that elucidates the user view, functional view, implementation view, and deployment view. The transformation between views is shown in Figure 8.1.

The user view describes the system parties, their roles and subroles, and cloud computing activities. Regarding the cloud computing roles, we can distinguish among three main roles in the cloud ecosystem (Figure 8.2) [3, 4]:

- Cloud service customer: The customer has a business relationship with the cloud service provider (as well as with the cloud service partner) for using cloud services. There are several defined subroles for the customer role, including cloud service user (an individual or an entity on the customer's side that uses cloud services), service administrator (for ensuring smooth operation of the service), service business manager (responsible for efficient acquisition and use of the cloud services), and service integrator (responsible for integration of the cloud service with the existing ICT services of the customer).
- Cloud service provider: This role provides the cloud computing services. Its activities are provided via several subroles: cloud service operations manager, deployment manager, service manager, business manager, customer care,

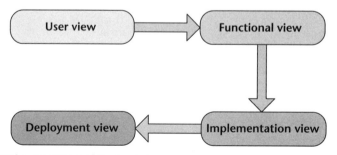

Figure 8.1 Cloud computing architectural views.

8.2 Cloud Systems and Architectures

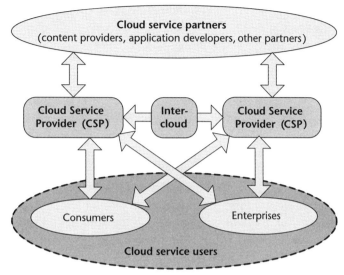

Figure 8.2 Cloud ecosystem.

intercloud provider services, service security and risk manager, and network provider.

- Cloud service partner: This is a party that is engaged in the support of either the cloud service customer or the provider. The type of cloud computing activities may vary depending on the type of partner. The partner may have several different subroles, including cloud service developer (responsible for design, developing, testing, and maintenance of the cloud service), cloud auditor (conducts audit of the provision and use of the cloud service), and cloud service broker (negotiates the relationship between the cloud service customer and provider, including assessment of the customers and setup of SLAs).

The functional view is related to functions that are needed to support a cloud computing service. The functional components form a so-called functional architecture for the cloud computing service. Specific functions are grouped into layers, thus forming a layering framework consisting of four layers [3]:

- User layer: This layer provides the user interface through which the user interacts with the cloud service and cloud service provider (e.g., web interface).
- Access layer: This layer provides access mechanisms for presenting the cloud service capabilities that are available in the service layer (e.g., a set of web pages). Where required, this layer is also responsible for handling encryption and checking the service integrity, as well as enforcing QoS policies (e.g., in cases when a telecom provider offers cloud services to its customers). So, this layer accepts the requests from customers for access a cloud computing service.
- Service layer: This layer contains the implementation of the cloud services offered by the cloud service provider and controls the software components. Such components, however, do not include the OSs in the hosts or the device

drivers. So, this layer depends on capabilities in the resource layer; sufficient resources are needed with the goal of meeting the requirements of the SLA between the cloud customer and service provider.

- Resource layer: This layer contains resources such as servers (placed in data centers), switches and routers for networking, storage devices, and noncloud software that runs on the cloud servers (includes host OSs, device drivers, hypervisors, and generic management software).

The implementation view describes the functions needed for implementation of the cloud service, including the service and the architecture parts). The deployment view describes how the cloud service functions are technically implemented within an existing infrastructure and its elements.

Based on the layering framework of the four layers just discussed, the cloud computing reference architecture is shown in Figure 8.3. It consists of the four layers and the functions within each of the layers. The functions that span multiple layers are called cross-layer functions, and they include the operational management function, cloud performance function, and security and privacy function.

8.3 Cloud Computing Service Models

Many cloud computing service models are possible. (Note that the cloud computing service model is used interchangeably with the cloud service category.) The end user, that is, the cloud service customers (CSCs), typically consumes cloud services in one of the following main five models [5]: IaaS, PaaS, SaaS, NaaS, and CaaS. In the following subsections we introduce the main characteristics of each of the cloud service models.

8.3.1 Infrastructure as a Service (IaaS)

IaaS is one of the representative cloud service categories. It provides an infrastructure capability type to the CSC end users [6]. The IaaS allows the CSCs to use the provided cloud infrastructure resources, which include processing, storage, or networking [7]. Hence, IaaS provides to CSCs the following cloud service functions:

- Computing service functions: These functions allow CSCs to provision and use processing resources including machine (physical or virtual) operations and functions such as virtual machine migration, snapshot, and backup.
- Storage service functions: These functions allow CSCs to use storage resources including life cycle operations and functions (e.g., snapshot, backup, load balancing, reservation).
- Network service functions: These functions allow the CSC to use the given networking resources, including IP addresses, load balance, firewalls, and virtual networking (e.g., virtual switch, VLAN).

In the IaaS model, cloud providers offer end users virtual machines, virtual storage, and virtual operating systems and applications software through the use of large

8.3 Cloud Computing Service Models

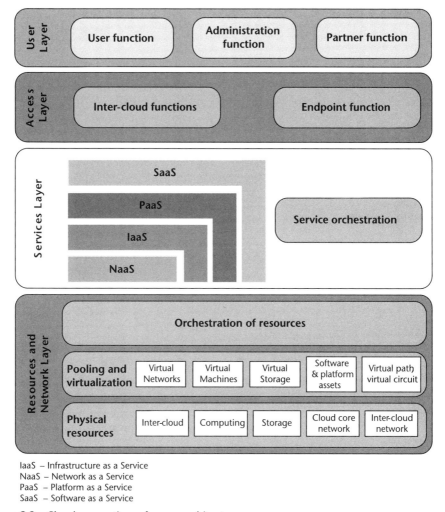

IaaS – Infrastructure as a Service
NaaS – Network as a Service
PaaS – Platform as a Service
SaaS – Software as a Service

Figure 8.3 Cloud computing reference architecture.

data centers (with large, physical computing resources). Connectivity to such clouds is performed either through the Internet (for residential users) or carrier clouds (via VPNs).

8.3.2 Platform as a Service (PaaS)

PaaS is a cloud service model in which a cloud provider provides to the end user a platform that includes operating systems, and execution environment for a given set of programming languages, databases, and web servers. In this model application developers can develop, test, and provide applications without needing to buy the underlying equipment for their provisioning. In the PaaS scenario the cloud user does not control or manage the underlying cloud infrastructure; instead, the end user has control over the applications deployed over the cloud infrastructure by using tools provided by the cloud service provider (CSP).

The initial idea of PaaS was to provide programming environments for developers (e.g., the start of Google's App engine in 2008 was limited to a specified number of developers). So, initially all PaaS deployments were public cloud services to speed up the development of applications and services (mainly OTT) on the Internet.

Three types of PaaS exist: public, private, and hybrid. The public PaaS is provided and fully maintained by the CSP. The private PaaS can be downloaded and installed on servers on the customer's side or in a public cloud. (It arranges different application and database components into a single hosting platform.) The hybrid PaaS refers to a private PaaS that uses the infrastructure of the CSP.

8.3.3 Software as a Service (SaaS)

SaaS is a cloud service model (and category) in which cloud providers install and manage application software in the cloud, thus reducing the need to run the software on the cloud user's own computers or devices (the user accesses the cloud applications by using cloud clients) [1, 2].

The SaaS architecture (including servers and their interconnection network toward the Internet) is used simultaneously by users of the CSP. The SaaS applications are typically hosted in the cloud, that is, on servers in data centers far away from the cloud users. A given SaaS application cannot access a given company's systems (e.g., databases), but many SaaS-based applications offer end users the ability to collaborate and share information, based on Web 2.0 functionalities. In that respect, most SaaS implementations can be accessed directly from a web browser by using web technologies such as HTTP, HTML, and JavaScript. That simplifies the use of SaaS applications since no additional installation of software is required on the customer's side (except some plug-in installation in certain SaaS cloud services). Many OTT services, including management and collaboration tools that run as OTT applications are based on SaaS. Popular SaaS offerings include email (e.g., Gmail, Yahoo mail), collaboration tools (e.g., WebEx from Cisco), Google Apps (which includes a dozen applications, such as Drive for managing documents, Maps, Gmail, Google+ as Google's social network service, YouTube, Calendar), social network services (e.g., Facebook), and so forth. The most common use of SaaS is to replace on-device (e.g., computer, smartphone) software and related data.

8.3.4 Network as a Service (NaaS)

NaaS is a cloud service model in which CSP provides transport connectivity and related network capabilities to end users.

General requirements for NaaS [8] are an on-demand network configuration, secure connectivity (between the end user and the CSP), QoS-guaranteed connectivity (if the cloud service is provided by telecom operators only, and not via the IAS, which is delivered using the best-effort principle), and heterogeneous network compatibility.

Regarding the high-level concept [9], NaaS offers three cloud capabilities:

- NaaS application: This refers to the case in which CSCs use network applications provided by the NaaS CSP. Examples of NaaS applications include a virtual content delivery network (CDN), virtual router, virtual EPC for 4G mobile networks, and virtual firewall.

- NaaS platform: In this case CSCs can use the network platform provided by the CSP. This type of NaaS platform provides software execution environments with different programming languages to deploy network applications (e.g., self-implemented network services by the CSC).
- NaaS connectivity: This is an infrastructure capability type of cloud service in which CSCs use network connectivity resources provided by the CSP. Examples of NaaS connectivity include VPNs, bandwidth-on-demand (BoD), and so forth. The CSP may choose to offer connectivity over logical, virtual, or logical networking functionalities. In some cases the CSP may make offers that extend over the IP networking provision, including control of optical networks, access to dark fiber via photonic switching, and so forth.

When cloud providers offer network and computing resources as a single product to end users, it is referred to as Network as a Service (NaaS) or cloud computing networking [7]. Examples of NaaS also include virtual network operators (VNOs) and mobile virtual network operators (MVNOs). Typically NaaS provides bearer connectivity without specific differentiation of the data that are carried through the network. So, NaaS is mainly targeted to best-effort Internet traffic, such as traffic from OTT services. However, services that are specific to a certain type of data carried through the network, such as VoIP, video, or instant messaging, are typically categorized as CaaS, as discussed next.

8.3.5 Communication as a Service (CaaS)

CaaS is a cloud service category in which the CSP offers the CSC capabilities for real-time interactions and collaborations [1]. It may be considered as a special case of the NaaS, that is, NaaS where individual service is important (e.g., VoIP, video telephony, instant messaging). On the other side, CaaS may be also regarded as part of the SaaS, that is, CaaS providers offer software-based products and services over the Internet. Similar to other SaaS solutions, cloud customers (e.g., business customers, such as enterprises and organizations) do not have to worry about maintenance of servers and connections or about administrative work to keep the communication platform functional.

The main attraction of CaaS is that it allows CSCs to take advantage of telecommunication operators' communication capabilities. One of the key characteristics of communication services offered by telecom operators is their high availability (e.g., 99.999% availability, known as "five nines," which is progressing toward "six nines" availability within the NGN framework), high performance, and highly secured environment, as well as reliable service monitoring and O&M capabilities 24 hours a day. From a telecom operator's point of view (as the CaaS provider), CaaS is about publishing APIs that give access to rich communication services. Considering that the all-IP transition of telecom operators is based on the NGN model, which uses IMS as its main system in the service stratum, practical CaaS implementation refers to provision of API (to the operator's IMS) to third parties, such as developers and partners (e.g., other operators and service providers, vendors, universities), with the goal of enhancing their own services. On the side of devices, CaaS is contributing with the provision of software development kits (SDKs), which may contain various communication building blocks such as codecs (for audio and video) and authentication, that are targeted for use in devices that are not natively built for communication (e.g., IP cameras, which have Internet connectivity capability, but not the communication capability).

A high-level view of the CaaS reference architecture is shown in Figure 8.4. It consists of several different components, including CaaS application, IMS services, a business support system (BSS) and operations support system (OSS), management application, and security services. The CaaS application typically consists of a management interface layer, which includes a configuration interface, billing interface, reporting interface; a communication service orchestration layer, which includes routing requirements based on keypad input by the user, such as call answer, prompt for input, processing the input, and routing call to appropriate destination; and a communication services layer, which includes services that perform a set of communication tasks, such as voicemail or call routing. Overall, CaaS is a cloud solution for real-time conversational services offered through a cloud, typically through the cloud interface offered by the telecom operator that acts as the CSP.

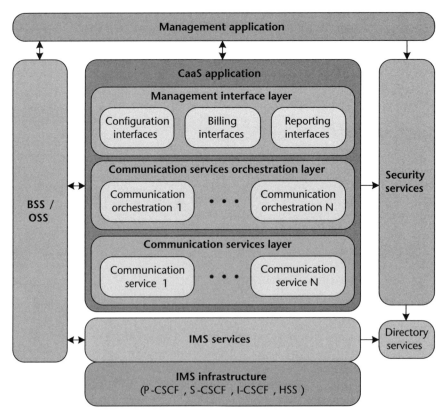

Figure 8.4 CaaS reference architecture.

8.3.6 Intercloud Computing

Intercloud computing refers to the interworking of CSPs in order to provide services to customers. Interrelationships among CSPs can be realized with direct communication among them or through an API.

In intercloud computing when one CSP (CSP-A) uses a certain service from another CSP (CSP-B), then CSP-A is considered to be the primary CSP, while CSP-B is the secondary one. It is possible, however, to have the reverse case in parallel; that is, in the given example, CSP-B may use a service provided by the other peer (i.e., CSP-A). In the case where an API is used, CSP-A uses services from CSP-B via an API provided by CSP-B, and vice versa.

Three main patterns are used for intercloud computing [10]:

- Intercloud peering: In this case CSPs interwork directly with each other.
- Intercloud federation: This pattern involves cloud services within a group of peer CSPs who combine their services to offer to customers. In this case multiple CSPs may share service-related policies, SLAs, and procedures related to service offerings and handling. However, there does not need to be a direct connection between each pair of CSPs in the federation.
- Intercloud intermediary: This refers to intermediation, aggregation, and arbitrage of cloud services offered by one or more peer CSPs by a given CSP. The intermediation refers to enhancing or conditioning the cloud service. Aggregation refers to the composition of a cloud service created from a set of services provided by the CSPs. Arbitrage refers to selection of a single service from a group of services offered by other CSPs.

For intercloud relations between CSPs, we can distinguish between two types of resources. The first type includes the underlying physical resources of the cloud infrastructure, which are controlled and managed by the CSP who owns them. The second type refers to resources that are abstracted from the underlying physical resources and then offered to CSPs as services. Based on the abstraction, the underlying physical resources become abstracted resources. So, intercloud service resource handling allows a given CSP to negotiate for use of abstracted resources from other peer CSPs, which are then provided as cloud services.

8.4 Cloud Security and Privacy

The perceived risks for cloud computing include confidentiality, integrity, and availability as cybersecurity objectives [2]. At the same time, cloud services are subject to local physical threats as well as external ones. Similar to other ICT applications, the possible threats to cloud computing services for both CSPs and CSCs include, but are not limited to, accidents, natural disasters, criminal organizations, hostile governments, and internal and external unauthorized and authorized cloud system access (including intruders, employees at the CSPs). The multitenancy characteristic of the cloud service implementations as well as various cloud service models increases the security and privacy risks of the end users and their data.

Different cloud security aspects are covered in standards from several standard developing organizations, including the European Network and Information Security Agency (ENISA) [11], the National Institute of Standards and Technology (NIST) [2], and the Cloud Security Alliance (CSA) [12]. However, the umbrella documents have been provided by the focus group on cloud computing of ITU-T [13].

In general, major cloud security objectives include the following [2]:

- Prevention of unauthorized access to a cloud computing infrastructure: This typically includes implementation of a logical separation of cloud resources, for example, a logical separation of the cloud user workload on the same server in the cloud by using hypervisors (i.e., virtual machine monitors) in a multitenant environment.
- Protection of customer data from unauthorized access: This includes supporting identity management (IdM), so the CSCs will have the ability to enforce policies on authorized access to their data and resources in the cloud.
- Protection from threats from hardware and software used on the CSP or CSC side, including trustworthiness and reliability of the software and hardware used.
- Implementation of security solutions into the design of web applications for access to cloud resources (e.g., use of SSL/TLS, certificates).
- Protection of web servers from attacks (e.g., installation of firewalls between the public Internet and cloud servers in data centers, applying patches to the software in use).
- Deployment of access control and intrusion detection systems at the CSP: This includes restricting physical access (of people, including unauthorized employees) to the network and devices, disabling unused ports and services, applying role-based access, minimizing the use of privileges, and making the use of antivirus software mandatory, as well as encryption of the communication end to end.
- A clear definition of responsibilities regarding the security measures between the CSPs and CSCs.
- Portability of cloud solutions, with the goal of providing to the CSC the ability to change the CSP when the provider fails to satisfy the confidentiality, integrity, and availability requirements.

Most of the problems in cloud computing are coming from loss of control, lack of trust, and multitenancy approaches. However, these problems typically are found in third-party cloud management models. (Self-managed clouds also have security threats, but there are different than these three.)

Loss of control refers to a CSP's ability to control data, applications, and resources and not lose control over them. In this respect, user identities are also stored in the cloud (e.g., email address, postal address, credit card number, cloud account password) and handled by the CSP. Different rules, security policies, and enforcement are also managed by the CSP, which is typically obliged to follow the legislation in place in the country in which their resources are located (e.g., the servers holding the platforms, applications, and data). So, CSCs rely on the CSP to ensure data security and privacy, cloud resource availability, and cloud O&M.

Another security issue is the level of trust between the CSC and the CSP. This issue typically appears in risky situations. Minimizing a CSC's lack of trust can be accomplished by means of certification (e.g., by trusted third parties) or with SLAs. However, SLAs define certain QoS parameters (e.g., uptime 99% of the time), but they do not allow customers to dictate their requirements. Minimizing a lack of trust can also be accomplished by creating a policy language that is understandable and processable by machines; for example, the policy statement "requires geographic isolation between virtual machines" requires physical and geographical separation between other tenants.

The multitenancy paradigm in the cloud may cause security issues due to possible conflict of interests between tenants and their opposing targets. These types of multitenancy security issues can be solved by the strong separation of different tenants.

Table 8.2 gives mappings for different security standards with regard to cloud services.

Considering the layered framework of cloud computing systems, security functions belong to cross-layer functions, which spanning all four layers—the user layer, access layer, services layer, and resources and network layer.

8.4.1 Application Layer Security for Access to the Cloud

Regarding the security solutions end to end between the CSP on one side and CSC on the other side, typical Internet security solutions are used, including SSL/TLS (between the application and transport layer, e.g., HTTPS) or VPN access to the cloud (with IPsec). Another possible application layer security mechanism is the use of Public Key Infrastructure (PKI) mechanisms for the cloud. However, most of the individual users do not have a commercial certificate. Typically, certificates have to be enterprise specific, which places a burden on enterprises to acquire certificates. This can be avoided if a CSP can act as the certificate and validation authority for its CSCs.

Note that there are several means of access control in the cloud, including access to the cloud, to services, to servers, to databases (including direct access and web queries), and to VMs and the objects within them. Regardless of the cloud deployment model, the CSP needs to manage the user authentication and access control procedures to the cloud. That can be a burdensome task for the CSP when many users access the cloud from different enterprises (organizations) with different access control policies. Such an approach requires user trust for the CSP in terms of security, and O&M of access control policies.

Finally, consumer-managed access control requires less trust on the part of the CSP. In this case a policy decision point (PDP) lies in customer's domain, whereas a policy enforcement point (PEP) belongs to the CSP domain (which is generally different than the first one), as shown in Figure 8.5. However, this approach requires

Table 8.2 Security Standards Mapping

Categorization of Security Aspect	Standard	SDO
Authentication and authorization	SSL/TLS (RFC 5246)	IETF
	X.509 Public Key Infrastructure (PKI) standards	ITU-T (X.509), IETF (RFC 3820, RFC 5280)
	Oauth (Open Authorization Protocol)	IETF (RFC 5849)
	OpenID authentication	OpenID Foundation
	eXtensible Access Control Markup Language (XACML); Security Assertion Markup Language (SAML)	OASIS
	Authentication standards	NIST
Confidentiality	SSL/TLS (RFC 5246)	IETF
	Key Management Interoperability Protocol (KMIP)	OASIS
	XML Encryption Syntax and Processing	W3C
	Escrowed Encryption Standard (EES); Advanced Encryption Standard (AES)	NIST
Integrity	XML Signature (XMLDSig)	W3C
	Secure Hash Standard (SHS); Digital Signature Standard (DSS); Keyed-Hash Message Authentication Code (HMAC)	NIST
Identity management	Requirement of IdM (Identity Management) in Cloud Computing	ITU-T
	Security Assertion Markup Language (SAML)	OASIS
	OpenID authentication	OpenID Foundation
Security, monitoring, and incident response	Security Content Automation Protocol (SCAP); Computer Security Incident Handling Guide; Guideline for the Analysis of Local Area Network Security	NIST
	Cybersecurity information exchange techniques (X.1500); common vulnerabilities and exposures (X.1520); Common Vulnerability Scoring System (X.1521)	ITU-T
Security policy management	eXtensible Access Control Markup Language (XACML)	OASIS
	Standards for Security Categorization of Federal Information and Information Systems	NIST
Availability	Guidelines for Incident Preparedness and Operational Continuity Management	ISO (22399:2007)

CSPs and CSCs to have trust relationship and a predefined way of describing users, resources, and access decisions. Also, the CSP should guarantee the support of the customer's access decisions. Examples of consumer-managed access control of its data can be found to a certain degree in Google Apps and Facebook (however, it is not full control). For example, Google Apps, as an SaaS cloud service, controls authentication and access to its applications, but the end users themselves can control access to their data (e.g., documents) via a defined interface (by the CSP, i.e., Google) to the access control mechanism. In the case of IaaS cloud services, customers can create user accounts on their VMs and then create user access lists for services located on the given VM.

8.4 Cloud Security and Privacy   319

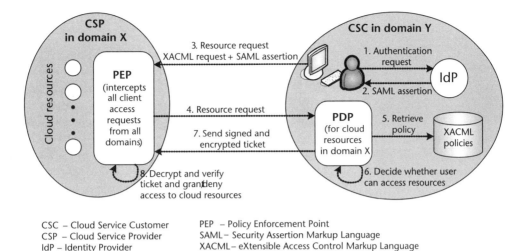

CSC – Cloud Service Customer   PEP   – Policy Enforcement Point
CSP – Cloud Service Provider   SAML – Security Assertion Markup Language
IdP – Identity Provider   XACML – eXtensible Access Control Markup Language
PDP – Policy Decision Point

Figure 8.5 Cloud access control.

8.4.2 Secure Cloud Access

For any cloud service (infrastructure, platform, or service), the CSP or the enterprise may control access to the services. However, in multitenant cloud solutions for enterprises there exist different legal entities (e.g., cloud provider, enterprise), so each one may need to be isolated in certain cases.

The three main types of secure access to the cloud [13] are as follows (Figure 8.6):

- Private authentication: In this case cloud users access the cloud via an enterprise VPN, while the CSP and the enterprise have a preconfigured secure tun-

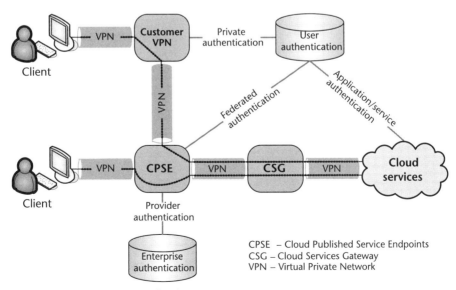

Figure 8.6 Scenarios for secure cloud access.

nel between them. In this scenario all control is on the side of the enterprise via its VPN authentication.

- Provider authentication: In this case the cloud provider provides the authentication via cloud published service end points (CPSEs), which refer to the points where the user's cloud APIs and VPN sessions are terminated and authenticated. Here are possible two subcases: (1) The user accesses the cloud without a separate enterprise VPN and authentication mechanism or (2) the user accesses the cloud through the enterprise and provider VPNs. The second case requires two distinct VPN login methods, one for the enterprise and the other for provider access.
- Federated authentication: In this case the enterprise agrees to federate its systems with the CSP, so the cloud users use enterprise VPN authentication mechanisms even when they are accessing the CSP's VPN. These systems may also be used for intercloud scenarios.

Note that authentication in cloud applications is separate from VPN authentication. In the case of public and federated authentication methods, certain parts of users' databases in enterprises may be used for both VPN and service authentication. On the other side, for a public authentication method, the CSP authentication may differ from service-specific authentication that is set up by the CSCs. In all cases, the cloud user may need to further authenticate itself through enterprise authentication [e.g., via HTTP or Secure Shell (SSH)].

Finally, it is important to manage the propagation of identity information in an orchestration network. For example, to satisfy QoS requirements (e.g., bit rate, delay) for a given cloud user, the CSP needs to associate the IP address of that user with its enterprise identity (the cloud user is identified either at the provider or enterprise edge). This can be achieved with tunnels between the CPSE and cloud services gateway (CSG) that is specific to cloud service user (see Figure 8.6). Such identity tunnels should be manually or automatically preset between the CPSEs and CSGs. Then, the CSGs are unaware of the cloud user's identity, but they can apply policies to the tunnel (e.g., SLA based).

8.5 Over the Top (OTT) Cloud Services

Cloud services are becoming part of many OTT services, such as web applications that offer storage, social networking services, and video-sharing sites. Each such OTT cloud-based service has a network infrastructure in the cloud ecosystem. It consists of several segments:

- Intracloud network: This is used to connect local cloud infrastructures, such as machines (hosts) that host protocols and applications from OSI protocol layers 4 to 7 (e.g., load balancers, application acceleration devices, firewalls).
- Intercloud network: This is the network (which may span across several domains) that connects cloud infrastructures that belong to a single or multiple CSPs placed at different locations.

8.5 Over the Top (OTT) Cloud Services

- Core network: This is typically a MAN or a WAN that is used by customers to access services from the CSP.

Some of the most well-known OTT cloud services are those deployed by Google, Amazon, and Microsoft. For example, Google's cloud is based on several layers, as shown in Figure 8.7. The lowest layer consists of Google global distributed computing, which includes servers, storage, and a network to interconnect them. The Google cloud OS, which includes Google File System (GFS), clusters, BigTable, MapReduce, and so forth, is deployed over the physical cloud infrastructure layer. Over the Google cloud OS are the Google AppEngine platform, Google Apps, as well as other Google services. On the top are third-party applications implemented through the Google AppEngine platform and also application marketing.

Another example for OTT cloud services is the Windows Azure platform, shown in Figure 8.8. The layering approach is similar to that of the Google cloud model. Physical data centers are located on the bottom, with hypervisors over them. The OS and databases are placed over the physical data centers. The platform includes computational services (VMs, web, Worker), storage services (drives, queues, tables), network services (connect, CDN), and databases (relational databases). The Windows Azure AppFabric (application services) and SQL Azure (data synchronization including client synchronization and database-to-database synchronization, as well as reporting including analytics) are placed over the OS and databases. Then, on top of Windows Azure, application runtimes, frameworks, and tools (e.g., Microsoft .NET framework, Java, PHP, Python) are placed.

Other functional cloud models have been deployed. However, from the two OTT cloud examples given above, we can derive a conclusion that is applicable in general. That is, OTT cloud deployments follow the layered cloud functional model, with a specific characteristic on a given layer depending on the technologies that are developed or used by the OTT cloud provider. On the other hand, all OTT cloud services lack end-to-end QoS support, because that can be provided only by

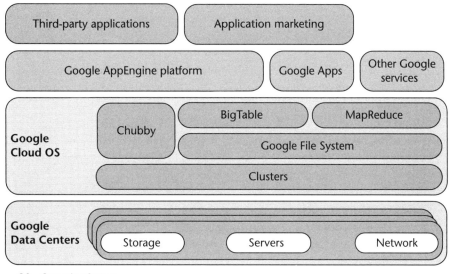

OS – Operating System

Figure 8.7 Example of a functional model for OTT services: Google cloud model.

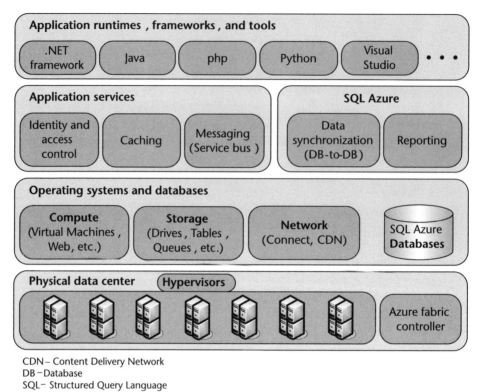

CDN – Content Delivery Network
DB – Database
SQL – Structured Query Language

Figure 8.8 Example of a functional model for OTT cloud services: Windows Azure platform.

telecom operators that give broadband Internet access to their customers. Overall, OTT cloud services cannot be differentiated by the telecom operators (e.g., to have guarantees on bit rates or delay) due to the network neutrality principle for IAS, which is based on the best-effort approach for OTT traffic. However, OTT cloud providers can partner with telecom cloud providers.

8.6 Telecom Cloud Implementations

Telecom operators look to cloud computing as a possible way to counter the loss of revenues from legacy services (e.g., telephony) and to increase their mediation role between customers and OTT cloud service providers. The telecom operators can contribute to enterprise cloud systems, because they can guarantee QoS end to end (between customers on one side and cloud resources on the other). In that respect, telecom operators may act as federators between different cloud entities, including public, private, community, and hybrid clouds.

Telecom operators provide access to the Internet to both individual users and enterprises/organizations. So, they have assets, including the networking infrastructure and service control entities (e.g., IMS). Further, telecom operators have SLAs in place with customers, so they have initially higher trust from their own subscribers. The NGN architecture in telecom operators is suitable for implementation of cloud services due to defined interfaces toward its service stratum, so third parties can contribute with the development of cloud applications and services.

Although telecom providers can offer services from cloud computing categories, their main differentiation from OTT CSPs is their ability to offer CaaS (for VoIP, video conferencing, instant messaging, over different end-user devices) and NaaS (e.g., flexible and extended VPN, BoD), which depends directly on the control of physical networking resources (e.g., routers, switches, and interconnection transport links), which can be accomplished by telecom operators. However, several other cloud computing services are also relevant targets for telecom operators, as shown in Table 8.3. Besides the cloud services given in Table 8.3, there is a plethora of many other different XaaS offerings (here the "X" replaces a letter that designates the type of offering), such as Security-aaS, Database-aaS, Storage-aaS, Content-aaS, and Integration-aaS.

The following sections give several different use cases for possible cloud services provided by telecom operators.

8.6.1 Desktop as a Service (DaaS)

DaaS is a cloud service in which users use virtualized desktops from the CSP in the form of outsourcing [14], so all data and used applications are run centrally in the cloud instead of locally on the machine of the cloud user. Hence, the CSCs can access an OS as well as applications through their user interface, which is typically in the location of the user, which is different from that of the application logic processing. In other words, it is desktop virtualization separation of the PC desktop environment by using a client–server model (the client is the user's PC, and the server is in the DaaS cloud). DaaS reduces the total costs of ownership, may provide a rich client experience, and provides separation of CSP and user responsibilities.

8.6.2 Cloud Communication Center

A cloud communication center provides enabling features for customer–enterprise interaction by using the communication and management capabilities provided by a cloud-based telecommunication infrastructure. Such capabilities include management of various resources in the cloud such as media storage, content, customer, enterprise agent, and transport and communication resources.

This service provides for the possibility of sharing of applications that are used by different enterprises, and the ability of the CSP to charge enterprises on a per-usage basis. These types of cloud services are convenient for customer relationship management (CRM) centers, support centers (e.g., for certain products), billing centers (e.g., third-party billing services), and enterprise resource planning (ERP) software, which typically consists of integrated applications that the enterprise can

Table 8.3 Relevant Cloud Services for Telecom Operators

	CaaS	NaaS	IaaS	PaaS	SaaS
Desktop as a Service (DaaS)	—	—	X	—	—
Cloud communication center	X	—	—	—	X
Service Delivery Platform as a Service (SDPaaS)	—	—	—	X	X
Flexible and extended VPN	—	X	—	—	—
Bandwidth-on-Demand (BoD)	—	X	—	—	—

use to collect, store, and manage data from various business activities (e.g., product planning, service delivery, marketing and sales, shipping and payment).

8.6.3 Service Delivery Platform as a Service (SDPaaP)

A service delivery platform (SDP) is a system architecture or environment that enables efficient creation, deployment, execution, orchestration, and management of different service classes. Hence, SDPaaS refers to capabilities in the cloud that are offered to users of SDP functionalities and applications provided by the CSP on one side, as well as capabilities to deploy, control, and manage SDP functionalities on the other side. The SDPaaS consists of SaaS and PaaS components. The SaaS component can be telecom SaaS, Internet SaaS, business SaaS, or consumer SaaS.

Cloud service users use APIs to access SDPaaS services in the cloud. The SDPaaS may offer traditional SDP services as well as new innovative convergent services across multiple operators' domains. Examples of relevant services provided with SDPaaS cloud solutions are shown in Figure 8.9. For example, telecom services provided through SDPaaS include VoD services, MMS), IPTV, news services, and so forth. Internet and mobile services offered through SDPaaS cloud computing include mobile search, video surf, map services, blogs, and address books. Other applications that can be provided via SDPaaS services include e-health, e-traffic, environment monitoring, agriculture monitoring, smart grids, and city emergency services.

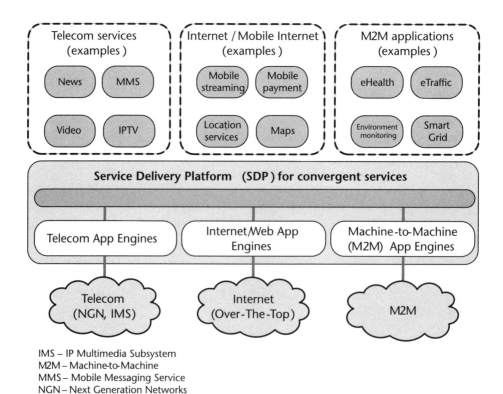

IMS – IP Multimedia Subsystem
M2M – Machine-to-Machine
MMS – Mobile Messaging Service
NGN – Next Generation Networks

Figure 8.9 Convergent services with SDPaaS.

8.6.4 End-to-End Service Management by Cloud Provider

For certain types of services, QoS support is needed end to end, which requires end-to-end management by the cloud provider (e.g., for VoIP services) [15]. We will illustrate end-to-end service management by describing a case in which CSPA is providing cloud VoIP service to CSP B, which is bundling the VoIP with other services and then reselling the package to the CSCs (Figure 8.10). CSP B provides connectivity services through its own core and transport network (e.g., based on IP/MPLS) and access networks (e.g., enterprise LAN/WAN, 3G or 4G mobile networks).

In the case illustrated in Figure 8.10, when a CSC has a problem with the offered VoIP services, it contacts CSP B, which sold the service to the customer, by using its CRM system. Then, the CSP B support agent should be able to see the health of the VoIP service from an end-to-end perspective, which requires visibility into the VoIP service elements and network management systems of both cloud providers, CSP A and CSP B. For that purpose there are two types of connection paths:

- Service delivery path: This path consists of combined cloud VoIP and IP/MPLS transport and core networks.
- Service management path: This path contains all logical management paths used for O&M functions, including service provisioning and assurance as well as charging to the CSCs.

To ensure the service works efficiently, all required functionalities end to end must work properly. For quick resolution of any problems with service provisioning, both providers should be able to see the system via CRM/dashboards and portals and to investigate its details. The customer service agent should be able to change the service configuration or initiate a new one. If useful management tools

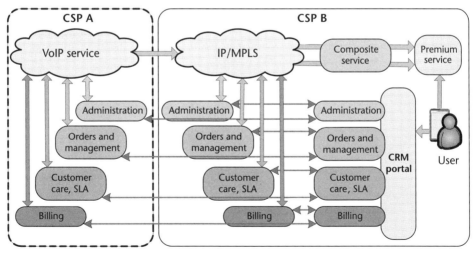

CRM – Customer Relationship Management
CSP – Cloud Service Provider
MPLS – MultiProtocol Label Switching
SLA – Service Level Agreement

Figure 8.10 End-to-end management of cloud services.

are lacking, the customers could become unsatisfied with the service, which can result in additional operational costs for the CSPs.

8.7 Mobile Cloud Computing (MCC)

Mobile technologies have been developing in parallel with Internet technologies during the past two decades. Their merging became a natural process that started with 2.5G mobile networks (e.g., GPRS) and continued via 3G (with its packet-switching part) toward the all-IP 4G mobile networks (e.g., LTE/LTE-Advanced). Internet technologies allowed for the development of cloud computing as the current emerging paradigm.

What makes cloud computing potentially valuable for mobile technologies? Well, the two most typical uses are storage of personal data in the cloud (through a mobile device) and the ability to perform computations on the stored data.

Can mobile cloud computing help in terms of the energy consumption in mobile terminals? Well, one solution is to make energy-efficient applications and services that take into account the number of computations required to complete a task. The other possible solution is to enable the mobile devices (where possible and needed for the application) to offload some energy-consuming tasks to some fixed servers in a cloud. In this way, cloud computing + mobile broadband access to cloud services gives us the mobile cloud computing (MCC) paradigm.

Can offloading of consuming computing applications to the cloud save energy for mobile devices? We can conduct a quantitative study to find the answer [16] by using the following formula:

$$MCC_{efficiency} = P_c \frac{C}{M} - P_i \frac{C}{S} - P_{tr} \frac{D}{B} \qquad (8.1)$$

where S is the speed of the cloud when computing C instructions, M is the speed of mobile when computing C instructions, D is the data that need to be transmitted, and B is the bandwidth of the wireless/mobile Internet (typically limited by the bit rates in the radio access network), P_c is the energy cost per second when the mobile phone is doing computing, P_i is the energy cost per second when the mobile phone is idle, and P_{tr} is the energy cost per second when the mobile is transmitting the data. Typically, clouds have a many times higher processing power than do mobile devices. In that respect, if we assume that a cloud is F times faster than a mobile device, that is, $S = F \times M$, then we can rewrite (8.1) as follows:

$$MCC_{efficiency} = \frac{C}{M}\left(P_c - \frac{P_i}{F}\right) - P_{tr} \frac{D}{B} \qquad (8.2)$$

If $MCC_{efficiency}$ is positive, then energy is saved by offloading computationally demanding tasks from a mobile device to the cloud; otherwise it is not. The formula gives a positive result (i.e., in favor of MCC) when D/B is sufficiently small compared with C/M while on the other side F is sufficiently large (i.e., the cloud can compute much faster than mobile devices).

A typical example where MCC use makes sense is a chess game. A chessboard has a total of 8 × 8 = 64 positions, and each of the two players controls 16 pieces at the beginning of the game. Then, each piece on the chessboard may be in 1 of the 64 possible locations and needs 6 bits to represent the location. If we need to represent the chess game's current state, it could be done with 6 bits × 32 pieces = 192 bits = 24 bytes, which is a size that can fit into a single packet over a mobile access network. However, the amount of computation required for the chess game can be very large (depending on its implementation), so offloading such a computational task (from the game) from a mobile device to the cloud can save energy and extend the battery life of the device.

On the other hand, one of the major limitations of today's mobile cloud services (i.e., MCC) is the energy consumption associated with the radio access technology, as well as the delay experienced in reaching the cloud provider through a mobile network.

Regarding macrocellular networks, mobile users located at the edge of macrocells are disadvantaged in terms of power consumption, because the B (bit rate) in the above equations can become very small, and cloud computing in such cases will not save energy. Therefore, it is hard to control battery power consumption in macrocellular mobile networks (this is another reason why small cells are needed in the 5G NGN mobile networks) [17]. The same analysis can be applied to the latency (i.e., the delay), which is very important for services such as MCC. Why? Well, as humans, we are physically sensitive to delay and jitter (i.e., average delay variation). As latencies increase, human interactive responses decrease. Whereas 4G has mobile network delays (including core and access) in the range of 10 ms, the delay in 5G mobile networks is expected to be reduced to 1 ms (that means less than 1-ms delays separately in the access and in the core parts). Because the interaction times foreseen in 5G systems in the so-called tactile Internet are quite small (on the order of milliseconds), the 5G systems can become an excellent network for the near-future MCC. [18]. However, the servers that host the cloud have to be as close as possible to the mobile network providers (e.g., located in the region or even in the mobile operator's network in some cases), because a limitation in the transfer of information is the speed of light; it is the maximal speed for transfer of information over any links. For example, if we take into account the length of the Equator, which is 40,000 km, then the longest distances (from a mobile handset to a server) will be more than 20,000 km (2 × 10^7 m), which, considering that the speed of light is approximately equal to 3 × 10^8 m/s), will give delays of around 70 ms due to propagation of the signals only (without different additional delays such as access delays, buffering delays, and processing delays, which further increase the total delay budget). So, where possible, clouds should be closer to the mobile devices they serve to provide an even better MCC use experience in 5G mobile networks.

With cloud computing the client–server paradigm (which is also used for MCC) is becoming more attractive. For typical client–server communication over the Internet, the server does not have the data until the client sends it to the server. In contrast, in cloud computing (including MCC) the server in the cloud already contains the data. For example, cloud services such as Amazon S3 can store data, while other cloud services such as Amazon EC2 (Elastic Compute Cloud) can be used to perform computation on the stored data by using S3 (Simple Storage Service).

To secure communication end to end between mobile devices and the cloud, encryption techniques require additional processing on both sides of the connection. If additional C_p instructions are required on the mobile device's side for encryption/decryption (for sending/receiving data to/from the cloud), then the equation for $MCC_{efficiency}$ becomes:

$$MCC_{efficiency} = \frac{C}{M}\left(P_c - \frac{P_i}{F}\right) - P_{tr}\frac{D}{B} - P_c\frac{C_p}{M} \tag{8.3}$$

So, MCC can potentially save energy for mobile devices, but not all mobile applications are energy efficient when migrated to the cloud.

Overall, MCC has similarities and differences compared to general cloud computing. The same cloud services offered via fixed broadband Internet access can be offered to mobile devices. All OTT services that use cloud services (e.g., video-sharing sites, social networking, data storage sites) are typically used through both fixed and mobile hosts. However, mobile devices have smaller screens and different navigation characteristics (e.g., one-hand navigation on a smartphone), therefore user interfaces (which are typically web-based or applications downloaded from the Web and installed on mobile devices) should be adapted to mobile device characteristics with regard to screen size, navigation, and the capabilities of the existing web browsers. Also, mobile services should consider the energy overhead incurred by privacy security and data communication tasks before offloading them to the cloud.

Looking toward the next generation of mobile systems, with the estimated capacities of 5G mobile networks and MCC-based services, people's work patterns and habits can be dramatically changed. We are converging to a point in time where most Internet users will work primarily through Internet-based applications in clouds that are accessed through various networked devices including fixed and mobile ones.

It is possible that future MCC applications in 5G will have a major impact on most of the activities in our personal and business lives. Humans will become connected to the network more than ever, and will become dependent on it more than ever. So, the "killer" applications and services in 5G will almost certainly be MCC-based ones.

8.8 Regulation and Business Aspects of Cloud Computing

Considering the impact of cloud computing in different sectors, including economic (e.g., banking) and social (e.g., social networking), policy makers are considering embracing and controlling the cloud. The overall approach is that what happens in the cloud (e.g., data, transactions) cannot be outside the existing legislation in each of the sectors that are concerned, including the electronic communications sector.

Regarding the business aspects, cloud computing together with mobile technologies is being considered as the most emerging technologies in near future, and combining them should result in plethora of business models and ecosystems in ICT industry on a global scale.

8.8.1 Regulation Aspects of Cloud Computing

Considering the dispersive characteristics of cloud services, the question is how to regulate it? The right answer, as in many regulation approaches, is "It depends." Several laws are applicable to cloud computing, which is country specific. They include telecommunication laws, consumer protection laws, competition laws, and environmental and jurisdictional concerns [19].

Considering the different players in the cloud ecosystem, it is certain that a CSP for cloud services will be covered by telecommunication laws on the national, regional, or international level. Cloud computing may be considered in certain aspects to be regulated similar to the way legacy telecommunication services (e.g., telephony, succeeded by carrier-grade VoIP) are regulated, but it does have its own specifics. So, certain cloud services (e.g., NaaS, CaaS) can fall directly into the regulated sphere of telecommunications regarding networks and services. Other cloud services, such as SaaS and PaaS provided through public access to the Internet (based on the network neutral principle), are not treated as separate telecommunication services. On the other hand, with the growing use of cloud-based social networking, especially by the "born-digital" generations, use of always-on connectivity in certain environments (e.g., mobile ones) may initiate certain regulatory concerns (e.g., for certain technologies).

With the rise of cloud computing services, there have been many cases involving the protection of cloud consumers from abusive practices. Also, many cloud services such as social networking (e.g., Facebook) bind their customers with standard terms in the offered online agreement that requires all legal disputes to be solved in a court located in the origin country in which the CSP is a legal entity. (Typically the CSP is not a legal entity in other countries, although it is present globally via the network neutral principle of Internet.) That may cause uncertainty to customers regarding the laws that are applicable to their activities in the cloud as well as their data and information. Another potential problem with having customer data in the cloud is data lock-in, a situation that makes it hard or impossible for a customer to move and port data to other cloud providers (i.e., to change his or her CSP). In that respect, in certain regions (e.g., Europe) the regional administration (e.g., European Commission) has made proposals that suggest the enforcement of data portability to other CSPs by recognizing data portability as an individual right per se [20].

The lack of user data portability between different CSPs may cause concerns regarding the competitiveness of the cloud markets, which may be a reason for intervention by NRAs. This can also happen due to a lack of standards in a given area, or attribution of standards to the market leader. However, competition issues may also arise regarding user access to cloud services, which may refer to networks and infrastructures on which the cloud services rely. Such competition issues may range from unbundling to the network neutrality principle.

Further, cloud computing is based on large data centers that can be placed in different locations. In that respect, environmental concerns may trigger deployment of energy-efficient data centers that will reduce energy consumption and costs. Such an example is placing data centers in locations where natural cooling is available. In certain countries such practices may be required through legislation.

Finally, cloud computing has a transnational nature, which means that multiple jurisdictions may govern certain activities in the cloud or the path between the CSCs and CSPs. In some cases, the transfer of user data out of a user's jurisdiction area (e.g., its country) may result in undesired effects and legal prosecution in certain cases. That is the case in certain regulated services such as financial services, where cloud transfers and storage outside the jurisdiction of the given entity may violate national rules. In general, national regulators in all sectors (e.g., telecommunication sector, financial sector) typically are not willing to transfer their jurisdiction to any foreign authority.

8.8.2 Business Aspects of Cloud Computing

The business aspects and models in the cloud computing sphere are dependent on the type of cloud service and cloud ecosystem being used. Hence, key business roles in cloud computing are the CSC, the CSP, the cloud service partner, and the intercloud. The CSP has three distinct business subroles: provider of cloud applications, provider of the cloud platform, and provider of the cloud infrastructure.

The business value chain in a cloud ecosystem is shown in Figure 8.11. Telecom operators and OTT cloud service providers buy hardware and software components for the purpose of building clouds. Such components are needed for large-scale virtualizations, databases, storage, and so forth. Telecom operators and OTT cloud service providers deploy clouds with IaaS, PaaS, or SaaS capabilities. The applications offered in the cloud can be bought by third parties, developed in house, or created by application developers (who may use the offered platform to develop applications, and then sell the applications to the CSPs). The CSCs buy services from the CSPs. The main subroles of CSPs (IaaS, PaaS, and SaaS) define their cloud business models [19].

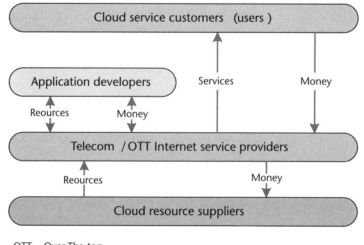

OTT – Over-The-top

Figure 8.11 Cloud ecosystem business value chain.

In the IaaS we can distinguish between two categories of business models: (1) storage provision (data offloaded to the cloud) and (2) provision of computing power (computing tasks offloaded to the cloud). For example, Amazon provides IaaS services based on their infrastructure as a computing service (EC2) and a storage service (S3). In this case, the pricing models are typically pay-per-use models or subscription based.

Regarding the business models of PaaS clouds, we can distinguish between two main types of platforms: (1) development platforms and (2) business platforms. Development platforms are targeted to developers to create their applications, and afterwards to upload their application code into the PaaS cloud where it can be run. A typical example of PaaS cloud development is the Google App Engine, which provides platforms for the deployment and management of applications in the cloud. There are also business platforms, such as SalesForce, which are targeted for development of business applications and services. The pricing models of the PaaS cloud are typically subscription based and per usage (price per user/month for a given offering).

Finally, applications are what most people get to know from cloud computing (it is in fact the interface for the customer). We can distinguish between SaaS applications and the provisioning of basic web services on demand. Typical examples of the SaaS are GoogleApps, which are completely accessible through a web browser. In the field of web service on-demand provisioning, the CSP offers web services hosted on a cloud with charging done on a pay-per-use basis.

In summary, cloud computing offers reduced costs to enterprises, increased storage, highly automated systems in the cloud, flexibility (e.g., easy change of a platform), higher mobility (e.g., work at a distance, service mobility), and possibilities for innovations (because cloud resources and applications are becoming instantly available to all cloud users). Finally, the cloud in its different "flavors" (SaaS, IaaS, and PaaS) has spread into all telecommunication/Internet market sectors, becoming the most disruptive Internet technology in both enterprise and consumer markets, including fixed and mobile environments.

References

[1] ITU-T Recommendation Y.3500, "Information Technology—Cloud Computing—Overview and Vocabulary," August 2014.

[2] National Institute of Standards and Technology, "NIST Cloud Computing Standards Roadmap," Special Publication 500-291, July 2011.

[3] ITU-T Recommendation Y.3502, "Information Technology—Cloud Computing—Reference Architecture," August 2014.

[4] ITU-T Focus Group on Cloud Computing, "Part 1: Introduction to the Cloud Ecosystem: Definitions, Taxonomies, Use Cases and High-Level Requirements," February 2012.

[5] ITU-T Focus Group on Cloud Computing, "Part 2: Functional Requirements and Reference Architecture," February 2012.

[6] ITU-T Recommendation Y.3513, "Cloud Computing—Functional Requirements of Infrastructure as a Service," August 2014.

[7] S. Azodolmolky, P. Wieder, R. Yahyapour, "Cloud Computing Networking: Challenges and Opportunities for Innovations," *IEEE Communications Magazine*, July 2013.

[8] ITU-T Recommendation Y.3501, "Cloud Computing Framework and High-Level Requirements," May 2013.

[9] ITU-T Recommendation Y.3512, "Cloud Computing—Functional Requirements of Network as a Service," August 2014.

[10] ITU-T Recommendation Y.3511, "Framework of Inter-Cloud Computing," March 2014.

[11] ENISA, "Cloud Computing—Benefits, Risks and Recommendations for Information Security," December 2012.

[12] Cloud Security Alliance, "Security Guidance for Critical Areas of Focus in Cloud Computing V3.0," 2011.

[13] ITU-T Focus Group on Cloud Computing, "Part 5: Cloud Security," February 2012.

[14] ITU-T Recommendation Y.3503, "Requirements for Desktop as a Service," May 2014.

[15] ITU-T Recommendation Y.3520, "Cloud Computing Framework for End-to-End Resource Management," June 2013.

[16] N. Fernando, S. W. Loke, W. Rahayu, "Mobile Cloud Computing: A Survey," *Future Generation Computer Systems*, Vol. 29, 2013.

[17] L. Lei et al., "Challenges on Wireless Heterogeneous Networks for Mobile Cloud Computing," *IEEE Wireless Communication*, June 2013.

[18] S. Barbarossa, S. Sardellitti, P. D. Lorenzo, "Communicating While Computing—Distributed Mobile Cloud Computing over 5G Heterogeneous Networks," *IEEE Signal Processing Magazine*, November 2014.

[19] C. Weinhardt et al., "Cloud Computing—A Classification, Business Models, and Research Directions," *Business & Information Systems Engineering*, Vol. 1, No. 5, 2009.

[20] ITU, "Demystifying Regulation in the Cloud: Opportunities and Challenges for Cloud Computing," GSR 2012 Discussion Paper, October 2012.

CHAPTER 9

Future Networks

With the evolution of telecommunications, network requirements and capabilities have changed over time. Some requirements, such as fair competition among different business players (e.g., network providers, service providers), remain important, but new requirements also arise in terms of network architectures. So far, the IP and related Internet technologies (e.g., DHCP, DNS, routing protocols, transport protocols) have succeeded in providing flexibility in telecommunication networks by hiding underlying link layer technologies from the network and upper layer protocols. Further, NGN architectures are being designed with functions that provide QoS support, identity management, and security in all-IP networks. However, the development of networks continues further with Future Networks paradigm, which is in the focus of this chapter.

9.1 Future Networks Framework by ITU

The evolution of the NGN framework within the ITU-T is called Future Networks. The reason for the Future Networks program is the increasing pace of development of new services, the continuous increase in access bit rates (in fixed and mobile networks), and the huge amount of network resources that have already been built. However, a fundamental change in telecommunications networks in a short period of time is less likely to happen due to the enormous amount of resources needed to build, operate, and maintain such networks. Also, the current IP-based network architectures are flexible enough to carry different types of services. Such flexibility is provided with the IP on the network layer (in all nodes and hosts in the telecommunications infrastructure), which hides all of the different underlying protocols below the OSI-3 layer. At the same time, the IP has high scalability in terms of the Internet as a group of interconnected ASs consisting of interconnected routers, and it provides the possibility to add QoS support and security where needed. However, we do not know if the networks can continue to fulfill service requirements in the future. Of course, this will depend on the markets for demanding services and applications and whether providers can cover the implementation and operation costs of an eventually changed network infrastructure. So the ongoing research efforts of different research groups are studying network virtualization, content-based networking, energy-efficient networks, and so forth. Hence, there is an expectation that some of the FN architectures can be put into trial deployments in the period from 2015 to 2020, with the possibility that they may become functional after 2020. For the purpose of standardization of future networks (FNs) the ITU-T has started a series of recommendations named "Future Networks."

Four objectives are defined for the Future Networks as shown in Figure 9.1 [1]:

- Service awareness: The services are expected to increase exponentially in the future, a trend that has already started with the worldwide deployment of broadband Internet access. FNs should provide services without a significant increase in network deployment and operational costs.
- Data awareness: FNs are expected to carry huge amounts of data in the access, core, and transit parts. However, users should be able to access different data (e.g., audio, video, web) easily, quickly, accurately, and safely, at the desired level of quality, and regardless of the access network and their location.
- Environmental awareness: It is recommended that FNs have the lowest possible consumption of materials and energy, which means they should be energy efficient and environmentally friendly.
- Social and economic awareness: FNs should be developed with awareness of the costs and competition, so their services should be accessible to all players in the Future Networks ecosystem, including end users, vendors, network operators, and service providers.

9.2 Network Virtualization for Future Networks

Currently, many different network infrastructures have been deployed. They consist of interconnected network nodes (e.g., switches, routers) and network hosts (e.g., servers, databases). All networks converge to Internet technologies that provide for the possibility of speedy applications/services innovation due to their openness for

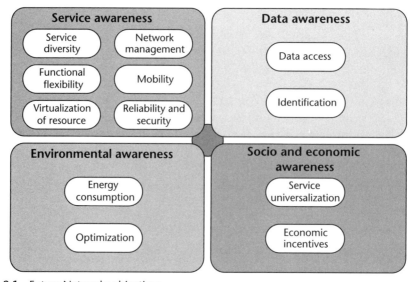

Figure 9.1 Future Networks objectives.

development of new applications without changing the underlying network architecture (best-effort Internet and NGN). In addition, new services can be added by using existing functions in the network for QoS, mobility, and security support. However, such network architectures provide for the possibility of changes only at the edges (e.g., at service providers).

Several questions could be asked about network development in the future: What types of nodes are used in the network? What type of information do they exchange? What are the nodes' power and bandwidth limitations? What are the specific requirements raised by different services from the network (e.g., online business services require high security, IPTV requires high bandwidth, voice and gaming require low delay and jitter end to end)?

The answer is that there is no single solution in the Internet that can fit all different requirements from heterogeneous services, users, and fixed and mobile network environments. That leads further to virtualization of the network resources (e.g., isolation of resources in a given physical node, such as the CPU, memory, or bandwidth). Then, multiple networks can be designed that will use different virtual components that exist in the physical network elements (e.g., switches, routers, servers) for different goals (e.g., different services, different customers). Such an approach to network design is referred to as network virtualization.

By ITU-T's definition [2], network virtualization is a method that provides for the possibility of multiple virtual networks, referred to as logically isolated network partitions (LINPs), coexisting in a single physical network. Virtual resources are created on physical objects in the network, such as switches, routers, and hosts. A given LINP uses virtual resources by means of their programmability features (which are needed for virtual network design). In terms of the end users (e.g., residential users, enterprises), LINPs can provide the same services to them as networks without virtualization.

A conceptual architecture for network virtualization is shown in Figure 9.2. It consists of multiple interconnected LINPs, where each LINP is built from multiple virtual resources. In this approach each physical resource may have multiple virtual instances, so it can be shared among multiple LINPs. To provide such sharing of physical resources among multiple LINPs, virtual resource managers (VRMs) are needed. VRMs interact with physical resource managers to provide tools for management and control of virtual resources, such as virtual switches and routers. Each LINP can have various characteristics regarding partitioning of the resources and networks, isolation (e.g., VPNs and VLANs), programmability (e.g., standardized interface through the VRM for access to virtual resources), AAA, and so forth. Overall, one of the goals of network virtualization is to provide higher utilization of the physical resources (by their sharing) and better adaptation of the networks to the requirements from services/applications and end users.

What network virtualization has appeared? Well, current IP-based networks are facing certain problems. One is the coexistence of multiple networks. IP-based transport networks typically use VPNs, and IP access networks use VLANs to provide isolated logical networks over shared physical networks (e.g., different offices of a given enterprise can be interconnected via VPNs across the operator's transport network). VPN security is based on IPsec, which is a well-established protocol, but it is less flexible for dynamic changes to an established VPN (e.g., upon request). In the access networks existing VLANs require configuration settings to

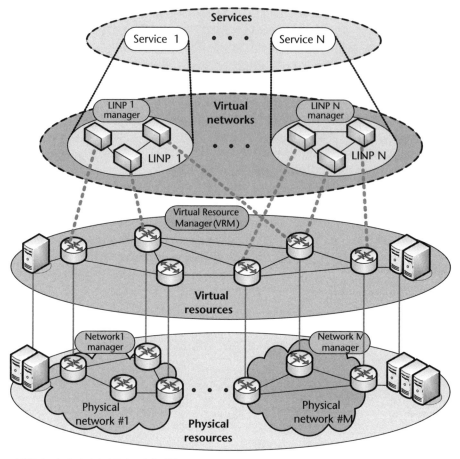

LINP – Logically Isolated Network Partitions

Figure 9.2 Network virtualization architecture.

be done manually by the network administrator. The introduction of LINPs, besides existing VPNs, VLANs, and agents (e.g., home/foreign agents in Mobile IP), should provide higher flexibility and dynamics to the networks (including access and core ones).

Another issue is the complexity of the practical management of network resources, such as switches and routers, due to the use of heterogeneous equipment. Network virtualization should provide simplified access to resources. However, for that purpose standardized interfaces are needed for access to virtual resources in FNs, and that requires changes in the existing equipment that is already installed in the network. The life cycle of the equipment in telecommunication networks is several years (e.g., 4 years), so longer use of a network's components is not a very important issue due to developments in processing power (e.g., according to Moore's law, the processing power of devices doubles every 18 months), as well as the continuous increase in the capacities of access and core networks with the deployment of broadband access to the Internet. But flexibility in provisioning, that is, network adaptation to the existing and newly deployed services by internal or external changes, is ab important advantage of network virtualization. In fact, network virtualization allows reuse of network resources by adding (or extracting) and aggregating logical resources to provide increased capabilities or capacities at significantly lower costs than an approach toward network change that requires new physical resources. Finally, network virtualization also makes it possible for research groups to use existing network resources that are used for commercial services to create testbed LINPs for experimental purposes, thus eliminating the need to create isolated network testbeds for testing new technologies and at the same time providing a real-world environment for the experiments.

Overall, LINPs have several design goals:

- Isolation: Network virtualization should provide isolation among different LINPs regarding power consumption and the capacities of each LINP.
- Network abstraction: Underlying network technologies should be hidden from the virtual resources that are used by LINPs.
- Reconfigurability: This is referred to as topology awareness of virtual resources and the associated physical resources used by the LINP, which is needed for interaction among virtual resources and for reconfiguration.
- Performance: Network virtualization introduces a so-called virtualization layer between the hardware and software in each physical resource for creation of isolated partitions in it (i.e., virtual resources). Such virtualization consumes additional CPU time in the physical device, thus leading to higher CPU utilization and lower bandwidth availability when compared to a non-virtualized network with the same physical resources.
- Programmability: The LINPs require new control schemes and mechanisms for control of the virtual resources (e.g., via a programmable control plane), but network virtualization must also support data plane programmability for control of routing and forwarding of data traffic within a given LINP.
- Management: Each LINP is isolated from others, so it has to be managed separately. Also, in network virtualization there are many complex mappings between physical resources and virtual resources, so there is a need for an integrated management system in a virtualization plane that can support monitoring, dynamic reconfiguration, topology awareness, and resource discovery and allocation.
- Mobility: In network virtualization mobility refers to movement of virtual resources, including end users (with their mobile terminals) and services. To provide service continuity to end users, the services must be moved together with users when they change the LINP or attachment point within the LINP (e.g., a single LINP can be a mobile network or a virtual mobile network).
- Wireless: Network virtualization also targets the wireless channels as virtual resources. Within a single wireless device with several active wireless connections to different wireless LINPs, scheduling is required that prevents the device from transmitting while receiving in a given frequency range, and vice versa.

Generally, several types of network virtualization are emerging, including cloud computing on one side and software-defined networks on the other, as part of the Future Networks concepts.

9.3 Software-Defined Networking (SDN)

Emerging implementation of network virtualization is called software-defined networking (SDN). The idea behind SDN is the existing situation in all networks in which different networks nodes (e.g., switches, routers) have their own OSs, which perform autonomous tasks within a given node. Each network node has hardware and software with specialized packet forwarding for many protocols (e.g., OSPF, RIP, BGP, multicast protocols, MPLS, differentiated services, integrated services, NAT, firewalls) and an OS with many features that are typically vendor specific. The SDN concept is the creation of a network OS that will perform the same tasks for the network as a host/node OS performs for the host/node. For example, ITU defines SDN as a set of techniques that enables network resources to be directly programmed, orchestrated, controlled, and managed, which facilitates the design, delivery and operation of network services in a dynamic and scalable manner [3].

Figure 9.3 shows the evolution of networks with autonomous operating systems in the network nodes toward SDN, where network nodes such as switches and routers are simplified to perform only packet forwarding and protocol-based communication between each other. At the same time, control of their processes is performed by a network operating system (NOS) that has an overall network view, because it controls many nodes in the network. With this approach, there is no longer the need for design of distributed control protocols. However, a standardized interface is needed between the network nodes and the NOS for the control traffic (something similar to the drivers used between the host OS and various interfaces). Such a standardized interface in SDN is OpenFlow [4].

OpenFlow is a standardized way to control flow tables in switches and routers. In some cases, already deployed equipment can opt to include OpenFlow by updating its firmware. So, if a node (switch or router) is to become part of an SDN, then it needs to have OpenFlow.

Figure 9.4 shows OpenFlow use in SDN. Each network node in SDN is required to have an OpenFlow client that communicates with an OpenFlow controller by using SSL/TCP secure communication.

OpenFlow has a flow table in the network node, where each entry in the table contains rules (it specifies source and destination IDs from the physical layer up to the transport layer, including the switch/router physical port, VLAN IDs, and source and destination pairs for MAC addresses, IP addresses, and port numbers), the action to be performed (e.g., forward packet to zero or more physical ports, encapsulate and forward to the controller, modify fields), and statistics (including packet and byte counters). In terms of the flow treatment, OpenFlow provides two main models: (1) a flow-based model and (2) an aggregated model. In a flow-based model, every flow is individually set up by the SDN controller, so the flow table in the network node contains one entry per flow. Hence, a flow-based model is appropriate for use in access networks. In contrast, the aggregated OpenFlow model has one entry in the flow table per aggregated traffic (consisting of a group of flows that belong to the same category), which is more convenient for use in backbone and transport networks.

Regarding the number of OpenFlow controller nodes, two main approaches are used: (1) centralized control with a single OpenFlow controller in the SDN and

9.4 Big Data

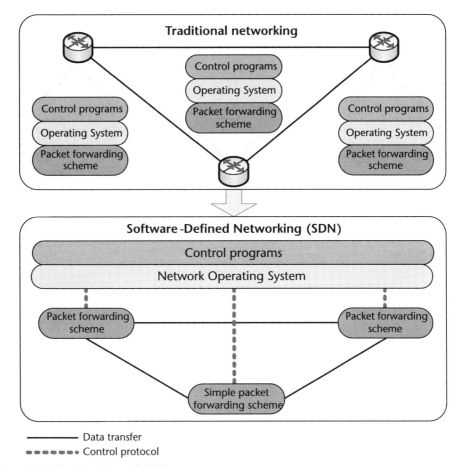

Figure 9.3 Evolution toward SDN.

(2) distributed control with two or more OpenFlow controllers in the SDN that provide load balancing and/or higher reliability.

In summary, SDN is a system-layered abstraction of the network that is programmable and flexible, and OpenFlow is the interface between network nodes (i.e., switches and routers) and controllers that enables SDN. With such a concept, SDN provides for the possibility of new innovations in deployed network architectures, not only in fixed-access and core networks, but also in optical transport networks [5], as well as software-defined mobile networks [6].

9.4 Big Data

With the development of cloud computing services, data can be stored and accessed through a cloud infrastructure. The data centers of cloud service providers and their resource suppliers are storing huge amounts of data (which are many times replicated via caches) in different storage servers around the world. Billions of individual and enterprise pieces of information are generated each day including supplier data, delivery slips, employment records in enterprises, customer complaints for different services (either public or private), as well as user-generated content such as

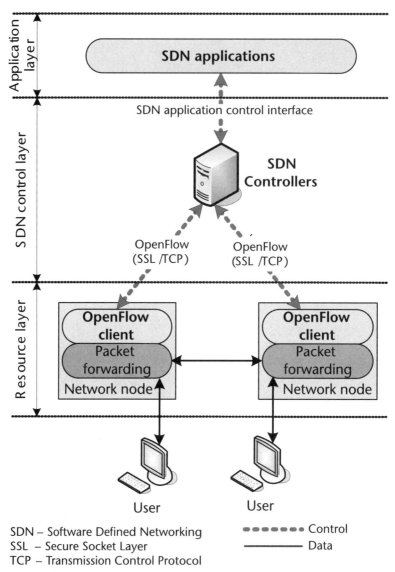

Figure 9.4 OpenFlow use in SDN.

messages, photos and videos posted on social networking or content-sharing websites, check-ins in different web portals, and so forth [7].

9.4.1 Big Data Definition

The telecom world based on Internet technologies is evolving from being an the Internet of contents in the past, to being the Internet of people in the present time, and then to the Internet of things in the near future. We can expect around 50 billion devices to be connected to Internet by 2020, which will generate much more data than we generate today in telecom and Internet infrastructures. Several standards developing organizations (SDOs), such as ISO/IEC [8] and NIST [9], define these accumulations of data as "Big Data."

> Big Data consists of extensive datasets with certain characteristics regarding the volume, velocity, variety, veracity, variability, and so on (also called Vs) that cannot be efficiently processed using existing technologies to provide value; therefore, they require a scalable architecture for efficient storage, manipulation, and analysis of the data.

Big Data is usually described [7] in terms of the four Vs [7, 9], five Vs, or more [8]. Below are the most common elements:

- Volume: This term refers to amount of data collected from all different sources of information, data collected anytime, anywhere, by anyone and everything.
- Velocity: This term refers to the speed of data processing, that is, the time needed to reach decision making from the time of data input.
- Variety: This term refers to the characteristics of Big Data. Big Data includes any type and any structure of data collected from diverse sources (e.g., sensor data, call records, maps, images, videos, audio files, log files, personal information, and many more).
- Veracity: This term refers to the accuracy of the data, which is essential for making crucial decisions based on the data (e.g., some data such as sensor data may be more trustworthy than data from social networking sites). So, Big Data systems need mechanisms to distinguish, evaluate, or rank different datasets.
- Variability: This term refers to changes in the data format, structure, type, and quality that have an impact on the supported applications, analytic mechanisms, and targeted problems [8].

With the goal of achieving scalability, Big Data consists of horizontally coupled independent data systems that can achieve the needed scalability for efficient processing of the massive amounts of information. Why horizontal coupling of data systems? Well, vertical scaling implies an increase in processing power, storage capacity, and speed to achieve greater performance. According to Moore's law, all such vertical scaling parameters double every 2 years. However, the volume of data is increasing much faster than the performance abilities of individual systems. Therefore, a horizontally scaled system is required that uses a loosely coupled set of resources in parallel (also referred to as massive parallel processing).

Why is Big Data happening now? Well, the main reason is that the development of cloud computing is making it possible to accomplish massive parallel computing on traditional computers. So, the costs for data acquisition are becoming significantly lower now than before.

9.4.2 Big Data Ecosystem and Reference Architecture

The Big Data ecosystem consists of the main players included in the Big Data services delivered to customers [10]. Its ecosystem consists of the following actors:

- Data provider: A data provider introduces data streams into the Big Data system. Hence, data provider actors include enterprises, public agencies, researchers, search engines such as Google, network operators, end users, websites, FTP sites, email services, and all other Internet applications and services.
- Big Data application provider: This type of provider executes the manipulations of the data life cycle to meet security, privacy, and system orchestrator requirements. This role in the ecosystem is used by application or platform specialists and consultants, which perform collection, preparation, analytics, visualization, and access processes on the data.
- Big Data framework provider: This type of provider establishes a computing framework based on algorithms that transform the input data while preserving its privacy and integrity. These actors are typically in-house clusters (for telecom operators), data centers, and cloud providers. They provide infrastructure, data platforms, and/or processing frameworks.
- System orchestrator: The system orchestrator defines and integrates the required data application activities over a given operational vertical system. Actors in this part are business leadership, consultants, data scientists, and information, software, security, privacy, and network architects.
- Data consumers: These are the end users who use the results offered by the Big Data application provider. This role is played by end users, researchers, applications (e.g., for user-oriented or context-based service provision), and systems.

The Big Data ecosystem reference architecture is shown in Figure 9.5. It depicts the Big Data flow from collection of the data to its usage by customers, with several data transformations along the way. Data can be collected from different sources and in different forms and types (e.g., public data, private data, social data, sensing data). Typically, similar data sources are obliged by similar policies. For example, data from cookies in web browsers are obliged to follow the formal policies set individually by the end user, while the charging records of telecom operator's customers follow more strict policies (in-house as well as legislation) regarding their processing as data. After collection of the data, initial metadata are created to facilitate subsequent aggregation or lookup. Further, smaller sets with easily correlated data are aggregated into a larger data collection (in such a case the datasets have similar security and privacy considerations). On the other hand, matching datasets with dissimilar metadata (i.e., keys) are also aggregated into larger collections. A typical example of matching is that of the advertising industry on the Internet; advertisers place different commercial ads on web pages that are targeted to the individual end user. Matching services may correlate HTTP cookies' values (obtained from browsing activities) with a person's real name (e.g., a real name is used for credit card payments over the Internet). Finally, data mining is performed to transform data and provide results from the data collected from different sources. Data mining can be defined as the process of extracting data and then analyzing it from various perspectives to produce summary information in a certain useful form that identifies existing relationships in the analyzed data. The two main types of data mining are

9.4 Big Data

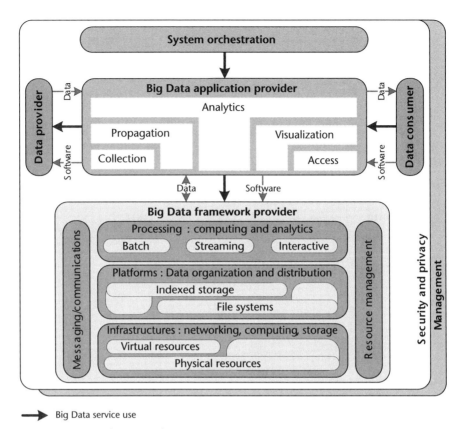

Figure 9.5 Big Data reference architecture.

(1) descriptive, which gives information about the analyzed data, and (2) forecasts, which are based on the data.

To provide Big Data services, an infrastructure is needed. The Big Data infrastructure consists of bundles of servers, storage, software (e.g., for databases, storage) and networking capabilities. These are needed to support data transformation processes and transfer and storage of the data. However, the infrastructure can be dedicated to Big Data (e.g., specialized data centers) or shared with other services such as cloud computing services (e.g., use of cloud resources for data storage or the use of core and transport IP networks for transfer of datasets and/or results from their analysis). Data storage and retrieval involved use of the Structured Query Language (SQL) and so-called noSQL databases, used for managing data in relational and nonrelational database management systems, respectively.

9.4.3 Big Data Technologies and Use Cases

Big Data is based on certain front-end Internet technologies for collection of data, transfer of datasets and results from analyses, and provision of Big Data services to customers. Also, there are back-end technologies, which involved aggregation, processing, and data mining of the Big Data in data centers (i.e., Big Data servers and databases). Existing Internet technologies related to Big Data are cloud computing (nowadays data are stored in clouds), web technologies (data have grown with the growth of the Internet and particularly the WWW, so many web technologies have to deal with Big Data), and databases (SQL and noSQL). These standards are part of the activities of ISO/IEC (e.g., for SQL and metadata, as well as cloud data management interfaces, open virtualization formats, and web services interoperability), ITU-T (e.g., cloud computing–based Big Data), W3C (e.g., XML and associated technologies, linked data, web ontology language, RDF, Sparse Query Language, provenance, and many others), and other SDOs.

Big Data is not a technology by itself, but its uptake has driven innovations in several different fields. Here are given several possible use cases for Big Data:

- Health: Analysis of large sets of data of diseases patterns is crucial for actors in the pharmaceutical and medical products sector.
- Solving the mysteries of the universe: The Large Hadron Collider at CERN is taking pictures of particle collisions at 40 millions of pictures per second, thus creating huge sets of unstructured data that have to be structured and analyzed by scientists.
- Movement of people: Data collected from different sources, such as RFID-based transport tickets, sensors and traffic cameras, GPS fleet tracking, and smartphone usage by drivers, can be used to create traffic routes for public transport, recommend routes to drivers to avoid congestion or to shorten travel time, and so forth.
- Monetization of network and service data assets: Telecom operators may monetize Big Data by utilization of large datasets obtained from network usage (calls, connections, messages, traffic volume dynamics), charging/billing records, and so forth. They can capitalize on Big Data during network planning, during deployment or operation of network infrastructures (e.g., balancing the traffic), and for marketing purposes. For example, by combining network insights and customer profiles, telecom operators and service providers can provide tailor-made offerings to increase revenue opportunities as well as attract and retain customers.

Typical present-day use cases of Big Data on the Internet are recommendations and marketing on popular global websites. For example, Amazon uses a patented recommendation engine, which aggregates data about users' browsing, purchasing, and reading habits, and then, based on a dataset of all the other customers with similar histories, makes extrapolations about what a given user would like to read next. A similar approach is used by Netflix for suggesting movies to its customers. Another example of the current use of Big Data on the Internet can be found in social networks (e.g., Facebook). The favorite social network site analyzes the social graph of a given user and then creates a mathematical structure used to model pairwise relations between objects to suggest friends and/or content to the user.

9.4.4 Challenges for Big Data

Big Data faces several challenges. One challenge is privacy and security [7]. Data collected from users can be used for tracking their movements, behaviors, and preferences, and to predict their future behavior with high accuracy. So, assurances are required that personal data will be used appropriately and according to accepted policies by the user and relevant laws. Another legal and regulatory issue is the ownership of the users' data (e.g., call records, Internet sessions). Some questions that need to be addressed by regulators and policy makers regarding Big Data include the following: Who has the rights to use data collected from user activities in the telecom networks or the Internet in general? What are the possible ways to use it?

Another challenge is a technical one. In the broadband Internet world, the largest part of the traffic is video and multimedia traffic. Knowing that, another Big Data challenge is analysis of real-time multimedia content (e.g., automatic extraction of metadata, events, in-video or in-image search).

Finally, the main drivers of Big Data are considered to be broadband, cloud computing, and the rise of social media. So, Big Data is a natural result of the ICT/telecom innovations and development in many subdisciplines.

9.5 Over-the-Top Versus Telecom Operator Service Models

With the deployment of broadband Internet access, OTT service developments have started to impact the traditional telecom service models. The main revenue generators for telecom operators in the pre-OTT era were fixed and mobile voice services, messaging in mobile networks, and leased lines for business users. Hence, the main battleground for competition between the telecom operators and OTT players was primarily for the following services:

- Fixed and mobile voice services: Broadband Internet access in both fixed and mobile environments spread the development of OTT voice services, starting with Skype for fixed hosts and continuing with a plethora of OTT voice services for mobile hosts, such as Viber and WhatsApp (covered in Chapter 7).
- Messaging: Besides voice, SMS has been a significant revenue generator for mobile operators since the deployment of GSM in the 1990s. However, almost all OTT voice services also included messaging integrated with OTT VoIP, so SMS is gradually becoming less important for many mobile users.
- Media: The emerging OTT video services (e.g., YouTube, Netflix) have changed the telecom game regarding television, VoD, music, advertising, and other integrated digital services. Telecom operators offer subscription-based models, whereas OTT services are mainly based on add-funded models (e.g., YouTube) although there are examples of subscription-based ones (e.g., Netflix).
- Cloud services: There are only a few OTT cloud players with compelling proportions: Google, Amazon, Microsoft, and Apple as the latest in the raw (with iCloud). Telecom operators, however, are well positioned regarding their network resources and customer relationships (e.g., SLAs). On the consumer side, OTT players have a global market, whereas telecom providers may benefit by just enabling cloud services (e.g., as value-added services) and getting customers to sign up for them by offering trustworthy and secure cloud services for storing personal digital valuables such as pictures, music,

videos, and documents. On the enterprise side, OTT cloud providers can provide full solutions to enterprises, whereas telecom operators can provide full cloud infrastructure by partnering with global cloud players (e.g., Amazon) and offering end-to-end QoS.

So, the traditional telecom operators, which are transiting to an all-IP infrastructure during the 2010s (e.g., with NGN deployments), are under pressure from the OTT global players. But they are not competing for profit, but for control of the value chain (Figure 9.6). For example, the mobile Internet's value chain includes mobile operators (i.e., telecom operators), OTT service providers, smartphone vendors (e.g., Samsung, Apple, LG), software vendors (e.g., Android from Google, iOS from Apple, Windows Mobile from Microsoft), and content providers (e.g., YouTube). In such a value chain, the competition is not symmetrical. Why? Well, OTT service providers do not bear the costs of providing the fixed and mobile broadband access infrastructure. In contrast, fixed and mobile telecom operators have costs related to deployment as well as O&M of the broadband infrastructure, including the transport and service stratums (if one uses NGN terminology). Connectivity is as important to OTT players as gas is to a car (they cannot provide services without broadband access to the Internet), but it is the telecom operators' role to provide access to the Internet (not the OTT themselves). Such asymmetry makes it difficult for telecom operators to protect certain traditional business models (e.g., time-based, volume-based, or message-based charging to its customers) for some legacy telecom services (e.g., telephony, messaging in mobile networks). At the same time, it gives more flexibility to the OTT players because Internet

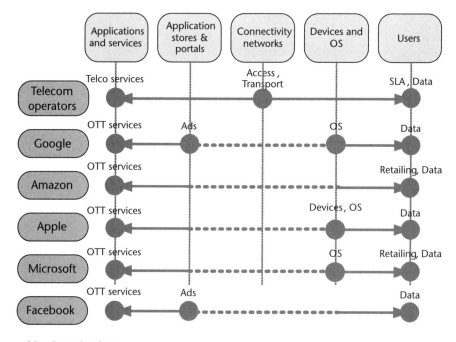

OS – Operating System
OTT – Over-The-Top
SLA – Service Level Agreement

Figure 9.6 Competition between telecom providers and OTT providers.

connectivity costs are paid to telecom operators, so many OTT services appear to end users to be free to use; they are not really free, however, because there are costs associated with broadband Internet access provided by telecom operators.

The core business models are different for mobile and fixed telecom operators and for OTT service providers (e.g., Google). Telecom business models are vertically integrated (all in one) and subscription based, with an SLA that specifies the agreed-on policies and certain QoS parameters (e.g., bit rates in the downlink and uplink directions). In contrast, OTT players can monetize from ads, downloads, acquisitions, and analytics (e.g., Big Data), so they can offer services for free (e.g., Skype-to-Skype calls, Google Docs) or even less than free (e.g., OTT service provider is sharing part of the revenues with the network operator). With such positioning, the vertically integrated (from the access network up to the provided services) business models of telecom operators make it very difficult for telecom operators to compete with OTT providers and their "free" innovative services and applications provided via the network neutrality principle.

Currently a status quo is in place regarding the telecom operator and OTT business models, if one does not want to sacrifice network neutrality and the ICT innovations that go along with it. The OTT providers see telecom operators as complements to their core business, rather than as competition. On the other hand, the telecom business models are targeted to user lock-in and subscriber acquisition (from other telecom operators in a given country). In fact, they have experienced declining revenues from certain strong revenue generators such as telephony and SMS, but broadband Internet access services are becoming a major contributor to their revenues, offered either individually or bundled with legacy services (e.g., with telephony as carrier-grade VoIP and television as IPTV). In such a manner, we can expect telecom operators to be less successful in copying the OTT service providers (this is almost impossible due to the asymmetry in the business models, as well as the higher flexibility and bigger global market for OTT services). So, telecom operators should leverage their advantages over OTT providers. Their advantages include end-to-end QoS support, user targeting, and privacy control, as well as provision of connectivity and network services to other players (e.g., fixed or mobile VNOs). Hence, future business models for telecom operators require subscriber-centric data consolidation with the goal of improving capital-to-operational expenditures, leveraging users' loyalties, improving the time to market for new services (although OTT providers have the advantage in terms of speedy global innovations), and engaging subscribers with their data (e.g., via cloud services provided by telecom operators). OTT providers and user device manufacturers, on the other hand, are also working in a high-risk environment that is changing very fast. Many OTT providers focus only on a few services, something that may create certain dependencies and result in failure on the global market if they accidentally lack needed support in the operating systems (as an example) or miss some new important trend for the services or user devices [11].

9.6 Cybersecurity

With the move of all telecommunication services onto the Internet, users are becoming increasingly dependent on the Internet for everyday communication (e.g., VoIP, email, messaging), commerce (banking, business, e-trading), different control systems (e.g., public utilities), and various kinds of information and entertainment. As a result, sensitive data are stored on the Internet. That increases security concerns, because the Internet is becoming a digital "mimicry" of the real world with all of the benefits and all of the threats, but in a digital form. In network and systems, design security is not always a priority, because other demands, such as cost, convenience, speed, and backward compatibility, may trump security. The other problem with security on the Internet lies in its transnational character, which makes it hard to implement the same security measures in all networks and systems globally because of the different legislations and capacities in different countries around the world.

What is cybersecurity? Cybersecurity refers mainly to information security on the Internet, which is protection of information against unauthorized modification, disclosure, transfer, or destruction. In other words, cybersecurity is security of cyberspace that includes all networks (e.g., routers and interconnection links) and all information systems (e.g., servers, databases).

9.6.1 Cybersecurity Technologies

Views on security and how to achieve it differ, but security is generally standardized in ITU-T X.805 [12], which defines the exact number of security dimensions for all end-to-end communication, which includes all Internet connections/sessions. These dimensions are access control, authentication, nonrepudiation, data confidentiality, communication security, data integrity, availability, and privacy, as shown in Figure 9.7. All of these dimensions are applied on three security layers:

- Infrastructure security layer: In this layer belong individual routers, switches, servers, and the communication links between any pair of them.
- Service security layer: This layer refers to security of service, which is provided by service providers to the customers. In this layer belong basic

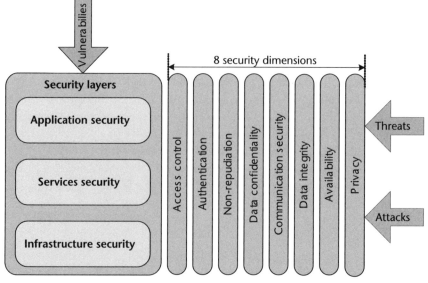

Figure 9.7 Security dimensions and security layers.

transport connectivity to service enablers (e.g., DHCP service, DNS service, AAA), as well as value-added services such as QoS support, VPN services, location services, telephony, and messaging. The service-layer security is targeted toward providing protection to service providers and their customers from any potential security threat (e.g., an attacker's attempt to deny a service provider the ability to offer services to its customers).

- Application security layer: This layer includes security of the network-based applications accessed by customers. Such applications include FTP and web browsing applications, messaging (e.g., voice messaging, email), customer relationship management, e-commerce, and video collaboration. There are four potential targets for security attacks in this layer: the user, the application provider, third-party middleware (e.g., web hosting), and the service provider.

Overall, an overview of cybersecurity is given in ITU-T X.1205 [13]. In general, the sophistication of the attacks is continuously advancing, but on the other hand the same statement can be made for the cybersecurity. The related cybersecurity technologies are outlined in Table 9.1, which provides mapping between cybersecurity techniques, categories, and technologies.

NGN security is defined in ITU-T Recommendations Y.2700 through Y.2799, which continue to evolve as part of the Future Networks security concepts.

9.6.2 Security in Future Networks

Future networks should be designed for the safety and privacy of their end users [1]. The rationale for this lies in the targeted use of FNs in human society including mission-critical services such as intelligent landlines, railway and air traffic management, e-health, emergency telecommunications, and reliable services in disaster conditions. So, FNs should provide an acceptable level of service even in cases of network faults; that is, they need to be trustworthy and challenge tolerant (to continue to provide service in the case of challenges to normal network operations).

Security for FNs can be provided by using multilevel access control, that is, assurance of user identification, authentication, and authorization (to use given services), which is in addition to the security requirements of the NGNs [14]. For example, network virtualization will bring many benefits to all actors, including providers and end users, but it also raises new security threats. In that manner, a malicious user can monitor or control virtual resources even in cases when they are not allocated to that user. Hence, access identification and AAA functions are considered to be essential to LINPs [15].

Overall, in FNs identification of different objects becomes crucial for the security and hence the implementation of the IdM framework [16] is mandatory in design and operation of FNs.

The identification framework for FNs consists of four components (Figure 9.8):

- ID discovery space: This components discovers various types of IDs (related to communication objects).

Table 9.1 Cybersecurity Technologies

Techniques	Categories	Technologies
Access control	Perimeter protection	Firewall
		Content management
	Authentication	N-factor (single factor uses user ID and password authentication, N-factor uses N elements for authentication including physical or smart tokens, biometrics)
	Authorization	Role based (user access based on its role)
		Rule based (user access based on certain rules)
Cryptography	Certificate and public key architecture	Digital signatures (based on issued certificates)
		Encryption
		Key exchange (based on transaction key or session key to secure communication)
	Assurance	Encryption
System integrity	Antivirus	Signature-based methods (protects against viruses, worms, and Trojan horses using signatures)
		Behavior-based methods (checks running programs for misbehavior)
	Integrity	Intrusion detection (used to warn network administrators of a security incident, such as compromised files)
Management	Network management	Configuration management (control and configuration of the network and its fault management)
		Patch management (installing updates and fixes)
	Policies	Enforcement (enforcing security policies by the network administrators)
Audit and monitoring	Detection	Intrusion detection (comparison of network traffic and log entries)
	Prevention	Intrusion prevention (alarming for suspicious network activities)
	Logging	Logging tools (monitor and compare log entries to find attackers)

- ID spaces: These are used to define and manage various kinds of identifiers, which include the following categories:
  - User ID: This refers to the user ID in a given network (e.g., username@example.com).
  - Data/content ID: This is assigned to certain data or content independently of the owner or the location. This approach leads toward information-centric networks, as well as easier mobility and caching of the contents.
  - Service ID: This category consists of two subcategories of service IDs: (1) a content service ID, which is used by client and server nodes to identify services, attributed with keys, sequence numbers, and states, and (2) a network service ID, which may, for example, specify a LINP, a VLAN, or a specific protocol used for handling the IP packets in the networks in terms of queuing and QoS support.

9.6 Cybersecurity

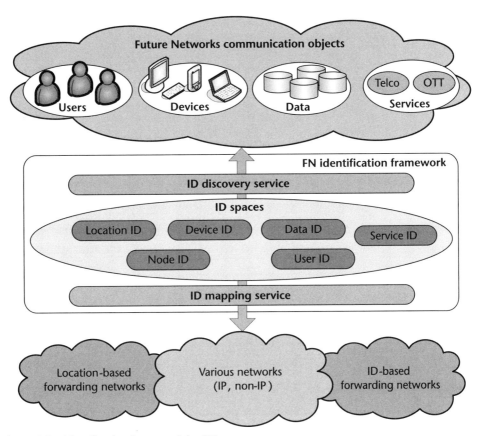

Figure 9.8 Identification framework for FNs.

- Node ID: This identifies physical or virtual objects/devices independent of its location in the network (e.g., one typical use is in mobile networks).
- Location ID: This is an identifier assigned to a user device or network node with the goal of locating it in the network (hence, it is dependent on network-layer protocols). In IP-based networks the network locator is the IP address assigned to a given network interface on a device or node.
- ID mapping registries: These maintain the relationship between different ID types.
- ID mapping service: This provides mapping of IDs from one category to IDs of another category.

Regarding the Internet and the NGN, the existing identifier approaches are very similar. For example, the basic NGN architecture does not define separate types of identifiers for users and devices, because they are both represented as URIs or URLs. Later, the ITU-T Y.2015 recommendation introduces the Node ID as to identifier to the transport layer (leaving the IP address as a location identifier at the network layer), thus separating the Node ID and location ID. However, that requires a mapping function between the IP address as locator and the Node ID as the node identifier. The approach of ID/location separation is also considered for FNs, which might be useful in particular for machine-to-machine (M2M) communication.

9.7 Impact of M2M

M2M technologies are used for communication between two or more machines without the need for human interaction. In general, the term M2M is used by different SDOs, including ITU-T, IEEE, and ETSI, but refers to the same communication paradigm. The M2M paradigm is directly related to the IoT, standardized by the ITU-T. The M2M technologies are considered to be a key enabler for the IoT (covered in Chapter 7 of this book). The NGN framework also uses the term machine-oriented communication (MOC), defined as form of data communication between entities where at least one entity does not necessarily need human interaction [17].

Many industrial developments of M2M solutions and applications are based on proprietary hardware (e.g., sensors, various devices) and/or software, something that results in a longer time to market, higher development costs, and higher prices for the solutions (due to their vertical integration into custom M2M systems, from physical devices up to applications). For that purpose the ITU-T established a focus group for M2M in 2012 with the goal of studying all needed requirements and developing specifications for a common M2M service layer that can be used by different industry sectors, across IoT vertical markets in a multivendor environment (instead of a customized solution). However, for certain M2M implementations (e.g., e-health), inclusion of additional communities (e.g., academies, SDOs, consortia) is also required. A comparison of the IoT reference architecture [18] and M2M service layer is shown in Figure 9.9.

M2M communication is also defined by the ITU-T as an ubiquitous sensor network (USN).

9.7.1 Ubiquitous Sensor Network (USN)

The USN is defined in NGNs as a conceptual network built over the existing physical network that makes structured use of the sensed data to provide knowledge services to anyone, anywhere, at any time [19]. In that manner, the term "ubiquitous" is derived from the Latin word "ubique" which means "everywhere."

The USN in NGN consists of wired or wireless sensor networks connected to the NGN functional entities. It opens up the possibility of providing the use of sensor networks and their data to different types of users globally, including individual users, enterprises, governments, and public organizations.

The USN has potential applications in civilian and military fields. The USN ecosystem with a NGN infrastructure is shown in Figure 9.10. It consists of the following parts:

- Physical sensor network: This consists of sensor devices with autonomous power (e.g., battery, solar energy) that are used for collecting information about the environment and transmitting it to a database.

9.7 Impact of M2M

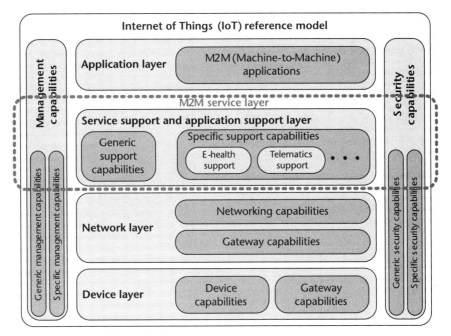

Figure 9.9 IoT versus M2M service layer.

- USN access network: This is a network consisting of sink nodes that collect information from the sensors on one side and provide communication with the external entities (e.g., servers for provision of the USN services) on the other side.
- NGN infrastructure: This includes NGN functional entities that provide necessary functions for the USN application and services, including both strata. The NGN transport stratum provides connection to heterogeneous sensor networks, address mapping for the objects (i.e., sensors) via the NACF, exchange of messages between different types of objects, and gateway functionalities between IP based and non–IP-based sensor networks. The NGN service stratum provides aggregation of information collected from the sensors.
- USN middleware: This refers to software that collects and processes large volumes of data collected from many USNs.
- USN applications and services: This refers to platforms and applications that enable the USN for particular use. In general, all USN applications can be grouped into three main groups:
  - Monitoring: This covers a wide range of monitoring goals, such as monitoring of human body parameters (e.g., blood pressure), presence (e.g., in classroom, in office), environment (e.g., weather forecast, hurricanes, earthquakes), structural parameters of building, behavior of animals, and so forth.
  - Detection: This is targeted to detection of changes for certain objects and associated parameters, such as a change of temperature in a room, field,

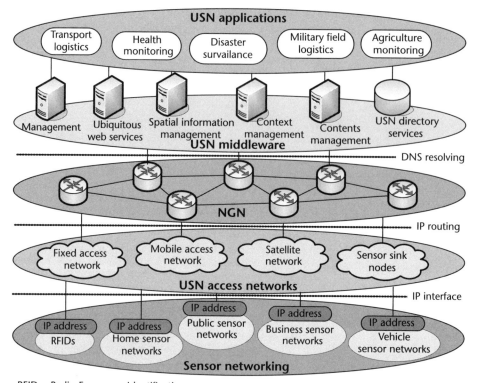

RFID – Radio-Frequency Identifications
NGN – Next Generation Networks
USN – Ubiquitous Sensor Network

Figure 9.10 USN ecosystem.

or human body, intruder detection, fire detection in a building, and so forth.

- *Tracking:* This includes tracking of humans and different objects, such as tracking of children, vehicle tracking in intelligent transportation systems, tracking animals in wildlife, and so forth.

With the availability of low-cost end devices (e.g., sensors, tags) and NGN deployments, we can expect USN services to emerge in different fields such as individual users, enterprises, governments, military, hospitals, police, and public services.

9.7.2 Common M2M Service Layer

Besides the ITU's work on IoT and USN as part of the IoT paradigm, different standardization activities have been conducted by SDOs in different regions in the world. That led to forming oneM2M in 2012 as a global standardization body for M2M and IoT (as a broader paradigm), which has been established by an alliance of SDOs to develop a single horizontal platform for exchange of data collected between different machines connected directly or indirectly to the Internet.

The oneM2M architectural model consists of three main layers [20]: the network services layer at the bottom, the common service layer, and the application layer at the top. So, M2M applications share a common service and network

9.7 Impact of M2M

infrastructure. The oneM2M functional architecture is shown in Figure 9.11. It has the following main functions:

- Application entity (AE): AEs implement the M2M application service logic. Each execution of such logic is termed an AE and it is uniquely identified with a so-called AE-ID. For example, AEs can be a remote blood pressure monitoring application or a fleet tracking application.
- Common services entity (CSE): This is the core of the oneM2M development, and contains a set of so-called common service functions (exposed to other entities via Mca and Mcc reference points, while Mcn is used for access to the network service entities, as shown in Figure 9.11). Each CSE has a unique identity CSE-ID.
- Network services entity (NSE): This provides services from the underlying network to the CSEs (e.g., device triggering, device management, and location services).

In practical implementations an M2M gateway may be needed between the local connectivity network and IP-based access network (fixed or mobile).

A resource-based information model is used in oneM2M ; that is, information is stored in systems as resources that can be identified with URIs. Such resources are organized in three structures and they can be created, read, updated or deleted, with the goal of manipulating the information.

Regarding the end-to-end communication protocols, the goal of oneM2M is not to reinvent "the wheel." So, the oneM2M service layer core protocols are as follows:

- Constrained Application Protocol (CoAP) binding: The CoAP is an IETF standard [21] that is targeted for use in simple low-power devices (e.g., sensors, switches) that need to be controlled or supervised remotely over the Internet.
- HTTP binding: This type of binding is based on HTTP client–server communication using the request–response principle, where AE has the capability

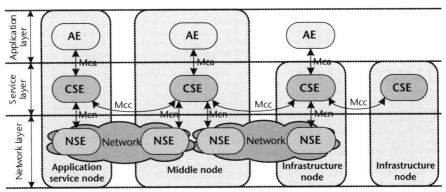

AE – Application Entity
CSE – Common Service Entity
NSE – Network Services Entity

Figure 9.11 OneM2M functional architecture.

of an HTTP client and CSE has the capability of both and HTTP client and server.

- MQ telemetry transport (MQTT) binding: This is a very simple and lightweight messaging protocol, designed for use on constrained devices attached to low-bandwidth, high-latency, or unreliable networks.

Many challenges remain, however, for oneM2M that have counterpart solutions. For example, secure communication is a solution for a large variety of scenarios, remote provisioning (various authentication approaches) is a solution for any device in any deployment challenge, and an access control policy is a solution for M2M privacy concerns (e.g., in a smart home).

Overall, the adoption and penetration of M2M depends on several actors in the value chain, including the telecom operators as infrastructure providers. In general, M2M communication offers new possibilities for fixed and mobile operators. In certain scenarios for M2M, older technologies, such as 2.5G in mobile environments, are used due to the lower price of the devices (e.g., for credit/debit card payment machines at shops). On the other hand, industries without an embedded communication infrastructure are developing M2M solutions in favor of service provider-centric solutions. In such cases M2M service/application providers provide solutions based on application platforms, standardized or customized devices, and the end-to-end transmission pipe (e.g., VPN over best-effort IAS).

Finally, we can easily predict that IoT with M2M will be one of the defining challenges in the following 10 years. However, the major challenge is to provide standardized interfaces and drive down device (or chipset) costs with the goal of providing mass deployments in many different industries including residential uses (e.g., smart home, smart cities).

9.8 Smart Networks and Services

With the speedy developments in the telecommunication/ICT world, the various networks, services, and end-user devices/terminals are requiring more complicated features and hence more complex capabilities. That brings into the spotlight the Smart Ubiquitous Networks (SUNs), which are defined as IP-based networks that can provide transport and delivery of a wide range of existing and emerging services to people and things [22].

Six capabilities should be accomplished with SUN deployment: context awareness, content awareness, programmable capabilities, smart resource management capabilities, autonomic network management, and ubiquitous capabilities. The context awareness capability refers to the ability to detect changes in the physical status of a given device. Such capability enables networks to dynamically obtain context information and monitor context changes. However, context awareness requires gathering context information (e.g., from end users and end devices), storing and retrieving of the contexts, context analysis (e.g., statistical analysis) and prediction, as well as context sharing (delivery and updating of context information) [22].

The content awareness capability refers to the ability to identify, retrieve, and deliver contents using content-related information and the location of the end user. The content awareness of SUN includes content discovery (e.g., based on metadata and user location), content caching (e.g., in local storage), and dynamic content distribution (e.g., based on the traffic load in the network, QoS capabilities at the user's location, and traffic optimization).

9.8.1 Smart Traffic Control

Many new smart devices and services increase the traffic burden on the network, especially with services that require video transmission over IP networks (e.g., smart TV). Those types of service also require smart networks that will support a wide range of services, including real-time and non–real-time as well as store-and-forward services. Therefore, new creative ways are required to ensure fair use of network resources [23].

There are many different QoS parameters, such as bandwidth (i.e., bit rates in the uplink and downlink), delay, jitter, losses, and so forth (described in Chapter 5, standardized with ITU-T Y.1541), that are used by SUN in smart traffic control and resource management functions (STCRMFs). The traditional traffic classification in IP-based networks regarding bandwidth is into two main categories—broadband and narrowband (currently bit rates above 2 Mbps are considered to be broadband [23])—although this definition is time dependent and it is moving up to higher speeds with time. However, SUN criteria are more complicated due to a variety of smart services; hence, a fine-grained bandwidth is needed. For that purpose five bandwidth types have been defined from Type 0 (up to 1 Kbps) to Type 4 (above 20 Mbps) (Figure 9.12).

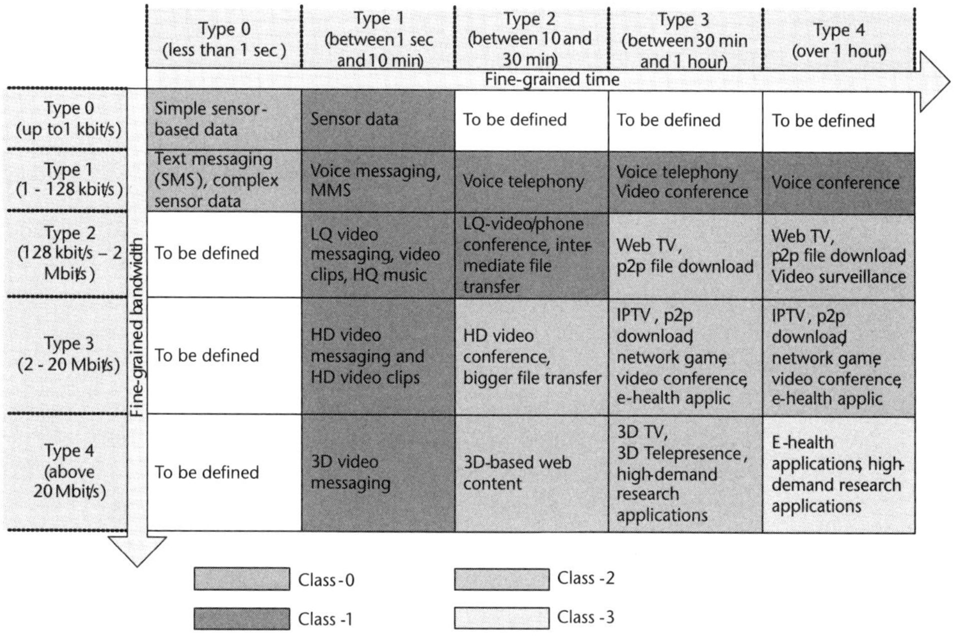

	Type 0 (less than 1 sec)	Type 1 (between 1 sec and 10 min)	Type 2 (between 10 and 30 min)	Type 3 (between 30 min and 1 hour)	Type 4 (over 1 hour)
Type 0 (up to 1 kbit/s)	Simple sensor-based data	Sensor data	To be defined	To be defined	To be defined
Type 1 (1 - 128 kbit/s)	Text messaging (SMS), complex sensor data	Voice messaging, MMS	Voice telephony	Voice telephony Video conference	Voice conference
Type 2 (128 kbit/s – 2 Mbit/s)	To be defined	LQ video messaging, video clips, HQ music	LQ-video/phone conference, intermediate file transfer	Web TV, p2p file download	Web TV, p2p file download Video surveillance
Type 3 (2 - 20 Mbit/s)	To be defined	HD video messaging and HD video clips	HD video conference, bigger file transfer	IPTV, p2p download network game video conference e-health applic	IPTV, p2p download network game video conference e-health applic
Type 4 (above 20 Mbit/s)	To be defined	3D video messaging	3D-based web content	3D TV, 3D Telepresence, high-demand research applications	E-health applications, high-demand research applications

Figure 9.12 Fine-grained traffic classes in smart networks

Besides the bandwidth, another important QoS parameter in SUN is service duration, which is also required to be fine grained due to inclusion of always-on and fixed bit rates for certain services (e.g., IPTV, network gaming, remote surgery). Hence, in SUN five service duration types are defined from Type 0 (less than 1 sec) to Type 4 (over 1 hr), as shown in Figure 9.12.

Combination of fine-grained bandwidth and fine-grained service duration for SUN further defines four traffic classes, from Class 0 to Class 3 (Figure 9.12). Class 0 includes sensor data transfers and SMS, so some services (e.g., critical sensor traffic) may require highest priority treatment (e.g., for detecting natural disasters). Class 1 is dedicated to most of the public services such as telephony (i.e., VoIP). Class 2 is targeted toward services with higher data volumes such as Web TV, video surveillance, P2P file download, network gaming, video conferencing, and telepresence. Hence, Class 2 services have a major impact on telecommunication business and OTT providers. Finally, Class 3 is defined for the highest bit rates (over 20 Mbps) and the longest time durations (over 1 hr). Therefore Class 3 seriously impacts telecom operators and service providers, and leased line connectivity is currently considered to be the best option to support it [23].

Regarding SUN functionalities the STCRMF includes several entities, such as a traffic monitoring and analysis function (TMAF), resource monitoring and analysis function (RMAF), and smart traffic and resource control function (STRCF). The TMAF monitors traffic to identify the heavy traffic. The RMAF monitors bandwidth usage per node and per flow using contextual information. Finally, STRCF receives context information (e.g., user behavior, location, device type, time), correlates traffic volume with the resource usage and context information, and afterward determines the optimal control mechanisms for smart traffic and resources control [23].

9.8.2 Smart Sustainable Cities

Smart networks and services may find endless use in everyday living and in society in general. With the increasingly smart capabilities of the ICTs, one of the most remarkable use cases is the development and future deployment of smart sustainable cities (SSCs).

According to the ITU-T Focus Group on SSCs [24], a smart sustainable city is an innovative city that uses ICTs and other means to improve quality of life, efficiency of urban operation and services, and competitiveness, while ensuring that it meets the needs of present and future generations with respect to economic, social, and environmental aspects.

The ICT architecture for the SSC aims to provide certain services to end users, including transportation services, e-government, e-business, smart health services, smart building services, smart energy services, smart water services, and so forth. With the goal of providing such services, the SSC is expected to be based on a multitier architecture from a communication point of view, as shown in Figure 9.13.

The SSC ICT architecture consists of the following layers (bottom up):

- Sensing layer: This layer consists of terminal nodes that sense the real world (e.g., sensors, RFID readers, cameras, GPS trackers, bar-code readers) and

9.8 Smart Networks and Services

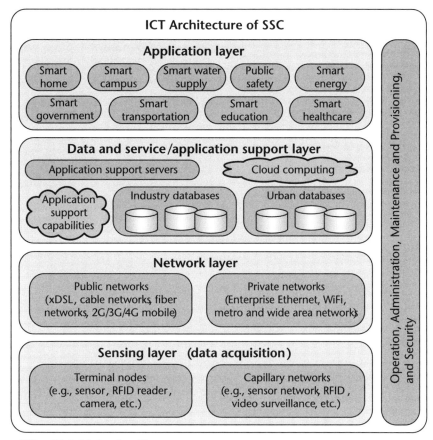

DSL – Digital Subscriber Line
ICT – Information and Communication Technologies
SSC – Smart Sustainable City

Figure 9.13 SSC architecture from a communication viewpoint.

capillary networks (e.g., RFID, video surveillance) that connect various terminals to the network layer.
- Network layer: This layer refers to various fixed and mobile networks provided by telecom operators (e.g., xDSL, cable networks, fiber networks, 2G/3G/4G mobile networks, Wi-Fi) including other metro networks owned by city stakeholders and enterprises.
- Data and support layer: This layer stores the collected data (transferred through the network layer) from the sensing layer into clouds, and provides the support capabilities (e.g., via application servers and databases) to different city-level applications and services. In fact, this layer makes the city "smarter."
- Application layer: This layer includes various applications that deliver SSC services (e.g., healthcare, urban governance, public safety, environmental protection).

All layers in the ICT architecture of the SSC are connected to the operation, administration, maintenance and provisioning, and security (OAM&P & Security) framework for higher reliability and sustainability of the provided services.

Overall, the SSC architecture is promising in terms of integrating many of the emerging ICT technologies including existing communication networks together with the IoT with its USN and M2M as well as WoT to create a sustainable living environment for citizens by using the ICTs based on the existing telecommunication infrastructure and Internet technologies.

9.9 Business and Regulation Challenges in Future Networks

Business and regulation approaches for traditional networks and services, based on the business models of telecom operators and OTT service providers and on regulation of the interrelations and interconnection of different actors on the ICT/telecommunication scene, are changing as the future networks change. This change is driven by the rapid availability of different and more complicated end-user devices and by the additional capabilities offered by networks and services.

A challenge in ICT infrastructure development is adaptation of business models and regulation (for fair use or sharing of resources) to keep up with the quick growth of the USN, M2M, and SDN cloud-based infrastructures.

With the development of smart networks with context and content awareness, another challenge is convergence of telecom businesses with initially nontelecom businesses (e.g., energy, transportation, environment, health, various public services). Such a convergence increases possible threats, which may be critical for certain services (e.g., health, public security). Hence, cybersecurity is a continuous challenge due to the parallel growth of illegal use of services and the ICT infrastructure. The main pillars for providing reliability and security are targeted to promotion of societal awareness regarding security issues and ethics.

Another business and regulation challenge in future networks is network neutrality sustainability. By some technopolitical definition, network neutrality (when speaking about Internet neutrality) is the principle that Internet service providers and governments should treat all data on the Internet equally. But what about network neutrality in future networks? Well, even the NGN framework allows all possible business relations to exist among end users, network providers, and third-party service providers. Also, there are some practical problems regarding the best-effort OTT services and network neutrality. For example, when one of the kids at home is playing online games (which have a low tolerance for delay end to end) and other one is watching HD videos on YouTube, and both are using the same broadband access to the Internet, the gaming traffic will suffer due to the delay caused by the higher volume of the video traffic, which is served in a best-effort manner together with the gaming traffic. In such a case, the end user may want to have the ability to choose which OTT service will have a higher priority or a certain preserved and shaped bandwidth. This could be accomplished by, for example, dragging that application to some networking management applications in a myHome gateway that will be accessible from all authorized devices (e.g., computers, laptops, smartphones) via a web interface (which is a default user interface on the Internet today). However, that function requires certain support from the telecom operators (e.g., provision of fine-grained traffic classes, with fine-grained bandwidth and service times).

Let's look at regulation efforts in some developed countries. For example, in May 2014, the Federal Communications Commission in the United States decided to consider two options regarding Internet services. The first option was to permit fast and slow broadband "lanes," and with that action to compromise network neutrality. The second option was to reclassify broadband as a telecommunication service, thereby preserving network neutrality. Currently network neutrality is still preserved, but FNs may lead to certain changes in that principle, especially with virtualization of network resources (e.g., with the use of SDN and LINP approaches).

Other challenges that will also be important for regulation in future networks are cybersecurity as well as Big Data issues. Cybersecurity is related to users' privacy on one side, to Internet "freedom" on the other side, and to national politics/policies on the third side. It is related to Internet governance as well, and sometimes the countries through which Internet traffic is routed are important, as is where the servers that store certain data (either user information or content) are located. This is also related to Big Data challenges, since the overall amount of data in the Internet is increasing exponentially, and most of it is stored somewhere (and the amount of storage will increase further with expected M2M use in upcoming years). However, such data belong to a certain person, company, or organization, and hence may be accessed and used now or in the future in different contexts, or even be traded (e.g., among companies/organizations). So, privacy is also an increasing challenge for FNs. Regarding the business challenges, new business models and actor roles are appearing with Big Data, such as data operators.

However, when most of the major service providers are on the global market (e.g., OTT providers), it becomes very hard to enforce certain national regulation policies over the globally accepted "freedom" of the Internet to which users and society have become accommodated, and no one can expect that to go back in future networks.

Finally, smart networks and services are expected to permeate all segments of everyday living, including nontelecom (i.e., non-ICT) sectors, such as smart sustainable cities, which use Internet technologies in fixed and mobile networks for the benefit of every citizen and society in general.

References

[1]   ITU-T Recommendation Y.3001, "Future Networks: Objectives and Design Goals," May 2011.
[2]   ITU-T Recommendation Y.3011, "Framework of Network Virtualization for Future Networks," January 2012.
[3]   ITU-T Recommendation Y.3300, "Framework of Software-Defined Networking," June 2014.
[4]   Open Networking Foundation, "OpenFlow Switch Specification," September 2012.
[5]   S. Gringeri, N. Bitar, T. J. Xia, "Extending Software Defined Network Principles to Include Optical Transport," *IEEE Communications Magazine*, March 2013.
[6]   K. Pentikousis, Y. Wang, W. Hu, "MobileFlow: Toward Software-Defined Mobile Networks," *IEEE Communications Magazine*, July 2013.

[7]   ITU-T Technology Watch Report, "Big Data: Big Today, Normal Tomorrow," November 2013.

[8]   ISO/IEC JTC 1, "Big Data," Preliminary Report 2014, Switzerland: ISO, 2015.

[9]   NIST Special Publication 1500-7, "DRAFT NIST Big Data Interoperability Framework: Volume 7, Standards Roadmap," April 2015.

[10] NIST Special Publication 1500-2, "DRAFT NIST Big Data Interoperability Framework: Volume 2, Big Data Taxonomies," April 2015.

[11] Knowledge@Detecon, "Future Telco Profitability in Telecommunications: Seven Levers Securing the Future," *Consulting Detecon*, 2014.

[12] ITU-T Recommendation X.805, "Security Architecture for System End-to-End Communications," October 2003.

[13] ITU-T Recommendation Y.1205, "Overview of Cybersecurity," April 2008.

[14] ITU-T Recommendation Y.2701, "Security Requirements for NGN Release 1," April 2007.

[15] ITU-T Recommendation Y.3012, "Requirements of Network Virtualization for Future Networks," April 2014.

[16] ITU-T Recommendation Y.3031, "Identification Framework in Future Networks," May 2012.

[17] ITU-T Recommendation Y.2061, "Requirements for the Support of Machine-Oriented Communication Applications in the Next Generation Network Environment ," June 2012.

[18] ITU-T Recommendation Y.2060, "Overview of the Internet of Things," June 2012.

[19] ITU-T Recommendation Y.2221, "Requirements for Support of Ubiquitous Sensor Network (USN) Applications and Services in the NGN Environment," January 2010.

[20] OneM2M Technical Specification, "Functional Architecture," January 2015.

[21] Z. Shelby, K. Hartke, C. Bormann, "The Constrained Application Protocol (CoAP)," June 2014.

[22] ITU-T Recommendation Y.3041, "Smart Ubiquitous Networks—Overview," April 2013.

[23] ITU-T Recommendation Y.3042, "Smart Ubiquitous Networks—Smart Traffic Control and Resource Management Functions," April 2013.

[24] ITU-T Focus Group on Smart Sustainable Cities, "Setting the Framework for an ICT Architecture of a Smart Sustainable City," May 2015.

CHAPTER 10

Conclusions

This book provided comprehensive, up-to-date coverage of the standardized Internet technologies including networking protocols and applications, as well as fundamental systems and architectures, from the Internet and telecom points of view by explaining the convergence and synergy between the standards developing organizations such as the IETF, ITU 3GPP, IEEE, and others, as well as the convergence of the global telecom and OTT sectors toward the Internet as a single networking platform for all ICT services. It included the legacy telecommunication services such as telephony (as VoIP) and TV (as IPTV); legacy Internet services such as email, web browsing, and P2P applications; and also emerging OTT services, cloud computing, and future networks. Besides the technologies, the book added value by covering the regulation and business aspects of the Internet and its services, something that is also important and influences the development of the technologies.

In summary, the book provided all-around coverage of the fundamental Internet technologies (Chapters 2 through 4) used in both the fixed and mobile environments, and then covered specific Internet standardization for the telecom sector in support of the transition of telephony and TV to all-IP environments with standardized signaling and QoS support. This is accomplished within the NGN framework of ITU (Chapter 5). We then discussed Internet technologies and implementations specific to the mobile broadband Internet (Chapter 6) and implementations of all existing broadband Internet services (Chapter 7), including carrier-grade services (i.e., QoS-enabled VoIP and IPTV) and OTT services for telephony (e.g., Skype, Viber), multimedia streaming (e.g., YouTube), social networks (e.g., Facebook), P2P services (e.g., BitTorrent). We then covered the emerging developments of cloud computing in fixed and mobile environments (Chapter 8) with different service models (e.g., IaaS, PaaS, SaaS, and combinations thereof) and different implementations (telecom operator–based and OTT-based clouds). Finally, the book covered the development of future networks based on ITU's framework, including network virtualization, SDN, as well as Big Data and smart networks and services (Chapter 9). Business and regulation aspects give additional "spice" to Chapters 6 through 9, and reflect the complexity of merging the initially separated telecom and Internet worlds in a broadband Big Data world.

Overall, this book is dedicated to the Internet technologies (technologies that provide connectivity and services end to end across IP-based networks) that have had a major impact on everyday living and in the development of our so-called "information society."

However, until the 1990s the dominant role in the world of telecommunications involved two services, telephony and television. Each of these services was delivered through completely separate networks built specifically for each service and adapted to each service's characteristics (e.g., adequate bandwidth required to transmit the signal from/to the end users). In the beginning these two services were provided by analog systems. Telephone networks were eventually digitized during the final decades of the 20th century, and then TV was digitized at the beginning of the 21st century. As all services were transitioned to the digital world, it became obsolete to have different networks for different types of services, because all types of information (voice, text, images, music, videos, and so on) were being transmitted across the network using the same digits or bits (e.g., ones and zeroes).

Simultaneously with this process of digitalization, the Internet was developing as a network for the transmission of multimedia content data, based on the simplicity and low cost of network elements (switches and routers) without built-in support for QoS, as opposed to what was provided in the traditional telephone and TV broadcast networks. The principle of a best-effort, network neutral Internet means that each IP packet will be received in any IP network connection and that any connection that may be required will be established. This principle is the main reason for the success of the IP, but it also became a major challenge when the time came for full integration of all telecommunication networks and services as well as the native and new emerging Internet services.

Parallel to this process of integration and convergence of the Internet and traditional telecommunications, Internet technologies were penetrating into nontelecom sectors, such as household appliances at home, items in the car, the environment, e-health, and smart cities, resulting in creation of smart ubiquitous networks. So, in the future we can expect everything—physical and virtual—to be connected to the Internet, and to provide data (e.g., from sensing networks), contents, or connectivity between humans, between humans and machines, and between machines. That imposes challenges, such as cybersecurity support for certain critical services, as well as QoS provisioning end to end. However, development of certain business models between different actors in the value chain is the most important for the mass deployment of certain emerging technologies (e.g., M2M and IoT in general), as well as an appropriate regulation framework regarding the access (e.g., frequency bands for mobile broadband access), core and transport networks (e.g., virtualized resources in the networks), and services/applications (e.g., telecom locked-in services, OTT services based on network neutrality, or hybrid ones based on partnering between different actors in the ICT ecosystem).

The future telecom/ICT developments that are now in the research phase (e.g., 5G mobile networks) will become a reality in the 2020s. Note, however, that all of them must maintain backward compatibility with existing IP-based networks and services (at least, in the short term). Broadband access in fixed and mobile networks is expected to continue further to higher bit rates (there is no foreseen limit) and that is expected to bring new emerging services that will use such an Internet "Information Highway." With mass deployments of the Internet of Things, including the Web of Things (for human access) and M2M for practical implementation of smart "things," the Internet will become omnipresent, just like the air we are breathing. Further, only our imagination is the limit.

About the Author

Toni Janevski, Ph.D., is a professor on the Faculty of Electrical Engineering and Information Technologies, Ss. Cyril and Methodius University, Skopje, Macedonia. He received his Dipl. Ing., M.Sc., and Ph.D. degrees in electrical engineering in 1996, 1999 and 2001, respectively, from the Faculty of Electrical Engineering and Information Technologies, Ss. Cyril and Methodius University. From 1996 to 1999 he worked for the Macedonian mobile operator Mobimak (currently T-Mobile, Macedonia), contributing to the planning, dimensioning and implementation of the first mobile network in Macedonia. Since 1999 he has been with the Faculty of Electrical Engineering and Information Technologies in Skopje. In 2001 he conducted research in optical communications at IBM T. J. Watson Research Center, New York. From 2005 to 2008 he was an elected member of the Commission of the Agency for Electronic Communications (AEC) of the Republic of Macedonia. From 2008 through 2016 he is an elected member of the Senate of the Ss. Cyril and Methodius University. In 2009 he established the Macedonian ITU (International Telecommunication Union) Centre of Excellence (CoE) as part of Europe's CoE network and has served as its head/coordinator since that time. He is the author of the book *Traffic Analysis and Design of Wireless IP Networks*, published in 2003 by Artech House Inc., USA. He also authored the book *Switching and Routing*, written in the Macedonian language and published in September 2011 by the Ss. Cyril and Methodius University. In 2012 he won the "Goce Delchev" award, the highest award for science in the Republic of Macedonia. Also, he received the Best Scientists Award from the Ss. Cyril and Methodius University in 2013. In April 2014 his second worldwide book, titled *NGN Architectures, Protocols and Services*, was published by John Wiley & Sons, UK. He has published numerous research papers and has led several research and application projects in the area of Internet technologies and mobile and wireless networks. Also, he has tutored and coordinated many international courses within the ITU Academy. He has been a Senior Member of IEEE since 2005. His research interests include Internet technologies; mobile, wireless, and multimedia networks and services; traffic engineering; quality of service; design and modeling of telecommunication networks, next-generation networks, and future networks.

Index

A

Access point name (APN), 239–240
Access Service Network (ASN), 200
Active document, 143
Active optical network (AON), 188, 190
Adaptive multirate (AMR), 243
Address conversion, 89
Address Resolution Protocol (ARP), 13, 34, 36
    standards, 98–100
Addressing, 199–201
Admission control, 207–8
Advanced Research Project Agency (ARPA), 2
Alliance for Telecommunications Industry Solutions (ATIS), 175
All-IP packet core, 230–38
Allocation and retention priority (ARP), 239, 242
Analog signal transfer, 2
Analog-to-digital (A/D) conversion, 2
Anycast address, 50
American National Standards Institute (ANSI), 9, 184
Application layer, 12, 321–22
Application server (AS), 215–16
Application Programming Interface (API), 80, 86
Application-specific functions (APP), 147
ARPANET, 3, 5, 127, 134
Asymmetric digital subscriber line (ADSL), 10, 183–86
Asynchronous Transfer Mode (ATM), 7, 25
Audiovisual object (AVO), 150
Authentication, authorization, and accounting (AAA), 162–70
Authentication, cloud computing, 323–24
Authentication header, 46, 159
Authoritative name servers, 128
Autonomous system (AS), 6, 18

Attribute-value pair (AVP), 167–168

B

Baseband unit (BU), 258
Basic service set (BSS), 198
Bearer service, 238–42
Beranek, Leo, 81
Berkeley socket, 80, 81
Berners-Lee, Tim, 6
Best-effort architecture, 18, 19, 22
    quality of service and, 206
    standardization and, 173
    TCP, 69–70
    WiMAX, 243
Bidirectional tunneling, 226–27
Big Data, 343–44
    challenges, 349
    definition, 344–46
    ecosystem, 346
    reference architecture, 346–47
    use cases, 348–49
Big-endian system, 85
Binary data interpretation, 85–86
Bind function, sockets, 87, 89
Bolt, Richard, 81
Bootstrap Protocol (BOOTP), 56
Border Gateway Protocol (BGP), 6
    discussion, 117–19
    operations, 115–17
    overview, 113–15
    message types, 17
    peers, 115–16
Bound end-to-end tunnel (BEET), 230
Breakout gateway control function (BGCF), 214–15
Bridge, 15, 97
Broadband Internet
    defined, 25, 182–83

367

Broadband Internet (continued)
    fixed access, 98–100, 183, 184–92
    growth, 8–11, 269
    mobile and wireless, 100–2, 183, 191–99
    next-generation, 183–99
    regulation, 302–6
    services requirements, 269
Broadband remote access server (BRAS), 186
Byte ordering, 89

C

Cable broadband networks, 186–88
Call session control function (CSCF), 214–16
Care-of-address (CoA), 222, 225, 226
Carrier Ethernet (metro Ethernet), 190–91
Carrier Sense Multiple Access with Collision Detection (CSMA-CD), 96
Carrier-grade mobile VoIP, 245–47
Carrier-grade NAT (CGN), 41
Carrier-grade services, 267
Central processing unit (CPU), 80
Cerf, V., 4
Circuit-switched (CS) network, 3, 102
Classful IPv4 addressing, 37–38
Classless interdomain routing (CIDR), 38–39
Client-server model
    networking architecture, 91–93
    TCP sockets, 84–85
    UDP sockets, 90–91
Cloud computing
    architecture, 312–14
    ITU framework, 309–12
    mobile technologies, 330–32
    OTT services, 324–26
    regulation, 332–35
    security, 319–24
    telecom and, 326–20
    service models, 314–19
Close function, 88–89
Cloud communication center, 327–28
Cloud published service end point (CPSEP), 323–24
Cloud RAN (C-RAN), 258–59
Cloud Security Alliance (CSA), 321
Cloud service customer (CSC), 312, 314, 317, 321–22, 334

Cloud service partner, 313
Cloud service provider (CSP), 312–13, 316, 318–19, 320–21, 322, 333, 334
Cloud services gateway (CSG), 324
Communication as a Service (CaaS), 311, 312, 317–18
Community cloud, 311
Compute as a Service, 311, 312
Congestion, defined, 69–70
Congestion control, 6, 12, 18, 70–73
Connection admission control (CAC), 282
Connect function, 90
Connectionless protocol, 31
Connection-oriented transport protocol, 65. *See also* Transmission Control Protocol (TCP)
Connectivity-service network (CSN), 198
Constrained Application Protocol (CoAP), 360
Contention-based service, 243
Contention-free service, 243
Convergence to all-IP, 23–27
Crocker, Steve, 3
Cybersecurity, 352–56, 364, 365

D

Data Storage as a Service (DSaS), 311, 312
Data traffic, 206
Datagram sockets. *See* User Datagram Protocol (UDP) sockets
Defense Advanced Research Project Agency (DARPA), 2, 75
Deployment view, 312, 314
Desktop as a Service (DaaS), 327
Destination address field, 34
Destination options header, 46
Device-to device (D2D) service, 259
Diameter protocol, 165–70
Differentiated services, 206
Differentiated services code point (DSCP), 283
DiffServ, 208–9
Digital communications, 2
Digital subscriber line (DSL), 9, 98, 184–86
Digital subscriber line access multiplier (DSLAM), 185–86
Digital Video Broadcasting (DVB), 248
Discrete multitone (DMT) modulation, 98

Distance-Vector Multicast Routing (DVMRP), 112
Distance-vector routing, 105, 106
Distributed RAN (D-RAN), 256–57
DOCSIS (Data-Over-Cable Service), 10, 99, 186–88
Domain, 36–37
Domain name space, 126, 127
Domain Name System (DNS), 5, 125–32, 200
Domain Name System Security Extensions (DNSSEC), 131
Double-crossing, 224
Duplicate ACK, 73
Dynamic DNS (DDNS), 131
Dynamic document, 143
Dynamic Host Configuration Protocol (DHCP), 7
    IPv4, 55–57
    overview, 55
    TFTP and, 133
    version 6, 57–60
Dynamic port, 60

E

E.164 standard, 200
Email (electronic mail), 3, 6, 134–40
Encapsulating Security Payload (ESP), 47, 159, 230
End of participation (BYE), 147
End-to-end identifier, 168
End-to-end performance, 210–13
End-to-end service management, 329–30
ENUM, 200–201
Ethernet, 13
    metro, 190–91
    network elements, 97–98
    standards, 95–96, 176
    Wi-Fi standards, 96–98
Ethernet virtual connection (EVC), 191
European Network and Information Security Agency (ENISA), 320
European Telecommunication Standardization Institute (ETSI), 175
Evolved packet core (EPC), 232–33
Evolved packet system (EPS), 195
Evolved UMTS Terrestrial Radio Access Network (E-UTRAN), 195–96
Equipment life cycle, 340
Extended service set (ESS), 198
Extended Unique Identifier (EUI-64), 53
Extension header, 46–48
External Border Gateway Protocol (EBGP), 115

F

Fast retransmit/recovery, 72–73, 74
Federal Networking Council (FNC), 8
Fiber-to-the-building (FTTB), 184–85, 188–89
Fiber-to-the-desk (FTTD), 188–89
Fiber-to-the-home (FTTH), 189
Fiber-to-the-premises (FTTP), 188
Fiber-to-the-x (FTTX), 188–89
Fifth Generation (5G) mobile broadband, 257
    architectures, 257–59
    regulation, 262–65
    services, 259–60
Fifth Generation (5G) network, 192
File Transfer Protocol (FTP), 6, 19, 85, 132–34
Finite state machine model (FSM), 116–17
Fixed broadband access networks, 98–100, 183, 184–91
Fixed-mobile convergence (FMC), 26, 213–14
Flags field, 33
Flow control, 62
Flow labeling, 44–46
Focus Group on NGN, 174
Foreign agent (FA), 222–23
Fourth Generation (4G) mobile networks, 11, 100, 192
    all-IP packet core, 231–38
    LTE/LTE-Advanced, 194–96
    QoS framework, 238–43
    TV, 248–54
    VoIP, 243–48
    Web, 254–57
Four-message exchange, 58
Four-way handshake, 230
Fragment header, 46
Fragment offset field, 33
Frequency division multiple access (FDMA), 192
Functional view, cloud computing, 312, 313–14

Future Networks, 337–38
　security, 353–56
　virtualization, 338–42

G

Gateway GPRS support node (GGSN), 100
Gateway router, 17–18
General Packet Radio System (GPRS), 10, 100, 192–93
Generalized Multi-Protocol Label Switching (GMPLS), 190
Generic access network (GAN), 213
Global System for Mobile (GSM) communications, 10, 11, 100, 192–94
Globally routable user agent URI (GRUU), 218
Google cloud services, 325, 326
Governance, Internet, 74–76
GPRS Tunneling Protocol (GTP), 234, 237–38
GSM Association (GSMA), 175
Guaranteed bit rate (GBR), 242
G.fast standard, 184–85

H

Header checksum field, 33
Hierarchical routing, 105, 108–9
High-speed packet access (HSPA), 194
High-speed uplink packet access (HSUPA), 194
Home address, 222
Home agent (HA), 222–24, 226–28
Home subscriber service (HSS), 215, 232–33
Hop-by-Hop Identifier, 167
Hop-by-hop options header, IPv6, 46
Host, 14, 97
Host Identity Protocol (HIP), 229–31
HTML (Hypertext Markup Language), 15, 15, 143–44
HTTP (Hypertext Transfer Protocol), 7, 8, 16, 85
　binding, 360
　fundamentals, 140–44
　protocol stack, 254–55
Hub, 14–15, 97
Hybrid cloud, 311

I

Identification field, 33

Implementation view, 312, 314
IMT-Advanced radio interface, 11
Information and communication technologies (ICTs), 1, 27, 362–64
Infrastructure as a Service (IaaS), 311, 312, 314–16
Institute of Electrical and Electronics Engineers (IEEE), 5, 173, 175, 176
　802.3 standard, 13, 14, 95, 96
　802.11 standard, 10, 95, 96, 197
　802.16 standard, 10, 11, 199
　optical broadband and, 190
Integrated services, 206
Integrated Services Digital Network (ISDN), 10
Intercloud computing, 318–19
Interconnection border control function (IBCF), 215–16
Internal Border Gateway Protocol (IBGP), 117
International Engineering Task Force (IETF), 173, 174, 175
International Mobile Telecommunications (IMT), 11, 103–4
International Standards Organization (ISO), 345
International Telecommunication Union (ITU), 9, 173–74
　broadband development and, 183–84
　cloud computing and, 309–12, 320
　future networks and, 337–38
　ITU-D, 183–84
　ITU-R, 11, 183, 261
　ITU-T, 172, 184, 189, 200, 202–4, 248–52, 309–12, 320, 337–38, 356, 362
International Telegraph Union (ITU), 173
International mobile subscriber identity (IMSI), 239
International Standardization Organization (ISO), 12
Internet, 1, 8. *See also* Internet architecture; Internet technologies
Internet access networks, 95
　Ethernet/Wi-Fi, 95–98
　fixed broadband, 98–100
　mobile broadband, 100–2
Internet as a service (IAS), 305
Internet Assigned Numbers Authority (IANA), 40, 75, 76

Internet architecture, 11–12
    dual role for addresses, 221–22
    Internet protocol layering model, 12–14
    Internet protocol networks, 14–18
Internet Architecture Board (IAB), 6
Internet Corporation for Assigned Names and Numbers (ICANN), 75
Internet Control Message Protocol (ICMP), 34–35
Internet Control Message Protocol v6 (ICMPv6), 48–49, 50
Internet Engineering Task Force (IETF), 3, 76
Internet Group Management Protocol (IGMP), 109–10
Internet layer, 13
Internet Message Access Protocol version 4 (IMAP4), 140
Internet Protocol (IP), 1, 4, 31–36
Internet Protocol Television (IPTV), 248–52
Internet Protocol v4 (IPv4), 7, 8
    addressing, 36–41
    DHCP, 55–57
    packet header, 32–34
    private addresses, 40–41
    transition to IPv6, 55–55
Internet Protocol v6 (IPv6), 7, 8, 39
    addressing, 49–54
    DHCPv6, 57–60
    extensions headers, 46–48
    flow labeling, 44–46
    fundamentals, 42–44
    ICMPv6, 48–49
    transition from IPv4, 54–55
Internet of Things (IoT), 25, 296–99
Internet Research Task Force (IRTF), 6, 75
Internet Security on Network Level (IPsec), 158–60
Internet Security on the Transport Layer (SSL/TLS), 160–62
Internet Service Provider (ISP), 36
Internet Society (ISOC), 75
Internet technologies
    architecture, 11–18
    broadband access, 8–11
    convergence, 23–27
    development, 2–11
    legacy telecommunications vs., 20–23
    new applications, 8–11
    security, 156–62
    signaling, 151–56
    standardization, 4–6
    traffic, 18–20, 205–6
    worldwide growth, 6–8
Internet Transmission Control Program, 4
Internet traffic, 18–20, 205–6
Interrogating-CSCF (I-CSCF), 215
IntServ, 208–9
IP addresses, 221–22
IP datagram/IP packet, 16
IP Multimedia Subsystem (IMS), 214–18
IP multimedia private user ID (IMPI), 217
IP multimedia public identity (IMPI), 217–18
IP television services (IPTV), 275–84
Iterative query, 130

J

Joint Rapporteur Group on NGN (JRG-NGN), 174

L

Label edge router (LER), 208
Label switched path (LSP), 208
Layer splitting, 14
Legacy telecommunications, 20–27
Library routine, 89
Licklider, J. C. R., 2–3
Link layer, 13, 14
Link-state routing, 105, 106–7
Little-endian system, 85
Local area network (LAN), 17
Local name server, 126
Location-based service (LBS), 25, 45
Location service, 154
Logically isolated network partition (LINP), 339, 341
Loopback address, 40
LTE/LTE-Advanced, 194–96, 231–32, 234
    MBMS, 253–54
    QoS in, 240–43
    PDP, 239

M

Machine-oriented communication (MOC), 356

Machine-to-machine (M2M) service, 257, 356–60
Macro-mobility, 221
Maximum bit rate (MBR), 242
Maximum datagram lifetime (MDL), 33
Maximum segment size (MSS), 65
Media gateway controller (MGC), 214–15
Medium access control (MAC) layer, 10
Memory management, 89
Message submission agent (MSA), 135
Message transfer agent (MSA), 135
Message transfer unit (MTU), 13
Message user agent (MUA), 135
Metcalfe, Robert, 5, 13
Metro Ethernet, 190–201
Metro Ethernet Forum (MEF), 190–201
Metropolitan area network (MAN), 17
Micro-mobility, 221
Mobile broadband network, 27, 221–22
    5G, 257–61
    regulation, 262–65
Mobile cloud computing (MCC)
    business aspects, 260–61
    development, 328–30
Mobile IP (MIP), 221–29
Mobile IPTV, 246–52
Mobile IPv4, 222–24
Mobile IPv6, 225–29
Mobile mobility management entity (MME), 232–34
Mobile QoS, 262–63
Mobile support messages, 49
Mobile technology, 10
    access networks, 97–98
    standardization, 100–2, 183, 191–99
Mobile TV, 248–54
Mobile Web, 254–57
Mobile WiMAX, 10, 11, 100–1, 199, 243
Mobility management and control functions (MMCF), 179, 245–47
Mockapetris, Paul, 5, 125
Moving Pictures Expert Group (MPEG), 144
    MPEG-2, 149–50
    MPEG-4, 150–51
    MPEG-7, 151
    MPEG-21, 151

MQ telemetry transport (MQTT), 360
Multicast address, IPv6, 50, 58
Multicast broadcast over a single frequency network (MBSFN), 253
Multicast extensions to OSPF (MOSPF), 112
Multicast Listener Discovery (MLD), 49, 110
Multicast IPTV, 277–79
Multicast routing, 109–13
Multimedia Broadcast Multicast Service (MBMS), 248, 253–54
Multimedia messaging service (MMS), 192, 239
Multimedia resource function controller (MRFC), 215–16
Multimedia streaming, 144–151, 284–87
Multi-Protocol Label Switching (MPLS), 14, 206, 208–9
Multipurpose Internet mail extension (MIME), 137–38
Multiplexing, 62–63

N

Name server, 126, 128–29
Naming, 199–201
National Institute of Standards and Technology (NIST), 320, 345
National regulatory agencies (NRAs), 265–66
National Science Foundation (NSF), 5–6
Neighbor discovery messages, 49
Network access identifier (NAI), 217
Network Address and Port Translation (NAPT), 42
Network Address Translation (NAT), 7–8, 41–42
Network as a Service (NaaS), 311, 312, 316–17
Network attachment and control functions (NACF), 179, 245
Network Control Protocol (NCP), 3, 4, 5, 79
Network File System (NFS), 61
Network identifier, 239–40
Network neutrality, 364–65
Network performance (NP), 203
Network Time Protocol (NTP), 58
Next-generation access network (NGAN), 183
Next-generation network (NGN), 25–26
    architectures, 180–82

broadband access, 182–99
cooperation by SDOs and, 175–76
network concept, 176–77
service stratum, 177, 180
standardization timeline, 174
transport stratum, 177–80
NGN Global Standardization Initiative (NGN–GSI), 174
Non-real-time services, 269
NSFNET, 5–6, 8

O

Open network architecture, 3
Open Shortest Path First (OSPF), 107
Open Systems Interconnection (OSI), 12
    basic architecture, 12–14
    network architecture, 14–18, 96
OpenFlow, 342–43, 344
Operator identifier, 240
Optical broadband access networks, 188–90
Options field, 34
Orthogonal frequency division multiple access (OFDMA), 192
Over-the-top (OTT) applications, 19–20, 26
    business aspects, 265–66
    cloud services, 324–26
    defined, 269
    mobile VoIP, 247–48
    multimedia streaming, 284–87
    peer-to-peer services, 287–91
    social networks, 291–96
    vs. telecom service models, 349–52

P

PacketCable, 187
Packet data network gateway (P-GW), 232–33
Packet data protocol (PDP), 239
Packet-switched network, 3
Passive optical network (PON), 99–100, 188–90
Peer-to-peer (P2P) systems, 8
    application types, 95
    Internet traffic and, 19
    networking architecture, 93–95
    OTT, 285–89

    structured architecture, 94
    unstructured architecture, 94
Personal computer (PC), 5
Platform as a Service (PaaS), 311, 312, 315–16, 335
Point-to-point (P2P) architecture, 188
Point-to-Point Protocol over Ethernet (PPPoE), 14
Policy and charging enforcement function (PCEF), 233
Policy and charging rules function (PCRF), 233
Policy decision point (PDP), 321
Policy enforcement point (PEP), 321
Post Office Protocol Version 3 (POP3), 138–40
Privacy, cloud computing, 319–24
Private cloud, 310–11
Private IP address, 40–41
Protocol field, 33
Protocol Independent Multicast (PIM), 111–13
Protocol layer, 1
Protocols, message, 15
Proxy agent, 169
Proxy server, 154
Proxy-CSCF (P-CSCF), 217
Public key infrastructure (PKI), 321
Public land mobile network (PLMN), 22, 24, 175
Public IP address, 40–41
Public cloud, 310
Public service identity (PSI), 218
Public Switched Telephone Network (PSTN)
    as legacy network, 20–23, 24
    sockets, 79, 83
    standardizations, 176–77, 205

Q

QoS class identifier (QCI), 242
Quality of experience (QoE), 204–5
Quality of Service (QoS)
    admission control and, 207–8
    convergence and, 74
    end-to-end, 210–13
    parameters and classes, 209–10
    introduction to, 201–203
    regulation, 262–63

Quality of Service (continued)
  standardization and, 203–13
  terminal device, 303

R

Radio access bearer, 239, 242
Radio access network (RAN), 246
Radio-frequency identifiers (RFIDs), 182
RADIUS, 163–70
Raw sockets, 82
Real-time services, 269
Real-Time Streaming Protocol (RTSP), 144, 147–48
Real-Time Transport Protocol (RTP), 14, 145–47
Received signal strength (RSS), 246
Receiver report (RR), 147
Recursive query, 130
Redirect agent, 169
Redirect server, 153–54
Regional Internet registry (RIR), 74
Registrar server, 154
Regulation
  cloud computing, 332–35
  future networks, 364–65
  mobile broadband, 262–65, 305–6
Relay agent, 169
Reliability, TCP, 62
Repeaters, 14–15, 97
Resolvers/resolution, 126, 131
Resource and admission control functions (RACF), 179, 180, 245
Resource record (RR), 131
Retransmission time-out (RTO), 72
Reverse Address Resolution Protocol (RARP), 36
Reverse path forwarding (RPF), 111
Roberts, Larry, 3
Root servers, 128
Round-trip time (RTT), 71–72
Route optimization, 226–27
Routers, 15, 17, 98
Routing
  defined, 102
  header, IPv6, 46
  multicast, 109–13
  overview, 102–4
  unicast, 104–9
Routing information base (RIB), 103
Routing Information Protocol (RIP), 106
RTP Control Protocol (RTCP), 145, 146–47

S

Security
  cloud computing, 319–24
  Internet, 156–57
  IPsec, 158–60
  SSL/TLS, 160–62
Secure Sockets Layer on Transport Layer (SSL/TLS), 160–62
Sender report (SR), 147
Service control functions (SCFs), 179
Service delivery platform (SDP), 328
Service Delivery Platform as a Service (SDPaaP), 328
Service stratum, NGN, 177, 180
Service-level agreements (SLAs), 179
Serving gateway (S-GW), 232–33
Serving GPRS support node (SGSN), 100
Serving-CSCF (S0CSCF), 216
Session Description Protocol (SDP), 151
Session Initiation Protocol (SIP), 151
  addressing, 152–53
  IMS, 214–16
  messages, 152
  network elements, 153–56
Shared tree, 111
Shortest path tree (SPT), 111
Signaling System number 7 (SSN7), 14, 21, 160–62
Simple Mail Transfer Protocol (SMTP), 12, 85
  content types, 137
  encoding types, 138
  POP3, 139–40
  fundamentals, 134, 135, 136–38
Simple Network Management Protocol (SNMP), 61
Single-carrier FDMA (SC-FDMA), 194–95
Slow start threshold, 70–71, 74
Smart sustainable cities (SSCs), 362–64
Smart traffic control, 361–62
Smart Ubiquitous Networks (SUNs), 360–64

Sockets, 79–83
  TCP, 83–89
  UDP, 89–91
Social networks, 289–94
Software as a Service (SaaS), 311, 312, 316
Software-defined networking (SDN), 259, 342–43, 344
Source address field, 34
Source description item (SDES), 147
Source tree, 111
Special purpose IP address, 40–41
Spectrum management, mobile broadband, 263–64
Standard Generalized Markup Language (SGML), 143
Standardization
  broadband access, 9–10
  DNS, 131
  FMC, 213–14
  IMS, 214–18
  multimedia streaming, 149
  naming and addressing, 199–201
  NGN, 174–77
  NGN architectures, 180–82
  NGN broadband, 182–99
  NGN service stratum, 177, 180
  NGN transport stratum, 177–80
  overview, 173–74, 218–19
  QoS framework, 201–213
  signaling, 151, 156
Standards developing organizations (SDO)s, 173–74, 345
Stateful addressing, 57–60
Stateful autoconfiguration, 53
Stateless autoconfiguration, 53–54
Static documents, 143
Stream sockets. *See* Transmission Control Protocol (TCP) sockets
Stream Transmission Control Protocol (SCTP), 234–37
Structured Query Language (SQL), 347–48
Subscriber location function (SLF), 215
Switches, 15, 97
System architecture evolution (SAE), 195, 231–32
System port, 60

T

Telecom cloud implementations, 326–30
Telecom service model, 349–52
Telecommunications and Internet Converged Services and Protocols for Advanced Networking (TISPAN), 175
Telecommunications, history of, 2, 14, 19
Telnet, 12, 84
Third-Generation (3G) mobile system, 11, 100–1, 102
3G Partnership Project (3GPP), 100
  mobile broadband standard, 192–94
  LTE/LTE–Advanced, 194–96
  standardization and, 173, 175–76
Three-way handshake, 66–68
Time division multiplexing (TDM), 190, 192
Time-to-live (TTL), 33, 129
Tomlinson, Ray, 3
Top-level domain (TLD), 127
Total length field, IPv4, 32–33
Traffic handling priority (THP), 239
Traffic management, IPTV, 282–84
Translation agent, 171
Transmission Control Protocol (TCP), 4, 5–6
  basic definition, 62
  congestion control, 70–73
  connection establishment, 65–68
  connection termination, 65–68
  operations, 62–63
  segment format, 63–65
  versions, 73–74
  windows, 68–69
Transmission Control Protocol (TCP) sockets
  binary data interpretation, 85–86
  interface, 86–89
  overview, 83–85
Transmission Control Protocol/Internet Protocol (TCP/IP), 4, 5–6, 19–20
Transport layer, 12, 14
Transport mode, 157
Transport protocols, 60. *See also* Transmission Control Protocol (TCP); User Datagram Protocol (UDP)
Transport stratum, 177–80
Triangle routing, 224
Trivial File Transfer Protocol (TFTP), 61, 133

Tunnel mode, 159
Two-message exchange, 58
Two-way simplex connection, 67
Type of Service (ToS) field, 32

U

Ubiquitous sensor network (USN), 356–58
UMTS/HSPA (universal mobile telecommunication systems/high-speed packet access), 11, 100–1, 238–40
Unicast address, 49, 51–53
Unicast IPTV, 279–82
Unicast routing, 104–9
   distance-vector, 106
   hierarchical, 108–9
   link-state, 106–8
Uniform resource identifier (URI), 141–42
Uniform resource locator (URL), 141–43
Uniform resource name (URN), 141–42
UNI-to-UNI performance, 210–13
UNIX operating system (UNIX OS), 80–81
Unlicensed mobile access (UMA), 213
Unsolicited grant service (UGS), 243
User agent (UA), 153–55
User agent client (UAC), 153–55
User agent server (UAS), 153–55
User Datagram Protocol (UDP), 5, 12, 14
   Internet traffic and, 18, 19
   overview, 60–61
   TFTP and, 133
User Datagram Protocol (UDP) sockets, 89–91
User experience, 256–57
User identities, IMS, 217–18
User ports, 60
User view, 312–13

V

Version field, 32
Very-high-bit-rate DSL (VDSL), 184–85

Video traffic, 206
Virtual local access network (VLAN), 339–40
Virtual Resource Manager (VRM), 339
Virtualization, Future Networks, 338–42
Virtual private network (VPN), 26, 160
Voice traffic, 205–6
Voice-over-IP (VoIP), 9, 26, 151, 156
   carrier-grade mobile, 245–47
   development, 270–71
   4G, 243–46
   OTT, 247–48, 271–75

W

Wavelength code division multiple access (WCDMA), 193
Wavelength division multiplexing (WDM) PON, 188–89
Web documents, 143–44
Web of Things (WoT), 25, 299–302
Wi-Fi, 10
   access points (APs), 97
   Ethernet standards, 96–98, 176
   standardization, 197–99
WiMAX, 10, 11, 176
   mobile standard, 199
   wireless access, 197–99
Windows Azure, 325
Windows, TCP, 68–69
Wireless access network (WAN), 10
Wireless application protocol (WAP), 239, 254
Wireless LAN (WLAN). *See* Wi-Wi
Wireless technology, 10, 100–2, 183, 191–99
World Wide Web (WWW), 6–7, 8, 19, 140–44
WWW Consortium (W3C), 75

X

X2 Application Protocol (X2AP), 234